A FESTSCHRIFT
for I. I. RABI

I. I. RABI

A FESTSCHRIFT
for I. I. RABI

Transactions of
The New York Academy
of Sciences

Series II Volume 38

The New York Academy of Sciences
New York, New York
1977

Copyright, 1977, by The New York Academy of Sciences. All rights reserved. Except for brief quotations by reviewers, reproduction of this publication in whole or in part by any means whatever is strictly prohibited without written permission from the publisher.

SP
Printed in the United States of America
ISBN 0-89072-060-6

TRANSACTIONS OF THE NEW YORK ACADEMY OF SCIENCES
SERIES II VOLUME 38
November 4, 1977

A FESTSCHRIFT FOR I. I. RABI

Editor
LLOYD MOTZ

CONTENTS

I. I. Rabi .... *Frontispiece*

Preface. *By* LLOYD MOTZ vii

Physical Interpretation of General-Relativistic Theories.
  By PETER G. BERGMANN 1

The Hyperfine Structure of the Ground State of Orthohelium in the
  Nonrelativistic Approximation. *By* G. BREIT, V. G. KAVEESHWAR,
  AND R. P. SINGH 10

The Predicted Infrared Spectrum of HeH$^+$ and Its Possible Astrophysical
  Importance. *By* I. DABROWSKI AND G. HERZBERG 14

Parity-Violating Electromagnetic Interactions of Nuclei.
  By G. FEINBERG 26

Quantization of the Damped Harmonic Oscillator.
  By HERMAN FESHBACH AND YOEL TIKOCHINSKY 44

A Diamond As Big As the Ritz. *By* ELLIOTT FLOWERS, ALAK RAY,
  MALVIN RUDERMAN, EDWARD SPIEGEL 54

The Fine Structure Constant α. *By* VERNON W. HUGHES 62

Vector and Tensor Gauge Particles in SL(6,c) Theory.
  By C. J. ISHAM, ABDUS SALAM, AND J. STRATHDEE 77

Some History of the Hydrogen Fine Structure Experiment.
  By W. E. LAMB, JR. 82

Recollections of a Rabi Student of the Early Years in the Molecular
  Beam Laboratory. *By* SIDNEY MILLMAN 87

Algebraic Incompatibilities Between Arnowitt-Nath Gauges and
  Supersymmetrized Gravity. *By* YUVAL NE'EMAN 106

Radioactivity's Two Early Puzzles. *By* A. PAIS 116

Notes on the Statistical Distribution of Single Population Level
  Spacings and Level Widths. *By* JAMES RAINWATER 137

The Electric and Magnetic Dipole Moments of the Neutron.
  By NORMAN F. RAMSEY 148

Superfluid Motions. *By* MARIO RASETTI AND TULLIO REGGE 161

The Majorana Formula. *By* JULIAN SCHWINGER 170
The Problem of Mass. *By* STEVEN WEINBERG 185
About Liquids. *By* VICTOR F. WEISSKOPF 202
A Few Specializations of the Generic Local Field in Electromagnetism
    and Gravitation. *By* JOHN ARCHIBALD WHEELER 219

PREFACE

This is the first in a series of Festschrifts that The New York Academy of Sciences is publishing to honor outstanding American scientists. In the past, particularly in Europe, the Festschrift was the highest tribute that a scientist's colleagues could pay him in recognition of his great accomplishments, but owing to the ever increasing publication costs of technical books publishers today are unwilling to undertake such an expensive project. Happily, The New York Academy of Sciences does not operate under the same constraints and limitations as does a commercial publishing firm and so it is publishing this Festschrift series not only to honor our great scientists but also as a service to our membership and to the science community as a whole.

When the members of the Board of Governors of The Academy voted to issue this series, they agreed unanimously that the first Festschrift should be dedicated to Dr. I. I. Rabi, not only because of his incomparable contributions to physics but also because of his ceaseless efforts on behalf of science and scientists in general and his great humanitarianism. This unanimous decision of the Board of Governors, a group of people working in many different sciences, may be taken as some measure of the esteem in which all scientists hold Rabi. That Dr. Rabi has been an active member of The Academy since 1949 and is now an honorary life member adds to our pleasure in dedicating to him this volume to which his former students and coworkers and his friends from various countries have contributed.

When Dr. Rabi, as assistant professor, returned to Columbia in 1930, after three years of postdoctoral studies at Munich, Leipzig, and Zürich, American physics was still struggling to make itself visible in the blinding radiance of the various European schools of physics. To be sure, there were accomplished experimentalists like the Nobel laureates Millikan and Compton, and outstanding theoreticians like Slater, Condon, the Nobel laureates Van Vleck and Wigner (a European American) and Oppenheimer who were beginning to forge American theoretical physics, but a school of American physics as such did not exist. Some nine years later things were quite different; a number of American schools of physics had been created, and American physics itself, if not superior to European physics, was at least its equal. If the eminence of a school is measured by its many disciples and the great influence it has had on the discipline it teaches, then the school of physics stemming from Rabi's work is pre-eminent, for there is hardly a branch of physics (some phases of chemistry and biology may also be included) that does not owe something to the molecular beam and magnetic resonance techniques that he developed in his laboratory at Columbia University. Casting about at random in the realm of past heroic deeds in physics that were a direct consequence of or strongly influenced by Rabi's work, we find, to name but a few, measurements of magnetic moments and spins of various nuclei (Rabi, Millman, Kusch, Zacharias); electric quadrupole moments of nuclei (Casimir, Feenberg, Wigner, Rarita, Schwinger); the fine structure of the hydrogen atom (Lamb, Retherford, Bethe, Schwinger, Feynman); anomalous magnetic moment of the electron (Kusch, Foley, Schwinger); quantum electrodynamics (Schwinger,

Feynman, Dyson); magnetic resonance spectroscopy (Bloch, Purcell, Alvarez, Dicke); nuclear structure (Bohr, Mottelson, Rainwater); microwaves, masers, and lasers (Townes, Zacharias, Schawlow); magnetic and electric moments of the neutron (Ramsey and students); electric resonances (Hughes and students); radio astronomy and the hydrogen 21-cm line (Purcell and colleagues); and interstellar molecules (Novick, Thaddeus).

That so much important and exciting research was generated by Rabi's work is remarkable, but then the period from 1930 to 1940 was a remarkable decade. With the discovery of the quantum statistics and with Dirac's discovery of the relativistic wave equation of the electron in 1928, quantum mechanics had become a beautiful mathematical system—a magnificent intellectual tool—for probing and understanding the structure of matter, and everyone—graduate student, postdoctoral fellow, professor—shared an exhilaration and excitement that was new and incredibly stimulating to physics in this country. In both the theoretical and experimental domains hundreds of problems were lying around to be solved by the new theories or to challenge and test them. With the discovery of the neutron, nuclear physics became part of this feverish intellectual game. At Columbia University, where Rabi was both experimentalist and theoretician (he gave the advanced courses in quantum and statistical mechanics) we all attended the weekly seminar conducted jointly by Gregory Breit and Rabi. We were, in a sense, all learning together but there were some in the postdoctoral group like John Wheeler (he seemed to have the correct answer to every question that arose) who were clearly marked for greatness. For a while Hans Bethe was with us, having come from Europe with an incredible reputation for being able to solve almost any theoretical problem. Later Julian Schwinger, while still a sophomore at CCNY, began attending the seminar, and Bob Marshak attended from the time of his senior year at Columbia until he went on to Cornell to do his graduate work with Bethe. Present also, among others, were Eugene Feenberg, Leo Szilard (a mysterious figure at the time, who seemed to know everyone and everything and seemed to have been everywhere; but then, all Hungarian physicists appeared somewhat superhuman to us), G. Placzek (a brilliant nuclear physicist who died prematurely), Henry Primakoff, William Rarita, Otto Halpern, Morton Hamermesh, Vernon Hughes, Norman Ramsey, Edward Teller, George Uhlenbeck, Franco Rasetti, Edoards Amaldi, and others. Fermi came near the end of the decade, as did Willis Lamb, and they both left indelible impressions on all of us. And so, with the emphasis in theoretical and experimental research shifting rapidly from atomic to nuclear physics, and with the outbreak of the Second World War, this extraordinarily creative decade, marked by Rabi's great experimental accomplishments and by the very rapid rise of American physics, ended.

The weekly events that occurred in the theoretical seminar touched only lightly on the work that was progressing so rapidly in Rabi's molecular beam–magnetic resonance laboratory, but we knew when something important was happening from the excitement he then communicated to us. Always ready to talk about his work or about anything else, Rabi was ever accessible in his large eighth-floor office in Pupin, which he still occupies. His door was invitingly open, and if he was not already in a heated discussion with some colleague or student, one had only to

Preface

step across the threshold to learn about his latest discoveries. The history of those discoveries is beautifully presented in this volume by Sidney Millman who, with Victor Cohen, Polykarp Kusch, Jerrold Zacharias, and Jerry Kellog, helped Rabi build his laboratory into a temple of science where, as Feynman stated, experiments and measurements of "fantastic precision" were performed.

Lloyd Motz
Columbia University, 1977

PHYSICAL INTERPRETATION OF GENERAL-RELATIVISTIC THEORIES*

Peter G. Bergmann

Department of Physics
Syracuse University
Syracuse, New York 13210

Introduction

The physical interpretation of general-relativistic theories differs in some significant respects from the corresponding task in Galilei (or Poincaré) invariant theories. With the latter kind of theory, knowledge of the coordinate values assigned to world points in *any* inertial-Cartesian frame of reference uniquely describes their relationship to each other; for instance, for two world points it would describe whether they are spacelike, timelike, or lightlike relative to each other, and what is the spacetime interval separating them. In a general-relativistic theory coordinate values serve almost exclusively to tell world points apart and to give us topological information about them, and even this information is suspect on occasion, as we all know from the history of our understanding of the geometry of the Schwarzschild radius, which was ultimately achieved by Kruskal.[1]

That is why the usual question asked of a physical theory, what can I expect to happen when and where, faces conceptual difficulties in a physical theory in which there are no inertial (or other rigid) frames of reference. That is why the problem of physical interpretation of general-relativistic theories gives rise to special problems.

Nevertheless, the question must be posed. A physical theory remains an empty shell until we have found a reasonable physical interpretation. And in our time, when Einstein's original theory of 1916 is being compared with alternative theories, such as Weyl's,[2] the so-called Einstein-Cartan theory incorporating torsion,[3] and the scalar-tensor theory,[4-8] it becomes particularly important to discuss to which extent these several theories are truly different. I shall endeavor to show that some of the differences between them are apparent, rather than substantive.

This paper will be devoted to the classical aspects of the question. The quantization of general-relativistic theories presents serious difficulties. Aside from their intrinsic nonlinearity, general-relativistic theories generally have vanishing Hamiltonian generators. Hence the "energy spectrum," which in other theories serves to define the vacuum state of the field, has no unambiguous meaning in general-relativistic theories. By the same token, the definitions of creation and annihilation operators, and with them the construction of a Fock space, are far

---

*This research was supported by National Science Foundation Grant PHY74-15246. This paper was originally presented at the 4th Soviet Gravitational Conference at the Belorussian University, Minsk, USSR, July 1–4, 1976.

from straightforward. It seems likely that more than one quantum theory correspond to the same classical field theory; without detracting from the importance of eventual quantization, I shall confine the discussion that follows to the classical problems.

Closed and Open Field Theories

In discussing the general theory of relativity, Einstein used to describe the two terms of the field equations,

$$G^{\mu\nu} + 8\pi\kappa P^{\mu\nu} = 0, \tag{1}$$

as the first consisting of marble, the second of plaster of Paris. Whereas the first term by itself describes in exemplary fashion the laws of the gravitational field alone, the second term represents all the (unspecified) contributions of other fields and of distributions of ponderable matter, to the sources of gravitation. A great deal of work has been done to explore the dynamics of the gravitational field outside its sources; more recently, Penrose, Hawking, and others have shown that even unspecified sources contribute essentially to the properties of the metric field if they satisfy certain requirements of nonnegativity of the nongravitational energy density.

I shall refer to a field theory as *closed* if all the fields of the theory are introduced explicitly. With this terminology the vacuum field equations of general relativity, $G^{\mu\nu} = 0$, represent a closed theory, but so do the Einstein-Maxwell field equations.

$$G^{\mu\nu} + \tfrac{1}{2}\kappa(g^{\mu\nu}\varphi^{\rho\sigma}\varphi_{\rho\sigma} - 4\varphi^{\mu\rho}\varphi^{\nu}{}_{\rho}) = 0, \varphi^{\mu\rho};{}_{\rho} = 0. \tag{2}$$

An *open* theory is one that permits the occurrence of extraneous sources, which might be visualized either as continuous or as isolated compact distributions. The latter type leads to ponderomotive laws.[9] In any event, source terms are subject to some restrictions, which are in effect integrability conditions for the existence of solutions of the field equations.

Closed and open field theories have different action integrals. Of course, either kind of Lagrangian will involve all those field variables that are to satisfy definite dynamical laws. In a closed theory these are indeed the only variables that will occur as arguments of the Lagrangian. In an open theory the Lagrangian will also contain terms with additional arguments, variables that are not to be considered as dynamical in the variational principle. These terms are referred to as matter terms, or *source terms*. Their presence is a formal admission that the theory is considered but an incomplete description of nature.

The integrability conditions on the sources mentioned above grow out of the invariance properties of the action integral of the closed theory that results if the source terms in the Lagrangian of the open theory are set to be zero. Because of the presence of arbitrary functions of space and time in the curvilinear coordinate transformations that leave the form of a general-relativistic theory unchanged, one can construct a set of four differential identities between the field equations resulting from the variation of the action integral, the generalization of the contracted

Bianchi identities. The right-hand sides of the field equations of the open theory, the source terms, must satisfy the same differential relations as the left-hand sides do identically if there are to be solutions.

Obviously, the source terms of an open-theory Lagrangian must depend not only on the "external" variables, the ones exempted from variation, but also on the dynamical variables proper; otherwise they would not contribute to the field equations. For instance, the (variational) derivatives of the matter terms with respect to the components of the metric tensor are interpreted as the components of the energy-stress tensor density, and the derivatives with respect to the electromagnetic potentials as the charge-current density. Brans and Dicke[4-8] have stipulated (in one formulation of their theory) that the matter terms contain no reference to the scalar variable, in other words, that the scalar source strength be zero (in that version), or algebraically related to the energy-stress density (in another formulation).

Observables

In a general-relativistic theory there is no natural way of separating the evolution in time from all the other coordinate transformations that form the invariance group. Hence the coordinate invariants are necessarily constants of the motion. It is these same variables that generate infinitesimal canonical transformations (in a Hamiltonian formulation) consistent with the preservation of the constraints, which in turn are nothing but those of the field equations free of highest-order derivatives off the local Cauchy hypersurface.

Only invariants can be determined entirely by Cauchy data in any theory in which the invariance group involves arbitrary functions that enter into the values of all noninvariant dynamical variables off the Cauchy hypersurface. Presumably, any act of physical measurement must deal with variables unaffected by coordinate transformations, as coordinates are not attributes of the real world but devices introduced for describing this real world conveniently. Hence, invariants are the only quantities that are observable in the ordinary sense of that word. Observables, then, in a general-relativistic theory are invariants, they are constants of the motion, they are the only variables that propagate from Cauchy data, and they are, in Dirac's terminology, the only first-class variables. Fundamentally, two physical theories should be considered equivalent to each other if their observables are in a one-to-one relationship to each other.

Unfortunately, this assertion is so simplistic as to be not only useless, but probably misleading as well. We know very little about observables, so little that the only known procedure for constructing a complete set yields a highly redundant set.[10] For the bread-and-butter analysis of alternative general-relativistic theories other methods must be invoked.

Actual observations and measurements of gravitational interactions have been of two kinds, laboratory experiments and astronomical measurements. Both have been in settings involving very weak gravitational fields. One may, of course, ask what is a weak field or, more precisely, compared with what is a gravitational field weak? All of the following criteria are equivalent to each other: (a) the local acceleration times the distance from the attracting center must be small compared to c^2; (b) the local curvature (in units of length) must be small compared to the square of the distance from the attracting center; (c) the kinetic energy that a test body must have to escape to infinity must be a small fraction of its rest energy; and (d) there exists a coordinate system that is Lorentz at infinity in which the (dimensionless) components of the metric tensor deviate locally but slightly ($\ll 1$) from their Lorentz-Minkowski values. None of these criteria is strictly local. Each formulation makes reference either to the distance from the attracting center or to the relationship to infinity. A strictly local criterion would have to involve at least first-order derivatives of the curvature, that is to say, third-order derivatives of the gravitational potentials, and would be very difficult to interpret in an intuitively satisfying manner.

If the field vanishes at infinity, it is possible to introduce asymptotic observables, variables that are invariant with respect to "localized" coordinate transformations without being constants of the motion. That is because in such a field the (pseudo-) group of curvilinear coordinate transformations may be decomposed into the semidirect product of transformations that are asymptotically (at infinity) identity, the normal subgroup, and of Poincaré or Bondi-Metzner-Sachs-type transformations, the factor group. There is, then, the possibility for the existence of variables that are invariant under local curvilinear transformations but sensitive to transformations and mappings that represent in some way the isometries of boundary conditions imposed at infinity. The local curvilinear transformations resemble non-Abelian gauge transformations, and especially in the case of weak fields the analogy can be carried quite far. Presumably, statements made about problems in celestial mechanics should be analyzed along these lines, but to my knowledge that analysis has not as yet been carried out in a comprehensive manner. In what follows I shall pursue a somewhat different line of approach.

Equivalent and Inequivalent Fields

Given any field theory, we may begin its analysis by decomposing all the geometrical and other building blocks into the irreducible representations of the theories' invariance groups. For general-relativistic theories the invariance group includes curvilinear coordinate transformations or, equivalently, the diffeomorphic mappings of space-time on itself. One can cast such a theory into the form of a fiber-bundle formalism, but I hesitate to do so, because the additional invariance seemingly introduced by the structure group of the fiber may do no more than compensate for the additional and redundant field variables. For instance, in conventional general relativity, if we choose to represent the metric tensor, which has but ten components, by tetrads, with 16 components, then the local Lorentz transformations, a six-parametric Lie group, merely serve to render the extra six components innocuous, without changing in any way the basic non-Lie invariance of the theory with respect to diffeomorphic space-time mappings. If instead of an orthonormal base at each world point one were to introduce arbitrary vectors, and with them a local metric, then the total number of components would increase to 26, the structure group GL(4) would be 16-parametric, and we should still have the very same theory. This is, of course, not to deny that there may be theories for which the bundle formalism is a necessity, and other theories, and some kinds of analyses, for which it is of distinct advantage to use such a formalism.

The metric tensor is not an irreducible representation of the local group GL(4), and the metric tensor field is not an irreducible representation of the group of space-time diffeomorphisms. Its irreducible parts are the conformal metric, a tensor density,

$$\tilde{g}_{\mu\nu} = |g|^{-1/4} g_{\mu\nu}, \qquad |\tilde{g}| \equiv -1, \qquad (3)$$

with but nine algebraically independent components, and the determinant of the metric, $|g|$, having a single component. The conformal metric is actually a faithful representation of SL(4), not of GL(4), whereas the metric determinant is a faithful representation of $R = GL(4)/SL(4)$.

In Riemannian geometry the components of the affine connection, the Christoffel symbols, do not represent an additional element, as they are obtained from the metric by purely local procedures. In a conformal geometry, which contains only the conformal part of the metric but not its determinant, the four components of the contracted affine connection represent a nonlinear representation of GL(4) that is independent of the metric. In a geometry with torsion the components of the torsion are not obtainable from the metric. Of the total of its 24 components, four can be split off by contraction on two indices. Finally, in a scalar-tensor theory the scalar obviously is a separate and irreducible degree of freedom of the field.

To summarize, without as yet having considered the field equations, theories based on Weyl-type geometry have different geometric building blocks from theories based on Riemannian geometry. Both scalar-tensor theories and theories with torsion contain elements that are independent of, and additional to, the metric; so does, of course, Einstein-Maxwell theory.

A given geometry does not by itself predetermine the field equations or the Lagrangian. If we concentrate our attention on theories with an action integral, the choice of Lagrangian is an important next stop. This choice is, of course, circumscribed by the requirement that the Lagrangian be a scalar density of weight one. It is also usual to select Lagrangians so as to avoid in the field equations derivatives of the field variables of unnecessarily high orders. This criterion is, however, somewhat ambiguous. For instance, the lowest-order Lagrangians that can be constructed within the framework of Weyl's geometry are quadratic in the components of the curvature tensor, and the resulting field equations involve fourth-order derivatives of the conformal metric. But if the components of the curvature tensor are introduced as field variables in their own right, then the field equations are reduced to second order, though the number of field variables and the number of field equations are increased by this maneuver.

For my purposes here I wish to consider the possibility that apparently diverse Lagrangians may result in equivalent theories. An early discovery is connected with the name of Palatini. If in the standard Lagrangian of general relativity the components of the affine connection are treated as variables to be varied independently of the metric, their variation leads to 40 field equations that reduce them to Christoffel symbols. That is to say, two apparently quite different variational problems with different numbers of field variables lead to equivalent systems of field equations. As long as we consider the closed version of the theory, the same holds even if we admit the possibility of torsion. That is to say, variation with respect to separate symmetric and antisymmetric components of the affine connection results in the former being equal to Christoffel symbols, and the latter being zero.

Next we come to the corresponding open theories. In the original open theory Einstein considered the components of the metric field to be the only independent field variables; accordingly in that theory the source is a symmetric tensor of rank two, usually known as the energy-stress tensor. It is assumed to represent the energy density, the linear momentum density, and the stress of all the nongravitational fields. In the presence of a gravitational field this tensor does not satisfy a conservation law (equation of continuity) but only a *covariant* divergence relation. Einstein already constructed the additional, purely gravitational terms that must be added to the source tensor to yield a conserved complex. This complex is not a geometric object; but such an object cannot exist in any case because of the principle of equivalence, which postulates that at any given world point it is possible to introduce coordinates in which all components of the gravitational field strength, the Christoffel symbols, vanish.

If we consider the affine connection as a set of field variables that are *a priori* independent of the metric, we can postulate additional source terms for these new field variables. In the versions of that theory proposed to date[3] the new field equations relate the source terms to the undifferentiated components of the affine connection so that its deviation from Christoffel symbols vanishes outside the sources. By a rather trivial manipulation we may arrange the field equations into Einstein's equations with an appropriately defined symmetric energy-stress tensor on the right, and into additional equations that relate the extra field components algebraically to their sources. That such an arrangement is at all possible proves that the supposedly new open theory does not differ from Einstein's original (open) theory, though new ways of looking at the original theory may have heuristic value.

Truly new theories might be obtained if the new third-rank tensors appear in the field equations in differentiated form, so that the field represented by them will propagate beyond the confines of the corresponding sources. For this to happen the Lagrangian would have to be of higher differential order than those considered heretofore; to take such ventures seriously one would have to have some strong physical motivation, as the formal or mathematical motivations appear rather weak.

In concluding this section, I should like to summarize a similar analysis performed on scalar-tensor theories.[11] By appropriate construction of the independent field variables, the Lagrangian can be given a form in which one term is identical with the gravitational term of conventional general relativity, and another term essentially quadratic in the first derivatives of the scalar field. This latter term leads to the possibility of a wave equation for the new scalar field. It is possible, further, to provide for nonremovable interaction between the electromagnetic and scalar fields, so that Maxwell's equations contain a nontrivial variable "dielectric constant of the vacuum." The ponderomotive equations of that theory involve a "scalar source strength" of particles, which does not obey any conservation law and thus permits test particles exposed to incident scalar fields to behave erratically. This theory clearly is not equivalent to one without a scalar field. The preponderance of the experimental evidence available to date appears to be against the scalar field.[12]

Ponderomotive Laws

Every general-relativistic theory with an action integral incorporates ponderomotive laws, in the following sense. If the sources of the (open-theory) field are confined to spatially compact regions, then these separable regions—bodies—are subject to laws of motion. These laws deal with two- and three-dimensional surface integrals whose domains of integration lie outside the sources but surround them.

If certain assumptions are made concerning the existence of a limit that might be interpreted as representing "test particles," then there result ordinary differential equations for these particles, such as the geodesic law in general relativity, or the Lorentz force for electrically charged test particles. Einstein never felt comfortable with the geodesic law as part of the fundamental postulates of the theory, because the geodesic law requires the separation of the total field into an "incident field" and a "self-field," a separation that is alien to the spirit of field theory, and doubly alien to a field theory that is intrinsically nonlinear. That is why he expended so much effort into what we all know today as the EIH theory. From the present status of our understanding of the ponderomotive laws we can more fully appreciate Einstein's misgivings. The only safe formulation of the geodesic law of motion is: "Test particles move along geodesics, except for those that move differently."

When we look at ponderomotive laws in alternative theories, we had better start from the integral formulations. In the scalar-tensor theory it is somewhat arbitrary which part of the field is to be considered the metric, and hence, which curves are geodesics. Similarly, in so-called Einstein-Cartan theories the identification of the affine connection, and with it the geodesics, are arbitrary. In all these cases the integral formulations are at least straightforward consequences of the postulated field equations.

Unfortunately, the integral form of the ponderomotive laws will not lead to unambiguous ordinary differential equations for the motions of test particles. The ambiguity is simply removed one step, and associated with the definition of test particles that are to have mass, but not spin, and possess zero electric charge, zero scalar source strength, and zero higher mass multipole moments. The equivalence, or inequivalence, of diverse general-relativistic physical theories can be established effectively by an analysis of their (closed) Lagrangians; consideration of their ponderomotive laws will point to possible kinematic effects, and thus facilitate the physical interpretation of the new degrees of freedom introduced.

It is remarkable how resistant Einstein's original theory is to most attempts at modification. Weyl's theory leads to a truly different theory, to be sure, but none of the others. The scalar-tensor theories essentially postulate the existence of a third classical (boson) field, of zero spin, in addition to the gravitational and electromagnetic fields, without modifying the gravitational field as such. The Einstein-Cartan theories are suggestive of couplings between the gravitational and the other fields that might not be thought of without the stimulus of the geometric pictures evoked by the notions of torsion; but these couplings can be accomodated within the framework of the original theory. Quite obviously, the decomposition of all fields into their irreducible components is a powerful formal device for analysis.

This is not to say that general relativity is the ultimate theory of the gravitational field, incapable of any further change. Throughout his later life Einstein was engaged in the search for a new and better "unitary" field theory, that might encompass all the fields met with in nature. He was working on the asymmetric theory at the time of his death. I believe that no physical theory ever will be "the" ultimate theory. But modifications of gravitational theory will have to be much more thoroughgoing than slight generalizations of general relativity to break truly new ground. They will require strong physical motivation, and they need not be classical field theories at all.

Summary

A set of formal field equations dealing with the dependence of the field(s) on the space-time coordinates (having the same form in all curvilinear coordinate systems) becomes a physical theory only if the fields, and their interrelations, are given a physical interpretation. That is to say, the totality of the structures occurring in the mathematical formulation must be identified with physical objects, and with the laws postulated for them.

The physical interpretation of general-relativistic theories presents a particular challenge, because one cannot determine by local means a global space-time framework in which measurements might be performed. A general-relativistic theory does not imply *a priori* the possibility of idealized clocks indicating proper time along arbitrary trajectories, nor even along geodesics in the presence of tidal forces. Nor will test particles travel on uniquely determined trajectories. The concept of observable is useful, but not all-powerful. These ideas have been illustrated by a comparison of Weyl's geometry, the Einstein-Cartan, and the Brans-Dicke theories with conventional general relativity.

References

1. KRUSKAL, M. D. 1960. Phys. Rev. **119**: 1743.
2. WEYL, H. 1952. Space-Time, Matter (Raum, Zeit, Materie). Dover. New York, N.Y.
3. TRAUTMAN, A. 1973. Ist. Naz. Alta Matematica Symposia Mathematica **12**: 139. This paper contains an extensive bibliography on the so-called Einstein-Cartan Theory.
4. BERGMANN, P. G. 1948. Ann. Math. (Princeton) **49**: 255.
5. JORDAN, P. 1956. Helv. Phys. Acta (Suppl. IV): 157.
6. THIRY, Y. R. 1951. J. Math. Pure Appl. **30**: 275.
7. BRANS, C. & R. H. DICKE. 1961. Phys. Rev. **124**: 925.

8. DICKE, R. H. 1964. *In* Gravitation and Relativity. Chiv & Hoffman, Eds. Benjamin. New York, N.Y.
9. An extensive bibliography of papers on ponderomotive laws will be found in INFELD, L. & J. PLEBAŃSKI. 1960. Motion and Relativity. Pergamon Press and Państwowe Wydawnictwo Naukowe. Warszawa, Poland.
10. BERGMANN, P. & A. KOMAR. 1960. Phys. Rev. Lett. **4**: 432.
11. BERGMANN, P. 1968. Int. J. Theoret. Phys. **1**: 25.
12. WILLIAMS, J. G., R. H. DICKE, *et al.* 1976. Phys. Rev. Lett. **36**: 551.

THE HYPERFINE STRUCTURE OF THE GROUND STATE OF ORTHOHELIUM IN THE NONRELATIVISTIC APPROXIMATION*

G. Breit, V. G. Kaveeshwar, and R. P. Singh

Department of Physics
State University of New York
Buffalo, New York 14214

Introduction

After the success of the Bohr-Sommerfeld Quantum Mechanics in various applications primarily, but not exclusively, to atomic spectra it was natural for physicists to inquire into the nature of the center of force having the properties of charge and mass available through the general progress of physics and chemistry by then. A few names and associated subject matter will make the meaning clear [Aston-isotopes, the Braggs-crystal structure, Mme Curie, Ernest Rutherford, existence of atomic nuclei of various types, and possibility of the transmutation of the elements]. The desirability of such an inquiry was well realized at the time by, e.g., the late Ernest O. Lawrence among experimentalists in the U.S., the Cambridge school in England, and W. Heisenberg and W. Pauli among continental theoretical physicists in Europe. One of the writers of this paper, M. A. Tuve, and a few other mavericks together with E. O. Lawrence were of the same opinion rather actively in some cases and periods.

It thus became natural for physicists as a group to pay more attention to properties of atomic nuclei *other* than their charge (electric, mass, etc.). The hyperfine structure of spectral lines with its dependence on nuclear magnetic moments was thus "in the air" as Werner Heisenberg would say. A paper on this subject by one of the authors of this present contribution to the Rabi Festschrift was published some time ago. One of the difficulties is that the very high precision of the experimental data attained at Yale under the leadership of Vernon W. Hughes makes such equations as are available questionable regarding sufficiency of their accuracy. It may be well to remind the reader at this point that *although* the *equations* of the problem are *separable,* the boundary condition

$$\psi = 0 \quad (r_1 = r_2)$$

makes *direct use of separability impossible.* A related difficulty is that at small distances between the two electrons the physically finite size of these particles matters from a strictly rigorous standpoint. As a matter of fact the equations of the problem are not known in an exact and well-defined sense. But it is generally recognized that progress is stimulated and furthered in such cases by first using an approximate representation of the actual situation obtained by supposing that

---

*This work was supported by the United States Energy Research and Development Administration.

$1/r_1$, $1/r_2$ and $1/r_{12}$ may be replaced by finite functions of each of the three quantities r_1, r_2, and r_{12}, respectively.

The motion of two electrons about an infinitely heavy nucleus may be treated by introducing the distances r_i and the angle between the direction in which the electrons lie. The two-particle wave function will be written in either of the three forms as below:

$$\psi^A(x_1,y_1,z_1,x_2,y_2,z_2) = \psi^B(r_1,r_2;\cos\theta) = \psi^C(r_1,r_2,r_{12})\ldots, \qquad (1)$$

and for brevity the superscripts A, B, C will be omitted when no misunderstanding arises by doing so. In this sense one may write:

$$\psi = \psi^A(x_1,y_1,z_1;x_2,y_2,z_2) = \psi^B(r_1,r_2;\cos\theta) = \psi^C(r_1,r_2;r_{12})\ldots \qquad (1')$$

where

$$r_i = (x_i^2 + y_i^2 + z_i^2)^{1/2}; i = (1,2),$$

$$\cos\theta = (\mathbf{r}_1\mathbf{r}_2/r_1r_2); r_{12} = (r_1^2 + r_2^2 - 2r_1r_2\cos\theta)^{1/2}$$

As is well known, the usual introduction of relative and center of mass coordinates leads in the case of one-electron spectra to the employment of an effective mass

$$\mu = mM/(m + M).$$

The difference between μ and m is responsible for the spectroscopic isotope shift in one electron spectra (Niels Bohr) and is thus responsible for the isotope shift observed in the sun's spectrum. It has led to the name Helium for the element. The main effect is the recoil of the nucleus caused by the moving electron.

In the case of the two-electron spectrum of the ^{3}He isotope a convenient first approximation is obtained by using an effective electron mass

$$\mu = m^3\text{He}/(m + {}^3\text{He}) \qquad (2)$$

in place of m and by temporarily forgetting about the complication of the nucleus recoiling simultaneously under the action of both electrons rather than just one. The necessity of doing so has been pointed out by D. S. Hughes and C. Eckart. The difference caused by this improvement is small but not negligible in view of the high accuracy desired. It is usually referred to as the Hughes-Eckart correction.

The Equations to Be Solved

The equation for relative motion will be written in terms of a function φ which when expressed in terms of the original ψ appears as below:

$$\varphi(r_1,r_2,\theta) \equiv r_1r_2\psi = \sum_0^\infty (2l + 1)^{1/2}\Phi_l(r_1,r_2)P_l(\cos\theta).$$

The Φ_l satsify a set of coupled partial differential equations

$$\left(\mathcal{L}_l + \frac{\lambda}{4}\right)\Phi_l = \mathcal{R}_l,$$

where

$$\mathcal{L}_l = \frac{\partial^2}{\partial r_1^2} + \frac{\partial^2}{\partial r_2^2} - l(l+1)\left(\frac{1}{r_1^2} + \frac{1}{r_2^2}\right) + \frac{1}{r_1} + \frac{1}{r_2} - \frac{M_{ll}}{2Z},$$

$$\mathcal{R}_l = \sum_m{}' M_{lm}\Phi_m/(2Z),$$

$$M_{lm} = (2l+1)^{1/2}(2m+1)^{1/2} \sum_0^{l+m} \frac{r_<^n}{r_>^{n+1}} \int_0^\pi P_l P_m P_n \sin\theta\, d\theta,$$

and $(r_>, r_<) = (r_1, r_2)(r_1 > r_2)$.

The antisymmetry to particle interchange requires

$$\Phi_l(x, y) = -\Phi_l(y, x).$$

It thus suffices to obtain Φ_l in the octant $x > y$.

The Panel Procedure

It being practically hopeless to solve the problem by usual mathematical methods, it was treated numerically. The octant $x > y$ was subdivided into a number of rectangles with sides parallel to the x and y axes. The procedure has some similarity to that used in Southwell's relaxation method but differs from it in that successive approximations can be well outlined, and are therefore especially well adapted for programming (alias coding) a calculation on a large digital computing machine.

Through *most* of the octant characterized by $x > y$ it is covered by regions that will be referred to here as *patches*. Within each patch the parameterization of the panel function is the same and it is the same for all l that are taken into account for that patch. As l increases, the term in $l(l+1)$ makes Φ_l small and for high enough l the corresponding Φ_l become negligible. (In the classical mechanics analogy the effect of these terms, being much like those of potential barriers, makes the corresponding $\Phi_l = 0$. Quantum mechanically there is some leakage through the barriers but in many cases the centrifugal effect overpowers the attractive effects caused by other parts of the $\mathcal{L}_l$.)

Each patch consists of cells. At the boundary of each cell the Φ_l is made to be continuous, but its derivative in a direction normal to the boundary (normal derivative for short) need not be continuous.

The Parameter $1 + \epsilon$ and the Virial Theorem Check

To the writers' knowledge the quantity $1 + \epsilon$ has been introduced into the subject of two-electron spectra by Breit and Doermann (BD).[†] In the notation of the BD paper and in angular momentum units $\hbar$ it has been shown (BD Eq. 36″) that the sublevels have the energies

[†] BREIT, G. & F. W. DOERMANN. 1930. Phys. Rev. **36**: 1732.

$$\mu\mu_0(1+\epsilon)\psi^2(0)\left(1,-\frac{1}{k},-\frac{k+1}{k}\right).$$

Here μ is the magnetic moment of the nucleus expressed in units

$$\mu_0 = e\hbar/2mc > 0.$$

In this formula e is positive *by definition* and μ_0 is the Bohr magneton; $k\hbar$ is the maximum angular momentum of an isolated nucleus, such as would be measured in a Stern-Gerlach experiment. The factor $1 + \epsilon$ is seen to correct for the presence of the second s-electron (i.e., the valence electron).

The term Virial Theorem is used here so as to correspond to the classical (Newtonian) mechanics in such a way that $-2KE/PE = 1 + x$. Here KE and PE are, respectively, abbreviations for kinetic and potential energy. The signs are chosen such that $x > 0$. A small positive x thus indicates that there is no objection to considering the calculations as "good" from the viewpoint of the Virial Theorem.

Some Numerical Results

	Late 1967	June 1972	Row Number
λ	$-1.087,614,679$	$-1.087,614,684,8$	1
$1 + \epsilon$	$1.037,004,0$	$1.037,004,544$	2
VTC	$1 + 2.0 \times 10^{-7}$	$1 + 1.0 \times 10^{-8}$	3
$\Delta^{\text{Hu. Ec.}}(\lambda)$	6.81×10^{-7}	6.76×10^{-7}	4
$\Delta^{\text{Hu. Ec.}}(1 + \epsilon)$	3.70×10^{-7}	3.49×10^{-7}	5

Notation

p, proton
n, neutron
ψ, wave function for relative motion of proton and neutron
r_1, r_2, distances of the two particles from their common center of mass
r_{12}, distance between the two particles
θ, angle between the two particles choosing order (1, 2) between them arbitrarily (with neglect of the Hughes-Eckart correction)
$\mu = mM/(m + M)$
$\varphi(r_1, r_2, \theta) = r_1 r_2 \psi$
P_l, Legendre Polynomial of order l $\left[P_0 = 1,\ P_n(x) = \frac{1}{2^n n!}\frac{d(x^2-1)^n}{dx^n} \cdots (n > 1)\right]$
$(x, y) = (r_1, r_2)\ (r_1 > r_2)$
e, charge of electron
m, mass of electron
c, velocity of light
h, Planck's constant
$\mu_0 = e\hbar/2mc > 0$
$\hbar = h/2\pi$
Z, nuclear charge
$\lambda = E/(Ry\, hc\, Z^2)$, Ry = Rydberg in cm^{-1}

THE PREDICTED INFRARED SPECTRUM OF HeH⁺ AND ITS POSSIBLE ASTROPHYSICAL IMPORTANCE*

I. Dabrowski and G. Herzberg

National Research Council of Canada
Ottawa, Canada

The helium hydride ion is a well-known ion in the mass spectrum of discharges through mixtures of helium and hydrogen, but no optical spectrum of it has yet been reported. In view of the high abundance of hydrogen and helium in the universe, and particularly the high abundance of H⁺ in H II regions, it appears likely that HeH⁺ is formed in some regions of stellar atmospheres or interstellar clouds. If its spectrum were known it would help in the detection of this molecular ion. The object of this paper is to report on the predicted infrared spectrum of HeH⁺.

Ab initio calculations of the electronic ground state of HeH⁺ ($X\ ^1\Sigma^+$) have been carried out by various investigators, starting with the early work of Beach (1936)[1] and culminating in the precise potential function of Wolniewicz (1965).[2] Very recently Kolos (1976)[3] and Kolos and Peek (1976)[4] have extended Wolniewicz' calculations to larger internuclear distances yielding a potential function that has a reliability comparable to that of the Born-Oppenheimer function for H_2 (Kolos and Wolniewicz[5]). It is therefore possible to predict, within an accuracy of a few cm⁻¹, the relative positions of all the vibrational and rotational levels of HeH⁺ in the electronic ground state.

The energies of electronically excited states of HeH⁺ have also been calculated[3,6,7] but with less accuracy. The first excited singlet state $A\ ^1\Sigma^+$ lies at 103243 cm⁻¹ above $X\ ^1\Sigma^+$. It has a very shallow potential function with a minimum at 2.9 Å, i.e., $r'_e \sim 3.8\ r''_e$. As a result, in absorption, only transitions to the continuum of $A\ ^1\Sigma^+$ (below 966 Å) have appreciable intensities, while in emission only transitions to the highest vibrational levels of the ground state are Franck–Condon allowed. These transitions will be difficult to observe in the laboratory and are unlikely to play a role in astronomical spectra.

The HeH⁺ ion does have a fairly large (r-dependent) dipole moment in the ground state, and as a result an ordinary rotation-vibration spectrum is expected in the infrared and may well play a role in astronomical spectra. The wavenumbers of these transitions are readily calculated from the potential function of Kolos and of Kolos and Peek. The prediction of the intensities requires in addition to the wave functions a knowledge of the electric dipole moment as a function of internuclear distance (r). Peyerimhoff[8] and Michels[6] have calculated from *ab initio* electronic wave functions the position of the center of the negative charges as a function of r. From their data the dipole moment referred to the center of mass was evaluated and is shown as a function of r (in Å) in FIGURE 1. A linear function

$$\mu = \mu_0 + \mu_1(r - r_e) = 1.66 + 4.29\,(r - r_e)\ \text{D} \qquad (1)$$

*Submitted September 8, 1976.

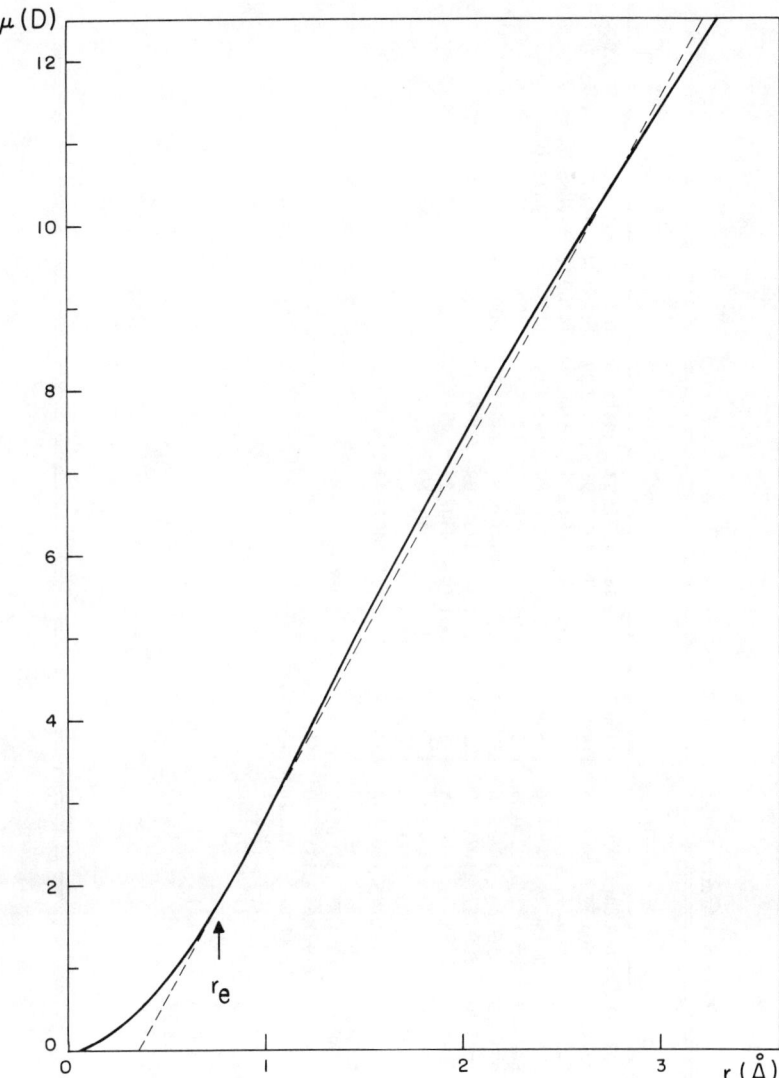

FIGURE 1. Dipole moment function of HeH⁺ obtained from the centers of the negative charges given by Peyerimhoff[8] and Michels.[6]

is a good approximation to the dipole moment function above $r = 0.4$ Å ($r_e =$ equilibrium internuclear distance).

From the potential function, by use of a program supplied by R. J. LeRoy, the bound energy levels given in TABLE 1 have been obtained. They agree within 0.5 cm^{-1} with those independently calculated by Kolos (unpublished). From these energy levels the positions of the lines in all possible bands in the infrared have

TABLE 1
Bound Energy Levels of HeH$^+$ in the $X\,^1\Sigma^+$ Ground State

J\v	0	1	2	3	4	5	6	7	8	9	10	11
0	0.00	2912.42	5517.88	7814.84	9798.35	11460.21	12789.46	13775.13	14415.18	14741.21	14855.37	14879.24
1	67.08	2974.06	5574.03	7865.39	9843.08	11498.81	12821.47	13799.98	14432.29	14750.74	14859.37	14880.08
2	200.85	3096.95	5685.96	7966.11	9932.17	11575.62	12885.08	13849.20	14465.95	14769.19	14866.73	
3	400.54	3280.36	5852.93	8116.27	10064.86	11689.85	12979.42	13921.80	14514.99	14795.23	14876.05	
4	665.00	3523.14	6073.82	8314.75	10240.03	11840.31	13103.18	14016.29	14577.59	14826.81		
5	992.72	3823.83	6347.18	8560.09	10456.16	12025.41	13254.63	14130.62	14651.21	14860.81		
6	1381.85	4180.60	6671.20	8850.47	10711.39	12243.17	13431.52	14262.09	14732.35			
7	1830.19	4591.31	7043.74	9183.74	11003.48	12491.16	13631.10	14407.29	14816.20			
8	2335.25	5053.49	7462.36	9557.39	11329.82	12766.52	13850.02	14561.80				
9	2894.24	5564.40	7924.31	9968.62	11687.43	13065.94	14084.24	14719.98				
10	3504.13	6121.02	8426.55	10414.29	12072.93	13385.54	14328.82	14873.46				
11	4161.62	6720.10	8965.77	10890.94	12482.53	13720.81	14577.54					
12	4863.22	7358.13	9538.38	11394.77	12911.97	14066.44	14821.96					
13	5605.23	8031.39	10140.56	11921.64	13356.40	14415.93						
14	6383.78	8735.96	10768.17	12466.97	13810.24	14760.83						
15	7194.84	9467.70	11416.80	13025.70	14266.86							
16	8034.21	10222.26	12081.69	13592.11	14717.86							
17	8897.56	10995.06	12757.64	14159.52								
18	9780.42	11781.25	13438.90	14719.71								
19	10678.14	12575.62	14118.88									
20	11585.87	13372.50	14789.66									
21	12498.51	14165.53										
22	13410.62											
23	14316.23											

been evaluated. From the vibrational wave functions (allowing for their dependence on J) and the linear approximation to the dipole moment function the line strengths have been obtained. The results, for the 1–0, 2–0, 2–1, 3–0, 3–1, 3–2 bands, are presented in TABLE 2. In FIGURES 2a and b the structure of the fundamental (1–0 band) is shown schematically at two different temperatures. The con-

TABLE 2
PREDICTED WAVE NUMBERS AND INTENSITY FACTORS FOR THE
1–0, 2–1, 3–2, 2–0, 3–1, AND 3–0 BANDS OF HeH$^+$

	J	$R(J)$	I	$P(J)$	I		J	$R(J)$	I	$P(J)$	I
1–0	0	2974.1	0.12			3–0	0	7865.4	0.000		
	1	3029.9	0.25	2845.3	0.13		1	7899.0	0.001	7747.8	0.000
	2	3079.5	0.37	2773.2	0.25		2	7915.4	0.001	7664.5	0.000
	3	3122.6	0.49	2696.4	0.38		3	7914.2	0.002	7565.6	0.001
	4	3158.8	0.61	2615.4	0.52		4	7895.0	0.002	7451.3	0.001
	5	3187.9	0.74	2530.4	0.66		5	7857.8	0.003	7322.0	0.001
	6	3209.5	0.86	2442.0	0.81		6	7801.9	0.004	7178.2	0.001
	7	3223.3	0.98	2350.4	0.97		7	7727.2	0.005	7020.3	0.001
	8	3229.2	1.10	2256.1	1.13		8	7633.4	0.006	6848.5	0.001
	9	3226.8	1.22	2159.2	1.31		9	7520.0	0.008	6663.2	0.001
	10	3216.0	1.35	2060.3	1.51		10	7386.8	0.010	6464.5	0.001
2–0	0	5574.0	0.004			2–1	0	2661.6	0.26		
	1	5618.9	0.008	5450.8	0.003		1	2711.9	0.51	2543.82	0.27
	2	5652.1	0.013	5373.2	0.006		2	2756.0	0.76	2477.08	0.54
	3	5673.3	0.018	5285.4	0.008		3	2793.5	1.00	2405.60	0.82
	4	5682.2	0.024	5187.9	0.011		4	2824.0	1.24	2329.79	1.12
	5	5678.5	0.031	5081.1	0.012		5	2847.4	1.48	2249.99	1.44
	6	5661.9	0.039	4965.3	0.014		6	2863.1	1.71	2166.58	1.77
	7	5632.2	0.047	4841.0	0.016		7	2871.0	1.94	2079.89	2.13
	8	5589.1	0.057	4708.5	0.017		8	2870.8	2.17	1990.25	2.52
	9	5532.3	0.068	4568.1	0.019		9	2862.2	2.39	1897.96	2.94
	10	5461.6	0.079	4420.2	0.019		10	2844.8	2.60	1803.29	3.41
3–1	0	4953.0	0.012			3–2	0	2347.5	0.40		
	1	4992.0	0.025	4840.8	0.016		1	2392.1	0.78	2240.8	0.41
	2	5019.3	0.040	4768.4	0.020		2	2430.3	1.15	2179.4	0.85
	3	5034.4	0.057	4685.8	0.028		3	2461.8	1.51	2113.2	1.31
	4	5037.0	0.076	4593.1	0.036		4	2486.3	1.89	2042.4	1.80
	5	5026.6	0.097	4490.9	0.043		5	2503.3	2.23	1967.6	2.33
	6	5003.1	0.120	4379.5	0.049		6	2512.5	2.55	1888.9	2.93
	7	4966.1	0.146	4259.2	0.055		7	2513.6	2.87	1806.7	3.56
	8	4915.1	0.175	4130.2	0.060		8	2506.3	3.16	1721.4	4.24
	9	4849.9	0.207	3993.0	0.066		9	2490.0	3.44	1633.1	5.00
	10	4769.9	0.242	3847.6	0.071		10	2464.4	3.69	1542.1	5.86

ventional spectroscopic constants in the ground state of HeH$^+$ derived from the energy levels in TABLE 1 are listed in TABLE 3.

In addition to the bound levels (TABLE 1) there are also a number of quasibound (predissociating) levels that lie above the asymptote of the potential function but below the maxima of the various centrifugal barriers. As an illustration we present in FIGURE 3 the effective potential functions for $J = 0$ and $J = 23$. The positions of the $v = 0$ and $v = 1$ levels are indicated. Some of the quasibound

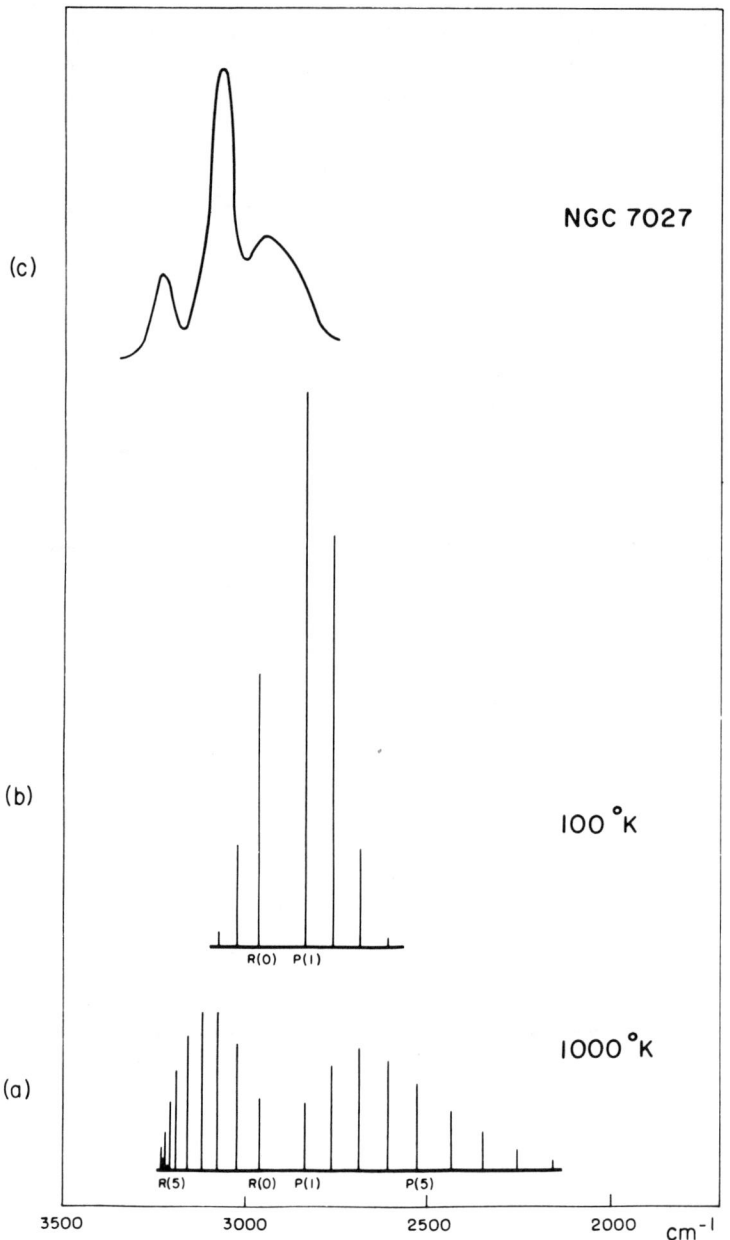

FIGURE 2. Predicted structure of fundamental of HeH$^+$ (a) at 1000°K (b) at 100°K, compared with (c) the observed spectrum of NGC 7027 after Merrill, Soifer, and Russell.[12]

TABLE 3
CONVENTIONAL MOLECULAR CONSTANTS IN THE GROUND STATE OF HeH$^+$

$B_e =$	34.887 cm^{-1}	$\omega_e x_e =$	157.71 cm^{-1}
$\alpha_e =$	2.636$_2$ cm^{-1}	$\omega_e y_e =$	0.454 cm^{-1}
$\gamma_e =$	-0.0305 cm^{-1}	$D_0^0 =$	14880.5 cm^{-1}
$D_e^{rot} =$	0.0161 cm^{-1}	$D_e =$	16455.6 cm^{-1}
$\omega_e =$	3228.4 cm^{-1}	$r_e =$	0.7743 Å

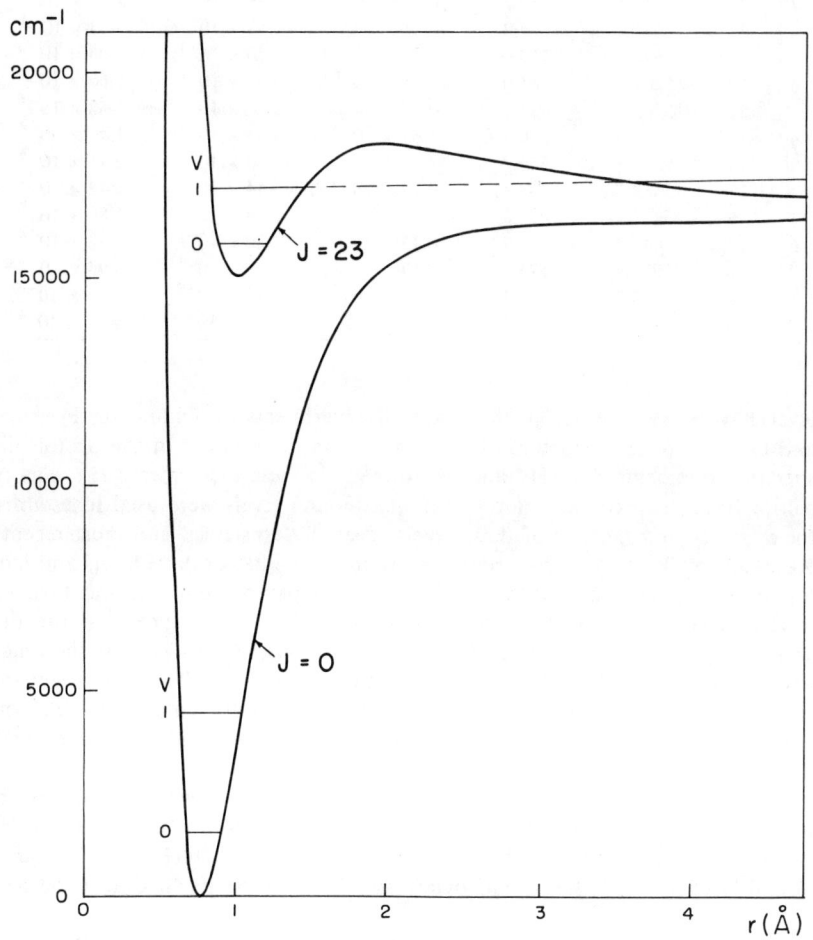

FIGURE 3. Effective potential functions of HeH$^+$ in the electronic ground state ($X^1\Sigma^+$) for $J = 0$ and $J = 23$.

TABLE 4
Energies (in cm^{-1}) Widths and Lifetimes of Quasibound Levels of HeH$^+$ in the $X^1\Sigma^+$ Ground State

v	J	$T_0(v, J)$	$T_0 - D_0^0$	$\Delta\nu$(cm^{-1})	τ_l(sec)	τ_r(sec)
0	24	15208.6	328.0	1.49×10^{-19}	3.55×10^7	—
	25	15751.2	870.7	4.11×10^{-7}	1.29×10^{-5}	—
	26	16917.0	2036.5	0.02	2.90×10^{-10}	—
	27	17699.0	2818.5	6.36	8.35×10^{-13}	—
1	22	14947.2	66.7	3.11×10^{-30}	1.71×10^{18}	3.38×10^{-4}
	23	15708.1	827.6	5.18×10^{-7}	1.02×10^{-5}	4.42×10^{-4}
	24	16433.6	1553.0	0.07	7.81×10^{-11}	6.25×10^{-4}
	25	17091.8	2211.2	19.8	2.69×10^{-13}	2.02×10^{-3}
2	21	15440.8	560.2	6.05×10^{-7}	8.78×10^{-6}	2.23×10^{-4}
	22	16055.4	1174.8	0.21	2.59×10^{-11}	3.33×10^{-4}
	23	16596.5	1716.0	40.8	1.30×10^{-13}	1.05×10^{-3}
3	19	15261.5	381.0	1.50×10^{-6}	3.55×10^{-6}	1.60×10^{-4}
	20	15765.5	885.0	0.73	7.25×10^{-12}	2.58×10^{-4}
4	17	15151.5	270.9	1.63×10^{-5}	3.26×10^{-7}	1.40×10^{-4}
	18	15544.7	664.2	3.32	1.60×10^{-12}	2.70×10^{-4}
5	15	15088.6	208.1	9.75×10^{-4}	5.44×10^{-9}	1.42×10^{-4}
	16	15372.4	491.8	15.5	3.44×10^{-13}	2.83×10^{-4}
6	13	15048.2	167.7	0.13	4.20×10^{-11}	1.52×10^{-4}
7	11	15005.9	125.3	5.17	1.30×10^{-12}	5.93×10^{-4}
8	8	14895.2	14.7	9.06×10^{-4}	5.86×10^{-9}	3.73×10^{-4}
9	6	14891.6	11.1	0.63	8.47×10^{-12}	4.72×10^{-4}

levels have been observed in the beautiful experiments of Schopman, Fournier, and Los[9] on the fine structure of the momentum distribution of the proton dissociation fragments of HeH$^+$ and its isotopes. In these experiments H$^+$ ions resulting from the predissociation of the quasibound levels were used to establish the existence and energies of these levels. Peek,[10] Bernstein,[11] and most recently Kolos and Peek,[4] have calculated the positions and widths of these levels and have found striking agreement with the results of Schopman, Fournier, and Los. We have independently calculated these levels using Kolos' new potential function and LeRoy's program. Our results for the positions, widths, and radiationless and radiative lifetimes of the levels are presented in TABLE 4. There are slight systematic differences in the energies from the values given by Kolos and Peek[4]: our values are up to 15 cm^{-1} higher. We have not been able to find the source of this discrepancy.*

In collisions between protons and He atoms, HeH$^+$ ions in the quasibound levels may be formed by inverse predissociation if the kinetic energy of He + H$^+$ is within the width of the levels. Following the formation of HeH$^+$ in such a quasibound level if may either predissociate again or during its lifetime undergo a

---

*If plotted as functions of J (after subtraction of $BJ(J + 1) - \bar{D}J^2(J + 1)^2$ with approximate B and D values), our energy values above the dissociation limit join smoothly on to those below it; the values of Kolos and Peek show a sudden jump of 15 cm^{-1} at the limit.

radiative (infrared) transition to a lower bound level. In the latter event a two-body radiative recombination of HeH⁺ results.

The infrared recombination spectrum is easily calculated from the energy levels (TABLES 1 & 4), the vibrational wave functions and the dipole moment function. In TABLE 5 the predicted wavenumbers and relative intensities of the recombination lines of HeH⁺ at 100°K and 1000°K are given. The intensities are

TABLE 5
PREDICTED INFRARED RECOMBINATION SPECTRUM OF HeH⁺ AT 100°K AND 1000°K

Transition $v'-v''$		ν	I_{100}	I_{1000}	Transition $v'-v''$		ν	I_{100}	I_{1000}
1–0	R(22)	2297.4	0	2.0	5–3	R(14)	2621.6	5.4	13.2
	R(23)	2117.4	0	0.3		R(15)	2346.8	0	1.5
2–0	R(20)	3854.9	0	4.8		P(16)	1496.5	0.5	1.2
	R(21)	3556.8	0	1.4		P(17)	1202.9	0	0.1
	P(22)	2030.2	0	0.1	5–4	R(14)	1278.4	0.8	1.9
	P(23)	1739.1	0	0.1		R(15)	1105.5	0	0.1
2–1	R(20)	2068.3	0	4.1		P(16)	370.8	0.1	0.2
	R(21)	1889.8	0	0.9	6–0	R(12)	10185.0	0.1	0.2
	P(22)	493.6	0	0.1	6–1	R(12)	7690.1	0.9	1.3
	P(23)	347.3	0	0.1	6–2	R(12)	5509.8	3.0	4.4
3–0	R(18)	5481.1	0.3	7.0	6–3	R(12)	3653.4	5.6	8.1
	R(19)	5087.4	0	1.6	6–4	R(12)	2136.2	5.5	7.8
3–1	R(18)	3480.3	0.7	17.4		P(14)	1238.0	0.2	0.2
	R(19)	3189.9	0	3.5	6–5	R(12)	981.8	2.4	3.5
	P(20)	1889.0	0	0.5		P(14)	287.4	0.1	0.2
3–2	R(18)	1822.6	0.3	6.9	7–1	R(10)	8884.8	0.2	0.2
	R(19)	1646.6	0	0.9	7–2	R(10)	6579.3	0.6	0.5
	P(20)	471.9	0	0.3	7–3	R(10)	4591.6	1.0	0.8
4–0	R(16)	7117.2	0.7	4.0		P(12)	3611.1	0.1	0.1
	R(17)	6647.1	0	0.9	7–4	R(10)	2932.9	0.8	0.7
4–1	R(16)	4929.2	2.8	15.6		P(12)	2093.9	0.1	0.1
	R(17)	4549.6	0	3.0	8–0	R(7)	13065.0	0.2	0
	P(18)	3370.2	0.1	0.4	8–1	R(7)	10303.9	1.1	0.2
4–2	R(16)	3059.8	3.5	19.8		P(9)	9330.8	0.1	0
	R(17)	2787.0	0	3.1	8–2	R(7)	7851.5	3.2	0.6
	P(18)	1702.6	0.2	1.1		P(9)	6970.9	0.5	0.1
	P(19)	1425.8	0	0.2	8–3	R(7)	5711.5	5.7	1.1
4–3	R(16)	1559.3	0.8	4.7		P(9)	4926.6	1.1	0.2
	R(17)	1385.2	0	0.4	8–4	R(7)	3891.8	5.9	1.1
	P(18)	431.7	0.1	0.3		P(9)	3207.8	1.6	0.3
5–0	R(14)	8704.8	1.6	4.0	8–5	R(7)	2404.1	3.5	0.7
	R(15)	8177.5	0	0.3		P(9)	1829.3	1.3	0.3
5–1	R(14)	6352.6	3.7	9.1	8–6	R(7)	1264.1	1.0	0
	R(15)	5904.7	0	1.5		P(9)	811.0	0.4	0
	P(16)	4866.4	0	0.2	9–2	R(5)	8544.4	0.4	0.1
	P(17)	4377.3	0	3.2	9–3	R(5)	6331.5	0.8	0.1
5–2	R(14)	4320.4	7.1	17.4	9–4	R(5)	4435.4	0.9	0.2
	R(15)	3955.6	0	2.5	9–5	R(5)	2866.2	8.2	1.6
	P(16)	3006.9	0.3	0.9	9–6	R(5)	1637.0	0.4	0.1
	P(17)	2614.7	0	0.1		P(7)	1260.5	0.2	0

derived from a combination of the Breit-Wigner formula with the usual intensity formula, i.e., from the expression

$$I \sim \frac{J' + J'' + 1}{Q_r(J')} \frac{\nu^4}{(\tau_l + \tau_r)} \exp\left\{-\frac{(T_0(v'J') - D_0^0)hc}{kT}\right\} R^{v'J'v''J''}, \quad (2)$$

where $R^{v'J'v''J''}$ is the vibrational transition moment calculated from J-dependent vibrational wave functions (based on the effective potential functions for the appropriate J-values); $Q_r(J')$ is the rotational state sum calculated from Boltzmann factors in which the energies $T_0(v'J') - D_0^0$ above the dissociation limit are substituted; τ_l and τ_r are the lifetimes for the radiationless transition and for radiative transitions to lower states, respectively. According to Equation 2 only those quasibound levels for which $\tau_l \ll \tau_r$ (that is, for which the total width is substantially larger than the radiative width) are important in the recombination process. In FIGURES 4a and b the two predicted spectra at 100°K and 1000°K in the region 1500–5000 cm^{-1} are plotted. It should be emphasized that the evaluation has been carried out with a linear dipole moment function (Equation 1). The small deviations from linearity may have a relatively large effect on the intensities of lines with $\Delta v > 1$.

Merrill, Soifer, and Russell[12] have recently reported the observation of four fairly strong unidentified features at 2.43, 3.09, 3.27, and 3.4 μm in the infrared spectrum of the planetary nebula NGC 7027. Still more recently Grasdalen and Joyce[13] have confirmed this observation and have obtained the two strongest features also in two other sources, the Orion nebula and M17. The possibility that these unidentified features are somehow related to the presence of HeH$^+$ was the motivation for the calculations reported above. Considering the abundance of H$^+$ in H II regions, which is confirmed by the presence of H I recombination lines in the spectrum of Merrill, Soifer and Russell, and in view of the presence of neutral He it seems not unlikely that HeH$^+$ may be formed in this source.

In FIGURE 2c, for comparison with the predicted structure of the fundamental, the spectrum of NGC 7027 from Merrill *et al.* is reproduced to the same scale. While the principal observed features are fairly close to the predicted position of the fundamental, the agreement is hardly convincing. It may however be significant that the predicted head of the R branch (which will be prominent only at temperatures above 1000°K) coincides closely with the observed feature at 3.09 μm. The $R(2)$ line agrees very closely with the strongest observed peak at 3.28 μm but in view of the absence of neighboring R and P lines this coincidence must be considered as accidental. Even if one accepts the identification of the peak at 3.09 μm with the R head of the 1–0 band one must conclude that something other than the 1–0 band gives rise to the principal peak at 3.28 μm and its longward companion. The R heads of the 2–1 and 3–2 bands would occur at 3.48 and 3.98 μm and cannot be responsible for the observed peaks.

Even if the stationary concentration of HeH$^+$ in planetary nebulae is not sufficient to lead to an observable emission of the fundamental and the overtones we must consider the possibility that the inverse predissociation of HeH$^+$ will lead to observable emission. Such a mechanism was also suggested by the observation of Townes[14] and Grasdalen and Joyce[13] that the 3.28 μm feature is substantially broader than corresponds to the instrumental line width. A clear-cut

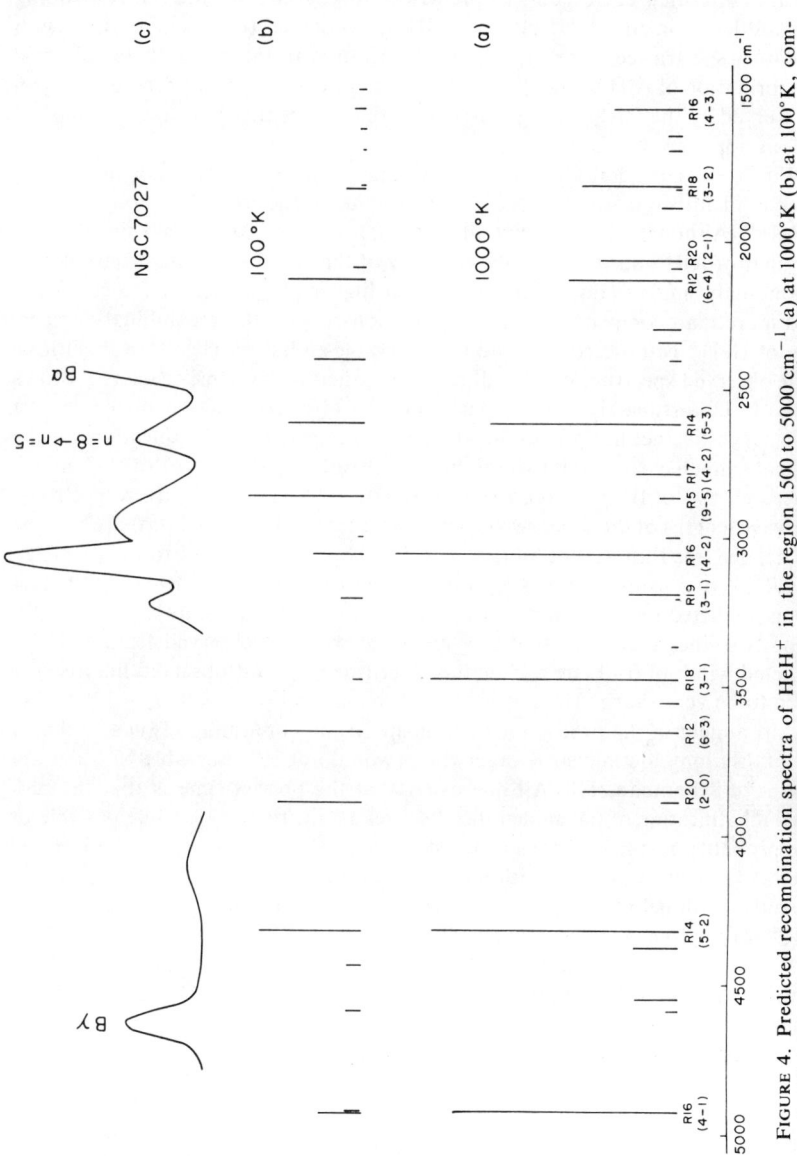

FIGURE 4. Predicted recombination spectra of HeH$^+$ in the region 1500 to 5000 cm^{-1} (a) at 1000°K (b) at 100°K, compared with (c) the observed spectrum of NGC 7027 after Merrill, Soifer, and Russell.[12]

example of the actual occurrence of an inverse predissociation has been established in the atmospheres of several long-period variables: near minimum light bright lines appear which were identified by Herbig[15] as those lines of the $^1\Pi - ^1\Sigma^+$ system of AlH that correspond to predissociating levels; no lines corresponding to bound levels in the $A\ ^1\Pi$ state of AlH are observed (for a comparison with laboratory spectra see Herzberg[16]). In other words in the atmospheres of these stars formation of AlH from Al + H by inverse predissociation proceeds and can be observed by the fairly strong emission of those lines that have the quasibound levels as upper levels.

With these considerations in mind we may inquire whether any of the observed unidentified features in NGC 7027 are due to the recombination spectrum of HeH$^+$. Although the radiative lifetime of HeH$^+$ is 100 to 1000 times larger than that of AlH and therefore the intensity of the recombination spectrum correspondingly smaller (Equation 2) the much higher abundance of He compared to Al more than compensates for this reduction so that the recombination spectrum of HeH$^+$ could certainly lead to observable emission. Therefore in FIGURE 4c the observed spectrum of Merrill et al. is plotted to the same scale as FIGURES 4a and b. Unfortunately this comparison of the observed and calculated spectra while suggestive neither proves nor disproves the proposed identification. In the spectrum predicted for 1000°K the lines $R(19)$ of 3–1 and $R(16)$ of 4–2 match rather well two of the four observed lines, viz., 3.09, and 3.27 μm, respectively. The wave lengths of the other two observed features (2.43 and 3.4 μm) are not too well defined (see FIGURE 1 of Merrill et al.[12]). They may be related to the predicted lines $R(14)$ of 5–2 and the pair $R(5)$ of 9–5 and $R(17)$ of 4–2. Most of the other strong predicted lines fall in regions in which no observations are available. One of the two lines that are tentatively identified with the observed features has a predicted width of 0.73 cm^{-1} while the other (the strongest observed line) is predicted to be very sharp. The upper level of this latter line ($v = 4, J = 17$) is also very prominent in the proton momentum spectrum, confirming (TABLE 4) that it has a fairly long lifetime since otherwise it would not be observable with the apparatus of Schopman et al. All one can say at the present time is that the proposed identification of the unidentified features as due to HeH$^+$, while possible or even probable, is not by any means established. Further studies of the infrared spectra of planetary nebulae with increased resolution and in spectral regions not yet studied will help to establish whether or not the recombination spectrum of HeH$^+$ plays a role in interstellar clouds.

We are greatly indebted to Dr. W. Kolos for advance information on his calculations and to Dr. R. J. LeRoy for the use of his program for the calculation of bound and quasibound levels. We are also grateful to Drs. P. R. Bunker and A. E. Douglas for many discussions and critical readings of the manuscript.

Summary

The infrared rotation-vibration spectrum of HeH$^+$ is predicted on the basis of Kolos' and Peek's *ab initio* potential function of the electronic ground state and the dipole moment function derived from Peyerimhoff's and Michel's electron density function. The existence of quasibound levels leads to the possibility of recombination of He + H$^+$ by inverse predissociation. The spectrum emitted in this process is also predicted. Two of the predicted lines agree with unidentified lines observed in a planetary nebula and the Orion nebula, but a conclusive identification is not yet possible.

References

1. BEACH, J. Y. 1936. J. Chem. Phys. **4**: 353.
2. WOLNIEWICZ, L. 1965. J. Chem. Phys. **43**: 1087.
3. KOLOS, W. 1976. Int. J. Quantum Chem. **10**: 217.
4. KOLOS, W. & J. M. PEEK. 1976. Chem. Phys. **12**: 381.
5. KOLOS, W. & L. WOLNIEWICZ. 1975. J. Mol. Spectrosc. **54**: 303.
6. MICHELS, H. H. 1966. J. Chem. Phys. **44**: 3834.
7. GREEN, T. A., H. H. MICHELS, J. C. BROWNE & M. M. MADSEN. 1974. J. Chem. Phys. **61**: 5186, 5198.
8. PEYERIMHOFF, S. 1965. J. Chem. Phys. **43**: 998.
9. SCHOPMAN, J., P. G. FOURNIER & J. LOS. 1973. Physica **63**: 518.
10. PEEK, J. M. 1973. Physica **64**: 93.
11. BERNSTEIN, R. B. 1974. Chem. Phys. Let. **25**: 1.
12. MERRILL, K. M., B. T. SOIFER & R. W. RUSSELL. 1975. Astrophys. J. **200**: L37.
13. GRASDALEN, G. L. & R. R. JOYCE. 1976. Astrophys. J. **205**: L11.
14. TOWNES, C. 1975. Private communication.
15. HERBIG, G. H. 1956. Publ. Astr. Soc. Pac. **68**: 204.
16. HERZBERG, G. 1965. J. Opt. Soc. Amer. **55**: 229.

PARITY-VIOLATING ELECTROMAGNETIC INTERACTIONS OF NUCLEI*

G. Feinberg

*Department of Physics
Columbia University
New York, New York 10027*

The interactions of nuclei with external electromagnetic fields have been extensively studied since around 1930. Much of the early work was carried out by I. I. Rabi and his collaborators.† In these experiments, they were able to observe such static electromagnetic nuclear properties as magnetic dipole moments,[1] electric quadrupole moments,[2] magnetic octupole moments,[3] and so forth. It is well known that this series of quantities is only one-half of the moments that characterize a classical system of moving charges. The other half, involving such quantities as electric dipole moments, magnetic quadrupole moments, electric octupole moments have not been observed for nuclei. I shall refer to these unobserved properties collectively as odd electromagnetic moments.

The reason for this is also well known. If it is assumed that the fundamental electromagnetic interaction is invariant under space reflections (P) or time reversal (T), then the unobserved odd electromagnetic moments will have the opposite transformation property under P or T than do quantities of the same tensorial rank constructed from the nuclear spin operator. However, under the assumption that the Hamiltonian describing the free nucleus is invariant under P or T, all static observables must transform in the same way under these operations as tensors constructed from the nuclear spin. Therefore, a nucleus can carry moments such as electric dipole, magnetic quadrupole, and so forth, only if parity and time reversal invariances both fail in some interactions involved in nuclei.[4,5] Since it is known that parity invariance holds to a high accuracy in strong and electromagnetic interactions, it is not surprising that the nuclear moments forbidden by P invariance should be highly suppressed in comparison with those allowed by P invariance.

Of course, we know that the weak interactions do not satisfy P invariance and the virtual effects of these interactions are expected to influence the properties of nuclei. Furthermore, observations of the decays of neutral K mesons[6] imply that some interaction fails to satisfy time reversal invariance, although it is not yet clear what interaction this is. The combined effect of the weak interactions and the unknown T noninvariant interactions will generate odd electromagnetic moments of nuclei, with some unknown strength. The precise values of these moments depends fairly critically on the ultimate source of T noninvariance. For example, estimates of the expected electric dipole moment of the proton and neutron have varied by ten orders of magnitude,[7] depending on the model used.

---

*This work was supported in part by the U.S. Energy Research and Development Administration.
†These include J. Kellogg, J. Zacharias, V. Cohen, P. Kusch, N. Ramsey, and others.

All of the electromagnetic interactions considered thus far are linear in the external electromagnetic fields. It is known that bodies composed of charges also undergo electromagnetic interactions involving higher powers of the external fields. Examples are the interactions with the electric and magnetic polarizabilities, which, for spinless systems takes the form

$$H_{\text{pol}} = -\tfrac{1}{2}\alpha_E \mathbf{E}^2 - \tfrac{1}{2}\alpha_H \mathbf{H}^2, \qquad (1.1)$$

where α_E and α_H are the polarizabilities, and $\mathbf{E}$ and $\mathbf{H}$ the external fields. These interactions are P and T invariant. However, there exist similar interactions, bilinear in external fields, which are odd under P, T, or both. For a spinless system, the only such interaction takes the form

$$H_M = -\tfrac{1}{2}\alpha_M \mathbf{E} \cdot \mathbf{H}. \qquad (1.2)$$

Here α_M is a new, "mixed" polarizability. By standard arguments, which will be reproduced below, it can be shown that if H_M is assumed to be P and T invariant, then α_M must be odd under both P and T. This implies that α_M must vanish for an undegenerate nuclear state, except to the extent that P and T noninvariant interactions contribute to the wave function for that state. Thus, α_M will originate in the same combination of interactions as the odd static moments.

For a nucleus with spin, other possibilities exist. For example, a bilinear interaction, which can occur for any system with spin greater than zero, has the form

$$H_J = \beta_J \mathbf{J} \cdot \mathbf{E} \times \mathbf{H}, \qquad (1.3)$$

where $\mathbf{J}$ is the nuclear spin, and β_J is another coefficient with the dimensions of polarizability. This spin-dependent mixed polarizability has the property that it is odd under P but even under T so that it can originate from weak interactions alone, without the need for the unknown time-reversal-violating interaction. It is therefore somewhat easier to calculate its magnitude, and this magnitude is expected to be somewhat larger than α_M. Some estimates for α_M and β_J will be given in the third section of this paper.

For spin zero or $\tfrac{1}{2}$ systems, no other polarizability interactions that are bilinear in constant external fields can occur. However, for a system of spin 1 or greater, yet another interaction, of the form

$$H_A = \gamma_A \{\mathbf{J} \cdot \mathbf{E}, \mathbf{J} \cdot \mathbf{H}\} \qquad (1.4)$$

can occur. This is analogous to the alignment-dependent electric and magnetic polarizabilities that are known to exist for such higher spin systems. The alignment-dependent mixed polarizability γ_A is again similar to α_M in that it is odd under both P and T, and therefore requires both parity-violating and time-reversal-violating interactions in order to occur. We would therefore expect γ_A to be similar in magnitude to α_M.

Still more possibilities for bilinear interactions exist if we consider variable external fields. However, in the interest of brevity, I shall confine myself to constant fields.

Similar odd polarizabilities can exist for other particles such as electrons. Much of the discussion given below can be applied to that case as well, although some of the numerical estimates will be different.

In view of the possibility that nuclei or atoms could have odd moments or odd polarizabilities, it is of interest to consider to what extent such quantities are observable. This has been done in detail for electric dipole moments,[8] but not to my knowledge for the higher odd moments, and not at all for the odd polarizabilities. I discuss this question in the fourth section of this paper.

Classification under P and T of Electromagnetic Interactions

The electric multipole operators for a system are defined in terms of coordinate operators by[9]

$$Q_{L0}^{(e)}(\mathbf{r}) = \frac{e}{\sqrt{2L+1}} r^L Y_{L0}^*(\theta, \phi). \tag{2.1}$$

It follows that under space reflection, these multipole operators transform as

$$P Q_{L0}^{(e)}(\mathbf{r}) P^{-1} = (-1)^L Q_{L0}^{(e)}(\mathbf{r}). \tag{2.2}$$

For any parity eigenstate $|\Psi\rangle$, we therefore have

$$\langle \Psi | Q_{L0}^{(e)}(\mathbf{r}) | \Psi \rangle = (-1)^L \langle \Psi | Q_{L0}^{(e)}(\mathbf{r}) | \Psi \rangle, \tag{2.3}$$

which implies that $\langle \Psi | Q_{L0}^{(e)}(\mathbf{r}) | \Psi \rangle = 0$ for L odd.

Under time reversal, we have instead

$$T Q_{L0}^{(e)}(\mathbf{r}) T^{-1} = \frac{e}{\sqrt{2L+1}} r^L Y_{L0}(\theta, \phi) = Q_{L0}^{(e)}(r)^* \tag{2.4}$$

$$= \frac{e}{\sqrt{2L+1}} r^L Y_{L0}^*(\theta, \phi), \tag{2.5}$$

so

$$T Q_{L0}^{(e)}(\mathbf{r}) T^{-1} = Q_{L0}^{(e)}(\mathbf{r}). \tag{2.6}$$

We suppose now that the state $|\Psi\rangle$ has a definite angular momentum J, m_J and that it is part of a rotationally degenerate multiplet that transforms as

$$T | \Psi, J, m_J \rangle = n_\Psi | \Psi, J, -m_J \rangle, \tag{2.7}$$

where n_Ψ is a phase factor. This is the assumption of time reversal invariance. Then

$$\langle \Psi, J, m_J | T^{-1} T Q_{L0}^{(e)} | \Psi, J, m_J \rangle$$

$$= \langle \Psi, J, -m_J | Q_{L0}^{(e)} | \Psi, J, -m_J \rangle \tag{2.8}$$

$$= \langle \Psi, J, m_J | Q_{L0}^{(e)} | \Psi, J, m_J \rangle. \tag{2.9}$$

The last two matrix elements are related by the Wigner–Eckart theorem,[10] so that

$$\langle \Psi, J, m_J | Q_{L0}^{(e)} | \Psi, J, m_J \rangle$$

$$= \frac{\langle J m_J L 0 | J m_J \rangle}{\langle J - m_J L 0 | J - m_J \rangle} \langle \Psi J - m_J | Q_{L0}^{(e)} | \Psi, J, -m_J \rangle. \tag{2.10}$$

By the properties of Clebsch–Gordan coefficients, this implies that

$$<\Psi, J, m_J | Q_{L0}^{(e)} | \Psi, J, m_J > = (-1)^L <\Psi, J, m_J | Q_{L0}^{(e)} | \Psi, J, m_J > \quad (2.11)$$

so that again the matrix element vanishes for odd L.

Therefore, we conclude that if P or T invariance is satisfied by the Hamiltonian of which $|\Psi>$ is an eigenstate, then all electric multipoles with odd L must vanish.

The magnetic multipoles differ from the electric ones by an additional factor proportional to $J \cdot \nabla$, where J is an angular momentum and ∇ is the gradient operator. Since this factor is odd under both P and T, we can repeat the above discussion, with the addition of a factor of (-1) in (2.3) and (2.11), and the conclusion that matrix elements of $Q_{L0}^{(m)}$ will vanish for even L.

The analysis given above uses a specific form for $Q_{L0}^{(e)}$ and $Q_{L0}^{(m)}$. However, all that is really necessary is the transformation properties (2.2) and (2.6), and the tensorial form of these operators. These relations can instead be derived as follows. We define the Q_{L0} as coefficients in the Taylor expansion of the linear interaction of the system with external electromagnetic fields. Specifically $Q_{L0}^{(e)}$ is the coefficient of an $(L-1)$th derivative with respect to a coordinate of an external electric field E_i, while $Q_{L0}^{(m)}$ is the coefficient of the $(L-1)$th derivative of an external magnetic field H_i. It follows that if the electromagnetic interaction is P invariant, that $Q_{L0}^{(e)}$, $Q_{L0}^{(m)}$ must satisfy

$$PQ_{L0}^{(e)}P^{-1} = (-1)^L Q_{L0}^{(e)},$$
$$PQ_{L0}^{(m)}P^{-1} = -(-1)^L Q_{L0}^{(m)}, \quad (2.12)$$

since the terms they multiply transform by $(-1)^L$ and $-(-1)^L$, respectively. Similarly, under time reversal, we must have

$$TQ_{L0}^{(e)}T^{-1} = Q_{L0}^{(e)*},$$
$$TQ_{L0}^{(m)}T^{-1} = -Q_{L0}^{(m)*}. \quad (2.13)$$

To see this, we note that an electric field, or any of its space derivatives, simply transforms into its complex conjugate under time reversal, while a magnetic field or its space derivatives transforms into minus its complex conjugate. If the electromagnetic interaction is to be invariant under time reversal, the coefficients of these field derivatives must transform in the same way as the field derivatives themselves, which leads to Equation 2.13. From Equation 2.12 or 2.13, we can immediately derive Equation 2.3 or 2.9, and the rest of the derivations given previously follow as above. The advantage of the latter approach is that it can more easily be extended to the case of the polarizabilities $\alpha_M, \beta_J, \gamma_A$, whose expression in terms of coordinate operators is not so apparent.

It is worth noting that the transformation properties in Equations 2.12 and 2.13 also apply if we multiply $Q_{L0}^{(e)}$, $Q_{L0}^{(m)}$ by arbitrary powers of r^2. Such operators determine the spatial distribution of the electromagnetic moments, and so we can conclude that not only must the integrated values of the odd electromagnetic moments vanish in a definite parity state, but also the spatial distribution of such moments must vanish identically as well. Similar conclusions hold for time reversal.

For the mixed polarizability interactions, it is simpler to avoid the angular momentum decomposition which requires the use of complex external fields. Instead, we use real fields E_i, H_i, which transform as

$$P E_i P^{-1} = -E_i,$$
$$P H_i P^{-1} = +H_i,$$
$$T E_i T^{-1} = +E_i,$$
$$T H_i T^{-1} = -H_i. \qquad (2.15)$$

It follows that in order to have P or T invariance of the electromagnetic interaction, we would need

$$P \alpha_M P^{-1} = -\alpha_M,$$
$$T \alpha_M T^{-1} = -\alpha_M, \qquad (2.16)$$
$$P \beta_J P^{-1} = -\beta_J,$$
$$T \beta_J T^{-1} = +\beta_J, \qquad (2.17)$$
$$P \gamma_A P^{-1} = -\gamma_A,$$
$$T \gamma_A T^{-1} = -\gamma_A. \qquad (2.18)$$

These equations can then be used to show that the matrix elements of α_M and γ_A vanish in eigenstates of an internal Hamiltonian that is invariant under P or T, as shown above for the odd Q_{L0}. On the other hand, while the matrix element of β_J must vanish in a parity eigenstate, it need not vanish in an eigenstate of a T invariant Hamiltonian, because of the plus sign in the second equation of Equation 2.17.

Since all electromagnetic interactions must be invariant under TCP, and the external fields E, H change sign under C, we can draw the further conclusion that α_M and γ_A are even under C, i.e., are the same for particle and antiparticle, whereas β_J is odd under C, and so is opposite for particle and antiparticle. Thus even a self conjugate system such as positronium can have nonzero values for α_M and γ_A, whereas such a system must have a nonzero value for all electromagnetic moments, odd or even, as well as for β_J, as those quantities all change sign under TCP.

Estimated Magnitudes of Parity-Violating Electromagnetic Interactions

Because they have spin $\frac{1}{2}$, individual nucleons can carry only an electric dipole moment of the odd electromagnetic moments. I shall not enter here into the question of estimating this electric dipole moment. A bibliography of papers dealing with this question is given in a recent experimental paper.[7] I instead assume that the neutron and proton may have intrinsic electric dipole moments d_N, d_P, and consider how these moments may generate various odd moments for complex nuclei.

In general, the odd moments of nuclei may have several distinct origins:

(1) From expectation values in nuclear states of the electric dipole moments of nucleons.

(2) From parity mixing in nuclear states due to the electric dipole moments of nucleons interacting with the internal electric fields in nuclei.

(3) From parity mixing in nuclear states due to other interactions in nuclei.

Let us consider first mechanism 1, in which the higher moments arise from a displacement of the nucleon electric dipole moments from the center of the nucleus. Assume that external fields $E(\mathbf{r})$, $H(\mathbf{r})$ act on the nucleons, and expand those fields in powers of $\mathbf{r} - \mathbf{r}_0$, with $\mathbf{r}_0$ the nuclear center of mass coordinate

$$E_i(\mathbf{r}) = E_i(\mathbf{r}_0) + (r - r_0)_j \frac{\partial E_i(\mathbf{r}_0)}{\partial r_j} + \cdots,$$

$$H_i(\mathbf{r}) = H_i(r_0^{-1}) + (r - r_0)_j \frac{\partial H_i}{\partial r_j} + \cdots. \quad (3.1)$$

The electric dipole moment of the nucleons, interacting with the external electric field, will induce electric moments for the nucleus as a whole, that are proportional to the expection values in the nuclear ground states, of the operators

$$Q_{r+1}^{(e)} \simeq \sum_n d_n S_{n,i} \rho_{n,k} \rho_{n,l} \cdots (r \text{ factors of } \rho_n), \quad (3.2)$$

where ρ_n are the center of mass coordinates of the nth nucleon, d_n its dipole moment and S_n its spin operator.

The sum is taken over all the nucleons. The term $Q_1^{(e)}$ is just the total electric dipole moment of the nucleus. The terms with r odd will vanish when evaluated in a nuclear state of definite parity, and so will be proportional to the square of the parity nonconserving interations, and are therefore uninteresting. However, a term such as $Q_3^{(e)}$ will not vanish in such a state, and will represent the electric octupole moment associated with the state. It can be seen from Equation 3.2 that the order of magnitude of this octupole moment will be $R_0^2 |d|$, where R_0 is the nuclear radius and $|d|$ is the magnitude of the electric dipole moment of a nucleon. From the existing limits on d, one can guess that the octupole moment arising from this source will be less than $e \times 10^{-50}$ cm^3. Higher odd electric moments such as $Q_5^{(e)}$ or arising from this source will be of order $R_0^4 |d|$, and so forth.

The essentially nonrelativistic electric dipole moment interaction considered thus far will not generate any odd magnetic moments of nuclei. However, associated with this nonrelativistic interaction is a relativistic interaction of the form

$$|d| \, \boldsymbol{\alpha} \cdot \mathbf{H}, \text{ where } \boldsymbol{\alpha} \text{ is the odd Dirac matrix} \quad (3.3)$$

When the expansion 3.1 is substituted into 3.3 there are generated higher magnetic moments, of the nucleus, proportional to

$$Q_{r+1}{}^{(m)} = \sum_n d_n \alpha_{n,i} \rho_n \cdots (r \text{ factors of } \rho), \tag{3.4}$$

In this case the terms with r even will vanish when evaluated in a parity eigenstate, while the terms with r odd will generate the odd magnetic moments. For example the magnetic quadrupole moment $Q_2{}^{(m)}$ will be proportional to $\sum_n d_n \alpha_n \rho_n$. The matrix element of this operator is proportional a product of large and small components of the nuclear wave function, and its order of magnitude in a nuclear state should be

$$d \frac{\hbar}{m_n c} \sim 10^{-38} e\,\text{cm}^2$$

Higher order odd magnetic moments will have extra powers of R, but not of $\hbar/m_n c$, e.g., the magnetic hexadecapole moment will be of order $d_n \hbar/m_n c\, R_0{}^2 \sim 10^{-64} e\,\text{cm}^4$.

I consider next the second mechanism for inducing odd electromagnetic moments in nuclei. For simplicity, take a nucleus with a single proton or neutron outside a closed shell. This nucleus will see an electric field produced by the protons in the nucleon core. If the core is spherically symmetric, this field will be the monopole Coulomb field

$$\mathbf{E}_{\text{core}} = \frac{Ze\hat{r}}{r^2} \tag{3.4}$$

where Z is the core charge, and $\mathbf{r}$ the coordinate of the odd nucleon relative to the center of the core. The electric dipole moment of the odd nucleon, interacting with E_{core}, will give an additional term in the internal nucleon Hamiltonian, of the form

$$H_{\text{e.d.}} = |d|\, eZ \frac{\mathbf{S}\cdot\hat{r}}{r^2} \tag{3.5}$$

This P odd, T odd interaction will produce a mixing of states with opposite parity, so that a state which in the absence of Equation 3.5 has the form $|\Psi_J^P\rangle$ where $P = +$ or $-$ for odd or even parity, will now take the form

$$|\Psi_J'\rangle = |\Psi_J^P\rangle + \sum_n |\Psi_{J,n}^{-P}\rangle \frac{\langle \Psi_{J,n}^{-P}|H_{\text{e.d.}}|\Psi_J^P\rangle}{E_0 - E_n}, \tag{3.6}$$

where the states $|\Psi_J^P\rangle$, $|\Psi_{J,n}^{-P}\rangle$ have energies E_0, E_n. This perturbed state $|\Psi_J'\rangle$ can have any of the odd electromagnetic moments allowed by its spin J, because the mixing terms in Equation 3.6 will have the correct parity and phase to generate such moments. Specifically if we evaluate the matrix elements in $|\Psi_J'\rangle$ of the odd terms in Q_{L0}, these will no longer vanish, but will instead be of the form

$$\sum \frac{\langle \Psi_J^P|Q_{L0}|\Psi_{J,n}^{-P}\rangle \langle \Psi_{J,n}^{-P}|H_{\text{e.d.}}|\Psi_J^P\rangle}{E_0 - E_n} + \text{c.c.} \tag{3.7}$$

The Q_{L0} involved in Equation 3.7 are those of the odd nucleon. If this is a proton, the form of Q_{L0} is that used at the beginning of the previous section. If however, the odd nucleon is a neutron, then the use of this form would give zero, as $e = 0$ for the neutron. However, it is well known that because of core recoil and core polarization effects, matrix elements such as $<\Psi_J^P | Q_{L0} | \Psi_{J,n}^{-P}>$ are not zero for a neutron, and are not given by taking $e = e_P$ for a proton. Instead, both P and N behave as if they had effective changes, of order e_P. Hence if the electric dipole moments of P and N are comparable, we would expect the induced moments we are considering, to be of similar magnitude whether the single nucleon is a neutron or a proton.

It is of interest to give a simple dimensional estimate of Equation 3.7. If there is not much suppression of matrix elements due to nuclear wave functions, then we may expect

$$<\Psi_J^P | Q_{L0}^{(e)} | \Psi_{J,n}^{-P}> \sim eR^L,$$

$$<\Psi_J^P | Q_{L0}^{(m)} | \Psi_{J,n}^{-P}> \sim \frac{e}{m_n} R^{L-1},$$

$$<\Psi_{J,n}^{-P} | H_{e.d.} | \Psi_J^P> \sim \frac{dZe}{R^2}, \quad (3.8)$$

where R is the size of the orbit of the odd nucleon, or approximately the nuclear radius R_0. We then obtain

$$Q_L^{(e)} \sim \frac{Z\alpha}{R_0} \frac{1}{E_0 - \overline{E}} d R_0^{L-1}, \quad (3.9)$$

where $\overline{E}$ is some mean excitation energy of the states with equal J but opposite parity to the ground state. Similarly, we find

$$Q_L^{(m)} \sim \frac{Z\alpha}{R_0} \frac{1}{E_0 - \overline{E}} \frac{d R_0^{L-2}}{m_n}. \quad (3.10)$$

By comparing with the previous discussion, we see that these induced moments differ from the moments that are direct matrix elements in the ground state of $H_{e.d.}$, by a factor of $(Z\alpha/R_0)/(E_0 - \overline{E})$, that is by the ratio of the Coulomb potential of the odd nucleon to the excitation energy. This factor can be expected to be large in heavy nuclei

$$\frac{Z\alpha}{R_0} \frac{Z}{A^{1/3}} \times 1 \text{ MeV},$$

whereas $E_0 - \overline{E}$ is typically 1 MeV or less, in nuclei with spin $\neq 0$. A case of special interest is the rare earth nuclei in which the Nilson levels of same spin and opposite parity almost are degenerate,[12] and $E_0 - \overline{E}$ may be 20 keV or less. Precise calculations of the induced moments in such nuclei will be presented elsewhere. However, it is clear that such induced moments must in general be considered together with the direct moments in estimating the values of odd nuclear moments.

Finally, we note that whatever the eventual explanation of T violation in the K° system, it is likely to generate a nonelectromagnetic interaction within nuclei that is P and T violating. This interaction, which we may label $H_{P \text{ odd}, T \text{ odd}}$ will lead to a mixing of nuclear levels just as $H_{e.d.}$, and in the presence of this mixing the odd nuclear moments will again have induced terms similar to those of Equation 3.7, except that $H_{e.d.}$ is replaced by $H_{P\,\text{odd},\,T\,\text{odd}}$. Repeating the discussion leading to Equations 3.9 and 3.10, we expect that these induced moments will now be of order

$$Q_L^{(e)} \sim \frac{\langle H_{P\,\text{odd},\,T\,\text{odd}} \rangle}{E_0 - \bar{E}} eR_0^L, \qquad (3.11)$$

$$Q_L^{(m)} \sim \frac{\langle H_{P\,\text{odd},\,T\,\text{odd}} \rangle}{E_0 - \bar{E}} \frac{e}{m_n} R_0^{L-1}. \qquad (3.12)$$

To give a more precise value, it is necessary to have a model of $H_{P\,\text{odd},\,T\,\text{odd}}$, and that will not be attempted here. However, it might be expected that the expressions 3.11 and 3.12 would be substantially larger than the corresponding expressions 3.9 and 3.10, as the latter involve an additional power of the fine structure constant. Some estimates of these induced moments involving specific models of $H_{P\,\text{odd},\,T\,\text{odd}}$ will be given elsewhere.

We consider next the estimate of the parity violating polarizabilities α_M, β_J, first for the nucleons; γ_A cannot occur for a spin $\frac{1}{2}$ particle. Such polarizabilities will arise from graphs involving weak interactions, that is virtual intermediate bosons, in addition to the external photons. An example of such graphs is given in FIGURES 1a and 1b. A detailed evaluation of such graphs is beyond the scope of the present article, and will be considered elsewhere. However, an order of magnitude estimate is possible, especially for $\beta_{J'}$, which, as indicated above, does not involve the poorly understood T violating interaction.

For β_J, we expect the order of magnitude to be, for a neutron or proton,

$$\beta_J \approx C_\beta \frac{\alpha}{2\pi} \frac{G_F}{m_N} \sim \frac{10^{-8}}{m_N^3} \sim 10^{-49} \text{ cm}^3, \qquad (3.13)$$

where C_β is a coefficient of order 1, α the fine structure constant, G_F the Fermi constant, and m_N the nucleon mass. The value of C_β could conceivably be much higher than this if states with virtual leptons play an important role in determining it.

For α_M, the graphs involved in calculating it would depend on the mechanism by which time reversal is violated. In model for which this violation is a definite small fraction ϵ of the ordinary weak interaction, we might expect that

$$\alpha_M \sim \epsilon \beta_J \sim 10^{-3} \beta_J \sim 10^{-52} \text{ cm}^3.$$

On the other hand, if some completely new interaction is involved in T violation, it is difficult to say in advance what value to expect for α_M. It would however be surprising if α_M were as large or larger than β_J for a nucleon.

When we turn to a complex nucleus, we find again that the parity violating polarizabilities can arise from the direct effects of the mixed polarizabilities of the nucleons and from indirect effects involving parity mixing within the nucleus. The direct effects of the nucleon polarizabilities themselves are obtained by simple or vector addition over all the nucleons in a nucleus. However, because of the different spin dependence of the terms with α_M and β_J, we expect that, for the direct terms

$$\alpha_{M\,\text{Direct}}^{(\text{nucl.})} \sim Z\alpha_M^{(P)} + N\alpha_M^{(N)}, \qquad (3.14)$$

so that unless α_M is an isovector quantity, a heavy nucleus will have a much larger

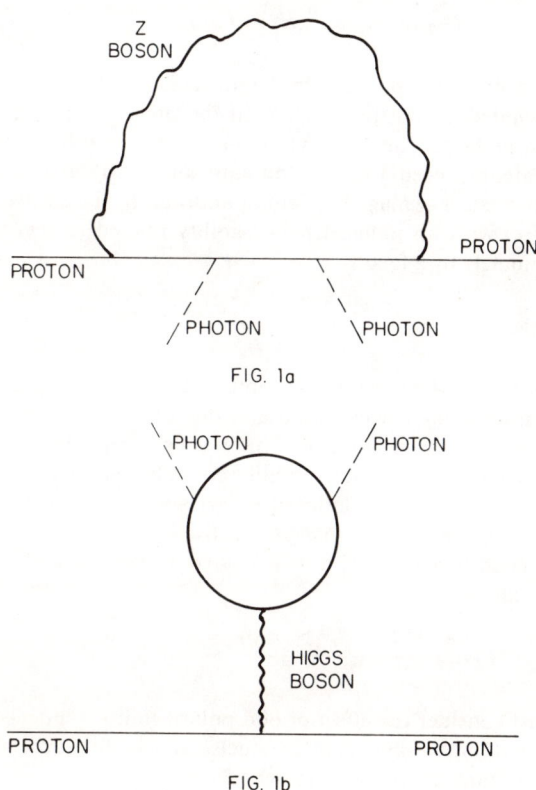

FIG. 1a

FIG. 1b

FIGURE 1. Two Feynman graphs that may contribute to odd polarizabilities of a nucleon. (Top) The source of the polarizability is the ordinary, T invariant weak interaction, here involving the neutral Z boson. This will contribute only to β_J. (Bottom) A hypothetical P and T violating interaction involving Higgs bosons is illustrated, which will contribute to α_M.

value for α_M than does an individual nucleon. On the other hand, for β_J we expect that

$$\beta_{J\,\text{Direct}}^{(\text{nucl.})} \sim \beta_J^{(P)} \text{ or } \beta_J^{(N)}. \qquad (3.15)$$

The induced polarizabilities involve a more complicated structure of effects. One such effect is analogous to Equations 3.7, with $H_{\text{e.d.}}$ replaced by H_2, where

H_2 is either of the Hamiltonians 1-2 or 1-3, in which **E** has been replaced by the internal nuclear electric field, given by Equation 3.4, or some modification thereof; i.e.,

$$H_{2,\alpha} = \sum_l \alpha_M^{(l)} H_i \cdot E_{i,l}^{(\text{internal})},$$

$$H_{2,\beta} = \sum_l \beta_J^{(l)} (H \times J)_i \cdot E_{i,l}^{(\text{internal})}. \quad (3.16)$$

Here the sum is over all the particles in the nucleus, and $E_{i,l}^{(\text{internal})}$ is the internal electric field *evaluated* at the lth nucleon. In the present case, the intermediate states that appear in the sum analogous to Equation 3.7 must have opposite parity to the ground state, but need not have the same angular momentum, because of the presence of the external magnetic field in addition to the electric field. Nevertheless, one still expects an induced polarizability related to the nucleon polarizability approximately by a factor

$$\frac{\langle E_i^{(\text{internal})} \rangle \langle e r_i \rangle}{E_0 - \bar{E}}. \quad (3.17)$$

This factor is essentially the same as that obtained for the induced odd moments, although it might differ somewhat because of the different intermediate states involved. So it is again possible that in some nuclei, where the excitation energy $E_0 - \bar{E}$ is small these induced polarizabilities could be especially large. It may be shown that contributions to the induced polarizabilities coming from replacing **H** by $\mathbf{H}^{(\text{internal})}$ are much smaller than those discussed here. The operator $H_{2,\alpha}$ also can, if evaluated in a state with $J > \frac{1}{2}$, give a contribution to γ_A, and this contribution will also be of order

$$\frac{\langle E^{(\text{internal})} \rangle \langle er \rangle}{E_0 - \bar{E}} \times \alpha_M.$$

Next, we must consider the effect of odd polarizabilities induced through the possible existence of odd moments of the nucleons. The first order interaction of a nucleon with constant external fields is written as

$$H_1 = dS \cdot \mathbf{E}_{\text{ext}} + \mu_L L \cdot \mathbf{H}_{\text{ext}} + \mu_S S \cdot \mathbf{H}_{\text{ext}}. \quad (3.18)$$

Taken in second order, this interaction will contribute both to α_M and γ_N, through terms such as

$$\sum \frac{\langle 0 | \mu S \cdot \mathbf{H}_{\text{ext}} | n \rangle \langle n | dS \cdot \mathbf{E}_{\text{ext}} | 0 \rangle}{E_0 - E_n} + \text{cross term}, \quad (3.19)$$

where $|0\rangle$ is the nuclear ground state, and $|n\rangle$ are *other* nuclear states of the same parity. Note that the ground state $|0\rangle$ must not be included in the sum. The sum 3.19 will contribute to α_M an amount

$$\alpha_M \sim \frac{\mu d}{E_0 - \bar{E}},$$

and similarly for γ_A if $J > \frac{1}{2}$ for the state $|0\rangle$. If we take

$$\mu \sim 10^{-13} \, e \text{ cm},$$
$$d \sim 10^{-24} \, e \text{ cm},$$
$$E_0 - \bar{E} \sim 10^{-6} \text{ ergs} \sim 1 \text{ MeV}. \quad (3.20)$$

Then $\alpha_M \sim 10^{-50}$ cm^3 and similarly for γ_A. This is somewhat larger than the direct nuclear polarizabilities coming from the sum of the nuclear polarizabilities, but it depends on the actual existence of a nucleon electric dipole moment as estimated in Equation 3.20. Note that this mechanism does not contribute to β_J, as such a term would have the wrong transformation under T. There are also other contributions coming from relativistic corrections to H_1 of Equation 3.18 but these appear to be smaller than the above estimate for α_M by several orders of magnitude.

Finally, we consider the odd polarizabilities induced by odd parity non-electromagnetic interactions in the nucleus. These will now occur in third order perturbations, such as

$$\sum_{n,m} \frac{\langle 0|H_{P\text{odd}, T\text{even}}|n\rangle \langle n|er \cdot \mathbf{E}_{\text{ext}}|m\rangle \langle m|\mu \mathbf{S} \cdot \mathbf{H}|0\rangle}{E_0 - E_n \quad\quad E_0 - E_m}. \quad (3.21)$$

This will contribute to β_J, an amount estimated as

$$eR_0 \times \frac{e}{m_N} \times \frac{\langle H_{P\text{odd}, T\text{even}}\rangle}{(E_0 - \bar{E})(E_0 - \bar{E})}.$$

The matrix element of $H_{P\text{odd}, T\text{even}}$ between nuclear states has been estimated[13.] to be 10^{-7} MeV. Thus we might expect this part of the induced polarizability β_J to be of order

$$\beta_J \sim \alpha R_0 \frac{\hbar}{m_N c} \times \frac{10^{-7}}{E_0 - \bar{E}} \times \frac{\hbar c}{E_0 - \bar{E}}$$

$$\sim 4 \times 10^{-47} \left(\frac{1 \text{ MeV}}{E_0 - \bar{E}}\right)^2 \text{cm}^3. \quad (3.22)$$

The corresponding contributions to α_M and γ_A will have a similar form to Equation 3.21 but involving $H_{P\text{odd}, T\text{odd}}$ instead of $H_{P\text{odd}, T\text{even}}$. Thus the contribution will depend on the matrix of $H_{P\text{odd}, T\text{odd}}$ in nuclear states, which is unknown, but likely to be smaller than this by several orders of magnitude.

These induced polarizabilities would appear to have the prospect of being substantially larger than the nucleon polarizabilities, even without any enhancement due to small energy denominators. It seems plausible that such induced effects will dominate the odd polarizabilities of nuclei.

Observability of Odd Parity Electromagnetic Interactions

We consider first the observability of odd electromagnetic moments. The difficulty in observing electric dipole moments of charged particles, including all nuclei other than neutrons is well known, and has been discussed in various places.[8] The difficulty stems from the fact that the average force acting on a charged particle in a stationary state must vanish. If only electric forces act on such a particle, the average electric field must then be zero, and so an electric dipole moment will have no first order effect on the energy of the particle. This is approximately the situation for a nucleus in an atom, although the relatively small spin-orbit and magnetic forces between nucleus and electrons do allow for a small nonvanishing value of the total electric field acting on the nucleus, and so permit some energy shift proportional to an applied electric field.[8] Furthermore, the electric field of the electrons in the atom, which is what tends to compensate the applied external electric field, is in general not homogeneous over the nucleus. As a result, even in the absence of magnetic interactions, a residual interaction occurs between the spatial distribution of the nuclear electric dipole moment and the derivatives of the atomic field. To see how these effects occur, we consider the Hamiltonian for the atom in an external electric field, neglecting magnetic and spin-orbit interactions

$$H = H_0 + H_1$$

$$H_0 = \frac{P^2}{2m_N} + \sum_i \frac{p_i^2}{2m_e} + Ze \int \rho_c(r' - r) V_c(r', r_i) d^3r' + \frac{1}{2} \sum_{i,j} \frac{e^2}{|r_i - r_j|}$$

$$H_1 = d_N \int S \cdot \nabla_{r'} V_c(r', r_i) \rho_D(r' - r) d^3r' + e \sum_i V_{\text{ext}}(r_i)$$

$$+ d_N \int S \cdot \nabla_{r'} V_{\text{ext}}(r') \rho_D(r' - r) d^3r'$$

$$- Z \int V_{\text{ext}}(r') \rho_c(r' - r) d^3r'. \quad (4.1)$$

Here P, r are the nucleus center of mass coordinates and momenta, r_i, p_i the electron coordinates and momenta, V_{ext} is the external electric potential, and $V_c(r', r_i)$ is the Coulomb potential of the electrons at the nucleus, given by

$$V_c(r', r_i) = - \sum_i \frac{e^2}{|r' - r_i|},$$

ρ_c, ρ_D are the spatial distributions of the charge and electric dipole moment of the nucleus, normalized by

$$\int \rho_c(r' - r) d^3r' = \int \rho_D(r' - r) d^3r' = 1.$$

For a constant external field, we may take $V_{\text{ext}}(r') = -r \cdot E_{\text{ext}}$, and obtain

$$H_1 = d_N \int S \cdot \nabla_{r'} V_c(r', r_i) d^3r' \rho_D(r' - r)$$

$$- e E_{\text{ext}} \cdot \Sigma r_i - d_N S \cdot E_{\text{ext}} + Z e r \cdot E_{\text{ext}}. \quad (4.2)$$

Next we write $\rho_D(r' - r) = \rho_c(r' - r) + \delta\rho(r' - r)$, where

$$\int \delta\rho(r' - r) d^3r' = 0.$$

We must now consider the first and second order energy shifts due to H_1, including only terms that are linear in both d_N and in E_{ext}. The first order shift is simply

$$W_1 = d_N S \cdot E_{ext}, \tag{4.3}$$

where I have not bothered to indicate the expectation value of S in the nuclear spin state.

$$W_2 = \sum_n$$
$$\frac{\langle 0 | \int d_N S \cdot \nabla_{r'} V_c [\rho_c + \delta\rho] d^3 r' | n \rangle \langle n | (Ze r \cdot E_{ext} - e E_{ext} \cdot \Sigma r_i) | 0 \rangle}{E_0 - E_n}. \tag{4.4}$$

Here $|0\rangle$ is the atomic ground state, $|n\rangle$ is a complete set of atomic excited states. Now

$$\int \nabla_{r'} V_c(r', -r) \rho_c(r' - r) d^3 r' = i [P, \int V_c(r', r_i) \rho_c(r' - r) d^3 r']. \tag{4.5}$$

However, the only term in H_0 that doesn't commute with P is the nucleus-atom Coulomb interaction term, so that

$$\int \nabla_{r'} V_c \rho_c d^3 r' = \frac{i}{Ze} [P, H_0]. \tag{4.6}$$

Hence

$$W_2 = W_{2,1} + W_{2,2},$$
$$W_{2,1} = \sum_n \frac{id_N}{Ze} S \cdot \frac{\langle 0 | [P, H_0] | n \rangle \langle n |}{E_0 - E_n}$$
$$(Ze r \cdot E_{ext} - e E_{ext} \cdot \Sigma r_i) | 0 \rangle + \text{cross term}$$
$$= -id_N S \langle 0 | [P, r \cdot E_{ext}] | 0 \rangle$$
$$= -d_N S \cdot E_{ext}. \tag{4.7}$$

Therefore $W_{2,1}$ exactly cancels W_1.

The remaining term in W_2 is

$$W_{2,2} = \sum_n d_N S \cdot \frac{\langle 0 | \int \nabla r' V_c \delta\rho d^3 r' | n \rangle \langle n | (Ze r \cdot E_{ext} - e E_{ext} \cdot \Sigma r_i) | 0 \rangle}{E_0 - E_n}$$
$$+ \text{cross term} \tag{4.8}$$

To evaluate this, we expand V_c in powers of $r_i - r$

$$V_c(r' - r_i) = V_c(r - r_i) - (r' - r)_l \nabla_{r,l} V_c + \frac{1}{2} (r' - r)_l (r' - r)_m \frac{\partial^2 V_c}{\partial r'_l \partial r'_m} + \cdots$$

$$\nabla_{r',j} V = -\nabla_{r,j} V_c(r - r_i) + (r' - r)_l \frac{\partial^2 V_c}{\partial r_l \partial r_j}$$
$$+ \frac{1}{2} (r' - r)_m (r' - r)_n \frac{\partial^3 V_c}{\partial r_m \partial r_n \partial r_j} + \cdots \tag{4.9}$$

Upon substituting Equation 4.9 into $\int \Delta_{r'} V_c \delta\rho(r' - r)d^3$, we note that the first term vanishes because of the definition of $\delta\rho$, while the second would be proportional to the nuclear electric dipole moment. Since $W_{2,2}$ is already proportional to one power of d_N, an extra factor of d_N would make it doubly small, and so this term is negligible. Therefore, we find that

$$\int \nabla_{r',u} V_c \delta\rho d^3 r' = \frac{1}{2} Q_{ml} \frac{\partial^3 V_c}{\partial r_m \partial r_l \partial r_j} \quad (4.10)$$

where $\quad Q_{ml} \equiv \int (r' - r)_m (r' - r)_n \delta_\rho(r' - r) d^3 r'$

and

$$W_{2,2} = d_N S_i Q_{ml} \sum_n \frac{\langle 0 | \frac{\partial^3 V_c}{\partial r_l \partial r_m \partial r_i} | n \rangle \langle n | (-e\Sigma \mathbf{R}_i \mathbf{E}) | 0 \rangle}{E_0 - E_N} + \text{cross term} \quad (4.11)$$

where $\quad \mathbf{R}_i = \mathbf{r}_i - \mathbf{r}.$

Thus a residual dipole interaction remains, which is proportional to the second moment of $\delta\rho$, the difference between the normalized electric charge and electric dipole distributions of the nucleus. If these distributions are spherically symmetric as for a spin $\frac{1}{2}$ nucleus, then $Q_{ml} \sim \delta_{ml} Q$, where Q, is essentially the mean square radius of the difference distribution.

The energy shift $W_{2,2}$ is smaller than what would be obtained for a nucleus that was held in place by a nonelectromagnetic force, by a factor which is essentially $(R_0/a_0)^2 \sim 10^{-10}$. It is the compensation of such a factor that makes the observation of the E.D.M. of a charge nucleus so difficult. An ingenious solution to this difficulty was found by Sandars and his collaborators,[14,15] who use the properties of polar molecules to enhance the effective external field acting on a nucleus. For these experiments, the use of a perturbation expansion in the external field is inappropriate. Instead, the observable effect is obtained by taking the expectation value of the electric dipole interaction in the eigenstate of the atom plus external field. To do this, one rewrites Equation 4.1 by including the interaction of the charges with the external field in H_0

$$H_0 = \sum_N \frac{P_N^2}{2m_N} + \sum_i \frac{P_i^2}{2m_e}$$

$$+ \sum_N Z_N e \int \rho_c(r' - r_N) V_c(r'_N, r_i) d^3 r' + \frac{1}{2} \sum_{i,J} \frac{e^2}{|r_i - r_j|}$$

$$+ eE_{\text{ext}} \cdot \Sigma Z_N r_N - eE_{\text{ext}} \cdot \Sigma r_i, \quad (4.12)$$

$$H_1 = -E_{\text{ext}} \cdot d_N S_N + \sum_N d_N S_N \cdot \int \nabla_{r'_N} V_c \rho_D(r' - r_N) d^3 r' \quad (4.13)$$

We have extended the Hamiltonian to include several nuclei in a molecule. The eigenstates $|\Psi\rangle$ of H_0 are not parity eigenstates because of the external field. The energy shift linear in d_N is given by

$$W_1 = <\Psi|H_1|\Psi>$$

$$= -\sum_N d_N S_N \cdot E_{\text{ext}}$$

$$= \sum_N d_N S_N \cdot <\Psi| \int \nabla_{r'_N} V_c \rho_c d^3 r'_N |\Psi>$$

$$+ \sum_N d_N S_N <\Psi| \int \nabla_{r'_N} V_c \delta\rho d^3 r_{N'} |\Psi>. \qquad (4.14)$$

By an argument similar to that leading to Equation 4.6, we can show that the sum of the first two is equal to

$$-i \sum_N \frac{d_N S_N}{Z_N e} \cdot <\Psi|[\mathbf{P}_N, H_0]|\Psi>$$

$= 0$ since $|\Psi>$ is an eigenstate of H_0.

The remaining term can again be expanded in powers of $(r' - r_N)$, leading to the result

$$W_1 = \sum_N d_N S_{n,j} \frac{Q_{lm}^{(N)}}{2} <\Psi| \frac{\partial^3 V_c(r_N)}{\partial r_{N,l} \partial r_{N,m} \partial r_{N,j}} |\Psi> \qquad (4.15)$$

For spherical distributions $Q_{lm}^{(N)} = Q^{(N)} \delta_{lm}$, this reduces to

$$W_1 = \sum_N d_N Q^{(N)} S_{N,j} <\Psi| \frac{\partial}{\partial r_{N,j}} \nabla^2 V_c |\Psi>. \qquad (4.16)$$

The atomic matrix element can, in some instances of polar molecules,[14,15] be much larger than $E_{\text{ext}} a_0^{-2}$ so that W_1 becomes much greater than $W_{2,2}$ given above, and indeed comparable to what would be obtained if these were no "shielding" of the applied electric field by the atomic electrons. This method has been used by Sandars[14,15] to obtain a limit for the E.D.M. of thallium nuclei.

The problem associated with constant external electric fields does not exist for the inhomogeneous electric and magnetic fields required to measure higher odd nuclear moments. No general theorem requires relevant derivatives of the fields to vanish even for a point nucleus. However, the value of the field at the nucleus is influenced by the presence of the atomic electrons in a nontrivial way. Furthermore, the magnitude of the interaction of higher odd nuclear moments is smaller than the E.D.M. by some factors of (R_0/a_0) or of (R_0/L) where L is the length over which the applied external fields varies. This tends to make the higher moments difficult to observe. The best one could hope for is to arrange that an externally applied field with the proper symmetry to interact with a given odd moment will induce a much larger internal field of the same symmetry, that will act directly on the nucleus. This is essentially the method used in References 14 and 15 to measure E.D.M. Supposing that this can be attained, we might take the effective field derivatives to be of order

$$H^{(n)} \sim e \left(\frac{h}{m_e c}\right) \frac{1}{a_0^{n+3}},$$

$$E^{(n)} \sim e \frac{1}{a_0^{n+2}},$$

where $E^{(n)}$, $H^{(n)}$ are the nth derivatives of the fields. In this case, the interaction energy with a nuclear magnetic quadrupole moment of magnitude $10^{-34} e$ cm^2 would be 10^{-20} eV. This is an order of magnitude smaller than the limiting value in existing searches for nuclear E.D.M.,[7,14,15] but it is not impossible that such a small interaction energy could be measured. The interaction energy of the electric field derivatives with a nuclear electric octupole moment of magnitude $10^{-46} e$ cm^3 would be 10^{-21} eV, or comparable to the magnetic quadrupole interaction, so that this coupling might also be eventually measurable. Higher odd couplings will contain extra factors of $R_0/a_0 \sim 10^{-5}$ and so appear more difficult to observe.

The observation of odd polarizabilities involves some of the same problems as does observation of odd electromagnetic moments. In particular, the shielding of an applied external electric field will tend to occur in this case as well. The arguments leading to Equation 4.7 can be repeated, providing that the external magnetic field is such that its interaction with the nucleus can be treated as a small perturbation. Under those circumstances, the odd polarizabilities of a charged nucleus would again be observable only to the extent to which this polarizability was distributed differently in space than the charge of the nucleus. This will again give a suppression factor or order $(R_0/a_0)^2$ in the interaction energy due to an odd polarizability interaction. It is presumably possible to compensate for this suppression to some extent by the Sandars trick of using the electric fields of polar molecules. If this is done, the nucleus might be subjected to an effective electric field, meaning $E(R_0/a_0)^2$ of up to 10^3 esu, and a magnetic field of 10^4 gauss. If the quantity β_J is as large as 10^{-45} cgs units, as suggested by Equation 3.22, the interaction energy due to the β_J term would be 10^{-26} eV, or much smaller than the interaction energies that have been searched for in the E.D.M. experiments. The interaction energy due to α_M is likely to be even smaller.

This problem would not exist for free neutrons, and it is perhaps not impossible to imagine exposing a neutron to the much larger atomic fields, and so obtaining much larger, and perhaps observable interaction energies from the odd electromagnetic polarizabilities.

It is also conceivable that such interactions could be measured for triplet positronium, or triple muonium. The former has the advantage of being neutral as well as having no magnetic moment to complicate the experiments. Of course, in these systems, one is sensitive to the quantities α_M, γ_A for leptons, and to parity violating interactions in atoms, which may have rather different magnitudes than for nuclei or hadrons.

In summary, it appears to be possible to search for magnetic quadrupole and electric octupole moments at nuclei by extension of techniques previously used for electric dipole moments. Detection of the odd nuclear polarizabilities apparently will require the development of significantly more sensitive methods.

I would like to thank Dr. C. K. Au, Dr. M. Y. Chen, and Dr. P. Sandars for helpful discussions.

References

1. RABI, I. I., J. M. B. KELLOGG & J. R. ZACHARIAS. 1934. Phys. Rev. **46**: 157.
2. KELLOGG, J. M. B., I. I. RABI, N. F. RAMSEY JR. & J. R. ZACHARIAS. 1939. Phys. Rev. **55**: 318.
3. ECK, T. G. & P. KUSCH. 1957. Phys. Rev. **106**: 958.
4. This was pointed out by LANDAU, L. 1957. Nucl. Phys. **3**: 127.
5. LEE, T. D. & C. N. YANG. 1957. Lectures at Brookhaven National Laboratory on Weak Interactions, January.
6. CHRISTENSON, J., et al. 1964. Phys. Rev. Lett. **13**: 138.
7. A survey of some theoretical productions for the electric dipole moments of nucleons is given in a paper by DRESS, W. B., et al. 1977. Phys. Rev. D **15**: 9.
8. SCHIFF, L. I. 1963. Phys. Rev. **132**: 2194.
9. For example, as defined by AKHIEZER, A. & V. BERESTETSKI. 1965. Quantum Electrodynamics. : 350–352. Interscience Publishers. New York, N.Y.
10. See, e.g., GOTTFRIED, K. 1966. Quantum Mechanics. : 302. Benjamin Books, New York, N.Y.
11. See the discussion by LANDAU, L. & M. LIFSHITZ. 1962. *In* Classical Theory of Fields. Revised 2nd edit. : 114. Addison-Wesley Publishing Company, Reading, Mass.
12. NILSSON, S. G. 1961. Mat. Fys. Medd. Dan. Vid. Selsk. **32**: 16.
13. HENLEY, E. 1969. Ann. Rev. Nucl. Sci. **19**: 367.
14. SANDARS, P. G. H. 1967. Phys. Rev. Lett. **19**: 1396.
15. HARRISON, G. E., P. G. H. SANDARS & S. J. WRIGHT. 1969. Phys. Rev. Lett. **22**: 1263.

QUANTIZATION OF THE DAMPED HARMONIC OSCILLATOR*

Herman Feshbach and Yoel Tikochinsky†

*Laboratory for Nuclear Science
and Department of Physics
Massachusetts Institute of Technology
Cambridge, Massachusetts 02139*

We hope Professor Rabi will enjoy reading this contribution which deals with a simple but fundamental problem.

Introduction

This study was stimulated by a recent and unexpected discovery of a phenomenon referred to as "deep inelastic scattering," which occurs when atomic nuclei collide. Deep inelastic scattering occurs when a large part of the kinetic energy of the incident nucleus is converted into internal energy of the nuclear system so that the kinetic energy of reaction products is a small fraction of that initial value. From a macroscopic point of view one can describe the collision as a dissipative process. In fact, such concepts as nuclear matter viscosity and friction have been employed in a phenomenological theory.[1] At a more fundamental level, one hopes to develop a microscopic theory in which the phenomenon is explained in terms of the underlying nuclear forces. A promising start on this approach has been made using the time-dependent Hartree-Fock method.[2]

The macroscopic theory must eventually develop a formulation in which quantum effects can be included in the description of processes in which friction or other dissipative processes play a role. One can presumably obtain such a formulation from a complete microscopic theory, that is, by describing in detail, including quantum effects, the processes responsible for dissipation. In another line of attack the details are not explicitly described. Rather their effects are included through parameters such as viscosity whose values can be determined from experiment and then compared with predictions from microscopic theory. In this paper a first step toward the latter approach is taken by developing a quantum theory of damped linear harmonic oscillators.

Classical Formalism

The equation of motion of the damped simple harmonic oscillator is:

$$m\ddot{x} + R\dot{x} + kx = 0, \tag{1}$$

---

*This work is supported in part through funds provided by ERDA under Contract EY-76-C-02-3069.

†Visiting MIT for academic year 1976–1977, on leave from Hebrew University of Jerusalem, Israel.

where x is the displacement of the oscillator from equilibrium. Canonical quantization requires the existence of a Lagrangian L from which Equation 1 can be derived as a Lagrange equation of motion. For this purpose the traditional formulation employing the Rayleigh dissipation function is not useful. The appropriate Lagrangian has in fact been known for some time.[3] It is given in terms of the variables x and an auxiliary variable y whose significance will shortly become clear. This Lagrangian is:

$$L = m\dot{x}\dot{y} + \tfrac{1}{2}R(x\dot{y} - \dot{x}y) - kxy. \tag{2}$$

It is easy to verify that the Lagrange equation obtained by varying y is Equation 1. The equation satisfied by y is obtained by varying x. It is:

$$m\ddot{y} - R\dot{y} + ky = 0. \tag{3}$$

This is the time reverse of Equation 1 for x so that solutions for y will grow with time as rapidly as those for x decay with time. It follows that bilinear forms such as the Hamiltonian H can be time independent.

The Hamiltonian is obtained in the usual manner. The canonical moments are

$$p_x = \frac{\partial L}{\partial \dot{x}} = m\dot{y} - \frac{1}{2}Ry,$$

$$p_y = \frac{\partial L}{\partial \dot{y}} = m\dot{x} + \frac{1}{2}Rx. \tag{4}$$

The Hamiltonian H is

$$H \equiv p_x\dot{x} + p_y\dot{y} - L = \frac{1}{m}p_xp_y + \frac{R}{2m}(yp_y - xp_x) + \left(k - \frac{R^2}{2m}\right)xy \tag{5}$$

It of course follows from the equations of motion that H is independent of time. This can be verified directly using the solutions to Equations 1 and 3 and the form for H:

$$H = m\dot{x}\dot{y} + kxy. \tag{6}$$

The value of H turns out to be $(k - R^2/4m)$ for the underdamped case for a unit initial amplitude.

Quantization[4]

The coordinates x and y and their canonical momenta are assumed to satisfy the Heisenberg commutation rules:

$$[x, y] = 0 = [p_x, p_y] \tag{7}$$

$$[p_x, x] = [p_y, y] = \hbar/i \tag{8}$$

When p_x and p_y are replaced according to Equation 4, we obtain:

$$[\dot{y}, x] = [\dot{x}, y] = \hbar/im, \tag{9}$$

which is consistent with Equation 7. It is a simple matter using the H given by Equation 5 to verify that the quantum Hamiltonian equations of motion are just Equations 1 and 3. We should once more emphasize that the reason one can write time-independent commutation rules such as Equations 7 and 8 is that they involve bilinearly the variables x or $\dot{x}$ and the time reverse variables so that the decay of x is compensated by the growth of y or p_x.

To see some of the consequences of Equations 7 and 8 it is useful to introduce the destruction and creation operators

$$a = \frac{1}{\sqrt{2\hbar\Omega}}\left(\frac{p_x}{\sqrt{m}} - i\sqrt{m}\Omega x\right) \qquad b = \frac{1}{\sqrt{2\hbar\Omega}}\left(\frac{p_y}{\sqrt{m}} - i\sqrt{m}\Omega y\right)$$

$$a^+ = \frac{1}{\sqrt{2\hbar\Omega}}\left(\frac{p_x}{\sqrt{m}} + i\sqrt{m}\Omega x\right) \qquad b^+ = \frac{1}{\sqrt{2\hbar\Omega}}\left(\frac{p_y}{\sqrt{m}} + i\sqrt{m}\Omega y\right) \quad (10)$$

where

$$\Omega^2 = \frac{k}{m} - \frac{R^2}{4m^2} \quad (11)$$

and

$$[a, a^+] = [b, b^+] = 1$$
$$[a, b] = [a^+, b^+] = 0. \quad (12)$$

In terms of a and b and their adjoints, H_0, where

$$H_0 \equiv \frac{1}{m}p_x p_y + m\Omega^2 xy \quad (13)$$

becomes

$$H_0 = \hbar\Omega[a^+ b + b^+ a]. \quad (14)$$

This suggests a further linear transformation:

$$a = \frac{1}{\sqrt{2}}(A + B), \qquad b = \frac{1}{\sqrt{2}}(A - B), \quad (15)$$

with the consequence that

$$H_0 = \hbar\Omega[A^+ A - B^+ B]. \quad (16)$$

Note that A and B satisfy the same commutation rules as a and b. In terms of the original variables

$$A = \frac{a + b}{\sqrt{2}}, \qquad B = \frac{a - b}{\sqrt{2}}, \quad (17)$$

which from Equation 10 involve symmetric and antisymmetric combinations of the x and y variables. For example:

$$A = \frac{1}{2\sqrt{\hbar\Omega}} \left[\frac{1}{\sqrt{m}} (p_x + p_y) - i\sqrt{m}\,\Omega(x + y) \right]. \tag{18}$$

A is even under the transformation $x \leftrightarrow y$ and B odd. It is possible to express H in terms of A and B

$$H = H_0 + H_1 \tag{19}$$

where

$$H_1 = i\frac{\Gamma}{2}(A^+B^+ - AB) \tag{20}$$

$$\Gamma \equiv \frac{\hbar R}{m}. \tag{21}$$

The decay constant of the classical variable x is just $(\Gamma/2\hbar)$.

In the limit, $R \to 0$,

$$H \to \hbar\omega(A^+A - B^+B), \tag{22}$$

where $\omega^2 = k/m$. We see that the states of this Hamiltonian become those of the simple undamped harmonic oscillator if only those states ψ, satisfying the condition $B\psi = 0$ are considered. That is the Hilbert space is restricted to those states generated by operating with A^+ which is even under the transformation $x \leftrightarrow y$. We refer to the basic phonon as the "even" phonon.

Since the eigenvalues of A^+A and B^+B are the positive integers and 0, the eigenvalues of H_0 are $\hbar\Omega(n_A - n_B)$ with eigenstates $|n_A, n_B\rangle$. Since only the difference $(n_A - n_B)$ enters, the eigenvalues of H_0 for all the states of the form $|n_A + n, n_B + n\rangle$, n an integer, will be identical. The perturbative operator H_1 commutes with H_0. Thus if the system initially is in an eigenstate of H_0 such as $|n_A, 0\rangle$ of the undamped variety the effect of H_1 will be to mix in states of the form $|n_A + n_B, n_B\rangle$, each of which are eigenstates of H_0 with identical eigenvalues. The original state is damped, the "sink" into which the energy goes are the states generated by B^+, which is odd under the transformation $x \leftrightarrow y$. The corresponding phonon is referred to as an "odd" phonon.

Eigenstates and Eigenvalues of H

In order to determine the time evolution of the states of the damped harmonic oscillator it is useful to obtain the eigenstates of the Hamiltonian. Toward this end note that H_1 together with two other operators, form an algebra. The operators are

$$X = \tfrac{1}{2}(A^+B^+ + AB), \tag{23a}$$

$$Y = \tfrac{1}{2}(A^+B^+ - AB), \tag{23b}$$

$$Z = \tfrac{1}{2}(A^+A + BB^+). \tag{23c}$$

H_1 is proportional to Y. Their commutators are

$$[X, Y] = iZ, \qquad (24)$$

$$[Z, Y] = iX, \qquad (25)$$

$$[X, Z] = iY. \qquad (26)$$

The operator H_0 commutes with X, Y, and Z and is the Casimir operator for this algebra. In fact

$$Z^2 - X^2 - Y^2 = h_0^2 - \tfrac{1}{4}, \qquad (27)$$

where

$$H_0 \equiv 2\hbar\Omega h_0,$$

so that

$$h_0 = \tfrac{1}{2}(A^+A - B^+B). \qquad (28)$$

The algebra of Equations 24–26 is sometimes designated as QU(2) or 0(2, 1). The angular momentum operators L_x, L_y, L_z satisfy Equation 24, but the analogs of (25) have a minus sign on the right hand side, e.g.,

$$[L_z, L_y] = -iL_x.$$

In analyzing the QU(2) algebra, it has been customary to label the states by the eigenvalues of $h_0, \tfrac{1}{2}(n_A - n_B) = j$ and by the eigenvalues of $Z - \tfrac{1}{2}, m = \tfrac{1}{2}(n_A + n_B)$. The noncompact nature of this algebra is indicated by the inequality

$$\tfrac{1}{2}(n_A + n_B) \geq \tfrac{1}{2}(n_A - n_B), \qquad m \geq |j|.$$

In the present context, however, we are interested in the eigenvalues of Y rather than those of Z. Toward this end we employ the relationship

$$e^{\mu X} Y e^{-\mu X} = Y \cos \mu + iZ \sin \mu$$

so that

$$e^{(\pi/2)X} Y e^{-(\pi/2)X} = iZ$$

or

$$Y = ie^{-(\pi/2)X} Z e^{(\pi/2)X}. \qquad (29)$$

Let the eigenstates of Z be denoted $|j, m\rangle$ with eigenvalue $\tfrac{1}{2}(n_A + n_B + 1) = m + \tfrac{1}{2}$. Then the eigenstates of Y are $e^{-(\pi/2)X}|j, m\rangle$:

$$Ye^{-(\pi/2)X}|j, m\rangle = i(m + \tfrac{1}{2})e^{-(\pi/2)X}|j, m\rangle. \qquad (30)$$

There is a second set of eigenstates and eigenvalues obtained from

$$e^{-(\pi/2)X} Y e^{(\pi/2)X} = -iZ. \qquad (31)$$

The eigenstates are $e^{(\pi/2)X}|j, m\rangle$:

$$Ye^{(\pi/2)X}|j, m\rangle = -i(m + \tfrac{1}{2})e^{(\pi/2)X}|j, m\rangle \qquad (32)$$

In both cases the eigenvalues are pure imaginary.

It can also be shown that m has a minimum value, which is identical for the eigenstates of both Equations 30 and 32. This can be shown in the traditional method as follows. The "lowering" operator is $(X + Z)$ as can be seen directly from Equations 24 and 25, from which one obtains

$$[(X + Z), Y] = i(X + Z).$$

If ψ_{im} is an eigenstate of Y with eigenvalue $i(m + \tfrac{1}{2})$, $m > 0$, it follows that

$$Y(X + Z)\psi_{im} = i(m - \tfrac{1}{2})(X + Z)\psi_{im}, \tag{33}$$

that is $(X + Z)\psi_{im}$ is an eigenstate of Y with m reduced by unity. Let the smallest value of m be M. Then

$$(X + Z)\psi_{iM} = 0. \tag{34}$$

Multiplying by $(Z - X)$ and using Equations 26 and 27 one finally obtains the equation

$$M^2 = j^2.$$

Hence the minimal value M is $|j|$. The "raising" operator is $X - Z$. Applying it to ψ_{-im}, $m > 0$, the set of eigenstates satisfying Equation 32 with eigenvalue $-i(m + \tfrac{1}{2})$ produces an eigenstate of Y with eigenvalue $-i(m + \tfrac{1}{2}) + i$ or $-i(m - \tfrac{1}{2})$. The minimum value of m for this set of negative imaginary eigenvalues is also $|j|$.

In summary, there are two sets of eigenstates, one with positive imaginary eigenvalues $i(m + \tfrac{1}{2})$, the other with negative imaginary eigenvalues, $-i(m + \tfrac{1}{2})$, where in both cases m equals $\tfrac{1}{2}(n_A + n_B)$. The positive imaginary eigenvalues for positive j are $i(j + \tfrac{1}{2})$, $i(j + \tfrac{3}{2})$, $i(j + \tfrac{5}{2})$, ..., where $j = \tfrac{1}{2}(n_A - n_B)$. Their negatives $-i(j + \tfrac{1}{2})$, and so forth, form the negative eigen-spectrum for positive j.

The eigenvalues of H are

$$2\hbar\Omega j \pm i\left(m + \tfrac{1}{2}\right)\Gamma = \hbar\Omega(n_A - n_B) \pm i\frac{\Gamma}{2}(n_A + n_B + 1). \tag{35}$$

The eigenstates with negative imaginary eigenvalues are decaying states while those with positive imaginary eigenvalues are growing states. As might be anticipated from this result, the set of states (32) is the time reverse of the states (30). To demonstrate this result note that under time reversal [5]

$$A \leftrightarrow -A^+$$
$$B \leftrightarrow -B^+$$

Hence H_0, X, and Z are even under time reversal while Y is odd. Relationship 32 is just the time reverse of the equation above Equation 29. Taking the time reverse of Equation 30 (indicated by the superscript T) yields

$$Y[e^{-(\pi/2)X}|j,m\rangle]^T = -i(m + \tfrac{1}{2})[e^{-(\pi/2)X}|j,m\rangle]^T.$$

Comparison with Equation 32 yields the

$$e^{(\pi/2)X}|j,m\rangle = [e^{-(\pi/2)X}|j,m\rangle]^T \tag{36}$$

Normalization

Although Y seems to be a Hermitian operator, the eigenvalues of Y are pure imaginary. This apparent contradiction is resolved if Y is not Hermitian. And indeed Y is not Hermitian because the normalization integral for the eigenstates of Y is infinite, a result that follows from the fact that the eigenvalues are imaginary. This infinity of the normalization prevents the immediate use of the metric of Hilbert space. It becomes impossible to expand a state vector in terms of ψ_{im} and $\psi_{im}{}^T$ using the usual methods to determine the coefficients. A further discussion would be necessary.

For the purposes of the subject under examination it is more convenient and appropriate to adopt another metric, that is to change the definition of the inner product to involve a state vector ψ and its time reverse ψ^T. Thus the "normalization" integral is given by

$$[\psi^T, \psi] \rightarrow \int \psi^T \psi \, dv, \tag{37}$$

where dv is the appropriate volume element. Note the absence of any complex conjugation of ψ^T. It can be verified that this normalization integral if convergent is independent of time when ψ is a solution of the Schrödinger equation.

The Green's theorem for an operator O is given, assuming convergence, by

$$[\chi^T, O\psi] = [O^T \chi^T, \psi] = [(O\chi)^T, \psi], \tag{38}$$

where O^T is the time reverse of O. Thus the Hermitian adjoint of the Hilbert metric is replaced by the time reverse. One can now immediately verify that if Equation 38 is valid the normalization integral of ψ_{im} is finite:

$$[\psi_{im}{}^T, \psi_{im}] = [e^{(\pi/2)X} | j, m \rangle, e^{-(\pi/2)X} | jm \rangle]$$

$$= \langle j, m | j, m \rangle \tag{39}$$

These equations follow from Equation 38, the evenness of X under time reversal and the normalization of the eigenstates of Z, which has real eigenvalues and whose normalization is well defined and chosen to be unity. Since Z is even under time reversal the difference between the two metrics disappears. It also immediately follows, using the same analysis, that

$$[\psi_{im}{}^T, \psi_{im}] = \delta_{m,m'} \tag{40}$$

so that $\psi_{im}{}^T$, and ψ_{im} form a biorthogonal set.

These considerations can be clarified through the use of a specific representation. It is easy to verify that the following satisfy commutation rules 24, 25, and 26:

$$Z = i \frac{\partial}{\partial \varphi}$$

$$X + iY = -e^{-i\varphi} \left[i \frac{\partial}{\partial \varphi} + j - \frac{1}{2} \right], \tag{41}$$

$$X - iY = -e^{-i\varphi} \left[i \frac{\partial}{\partial \varphi} - j + \frac{1}{2} \right],$$

where j is arbitrary, $0 \leq \varphi \leq 2\pi$. However, by computing $Z^2 - X^2 - Y^2$, one can identify j with $\frac{1}{2}[n_A - n_B]$. The eigenvalue equation

$$Y\psi_{jm} = i(m + \tfrac{1}{2})\psi_{jm}$$

can now be easily solved

$$\psi_{jm} = \left[\cos\frac{\varphi}{2}\right]^{j+m}\left[\sin\frac{\varphi}{2}\right]^{j-m-1} \tag{42}$$

The time reverse $\psi_{jm}{}^T$ is given by $m + \tfrac{1}{2} \rightarrow -(m + \tfrac{1}{2})$ or $m \rightarrow -(m+1)$

$$\psi_{jm}{}^T = \left[\cos\frac{\varphi}{2}\right]^{j-m-1}\left[\sin\frac{\varphi}{2}\right]^{j+m} \tag{43}$$

In order for ψ_{jm} to be single valued, j must be a half integer. The bounds on m, namely $m > |j|$ and the unit spacing for the eigenvalues, m, can be obtained using the arguments following Equation 32. If one introduces the values for m and j in terms of n_A and n_B, $j + m = n_A$, while $j - m - 1 = -(n_B + 1)$.

We see immediately shown that the normalization integral

$$\int_0^{2\pi} |\psi_{jm}|^2 d\varphi = \int_0^{2\pi}\left(\cos\frac{\varphi}{2}\right)^{2n_A}\left(\sin\frac{\varphi}{2}\right)^{-2(n_B+1)} d\varphi$$

is not defined. On the other hand

$$\int_0^{2\pi} \psi_{jm}{}^T\psi_{jm} d\varphi = \int_0^{2\pi}\left(\cos\frac{\varphi}{2}\right)^{2j-1}\left(\sin\frac{\varphi}{2}\right)^{2j-1} d\varphi,$$

$$= \frac{1}{2^{2(2j-1)}}\int_0^{2\pi} (\sin\varphi)^{n_A - n_B - 1} d\varphi,$$

which is well defined for odd positive values of $n_A - n_B$. This representation is valid only for these values of j, half odd positive integers. Orthogonality for a given j but differing values of m and m' can be readily proved verifying Equation 40 for this case.

Time Dependence

The eigenstates of the Hamiltonian H are these

$$\psi_{j,m}{}^{(\pm)} = e^{\mp(\pi/2)X}|j,m\rangle e^{-2i\Omega jt \pm (\Gamma/2\hbar)(2m+1)t}, \tag{44}$$

where the minus superscript refers to the decaying solution, the plus to its time reverse and therefore growing solution. With the aid of the orthogonality relationship 40 it is possible to solve the time-dependent Schrödinger equation for a given initial state.

As an example consider the initial condition

$$\psi(t = 0) = |j, m_0\rangle.$$

The solution of the Schrödinger which decays with time is

$$\psi(t) = e^{-[2i\Omega j + (\Gamma/2\hbar)]t} \sum_{m=j}^{\infty} e^{-m\Gamma t/\hbar} [e^{-(\pi/2)X} | j, m] a_{m,m_0}^{(j)}, \quad (45)$$

where

$$a_{m,m_0}^{(j)} = \langle j, m | e^{-(\pi/2)X} | j, m_0 \rangle. \quad (46)$$

The value of this coefficient can be calculated using the methods of Racah, which he employed in the evaluation of the Wigner D functions. The result is

$$a_{m,m_0}^{(j)}(\mu) \equiv \langle j, m | e^{\mu X} | j, m_0 \rangle$$

$$= \sqrt{\frac{(m-j)!(m_0-j)!}{(m+j)!(m_0+j)!}} \frac{\left(\sin \frac{\mu}{2}\right)^{m+m_0-2j}}{\left(\cos \frac{\mu}{2}\right)^{m+m_0+1}}$$

$$\times \sum_{l} \frac{(m+m_0-l)!}{l!(m-j-l)!(m_0-j-l)!} \left(\cot^2 \frac{\mu}{2}\right)^{l}$$

so that

$$a_{m,m_0}^{(j)} = (-1)^{m+m_0-2j} \sqrt{\frac{(m-j)!(m_0-j)!}{(m+j)!(m_0+j)!}} 2^{j+1/2}$$

$$\times \sum_{l} \frac{(m+m_0-l)!}{l!(m-j-l)!(m_0-j-l)!}. \quad (48)$$

For large t

$$\psi(t) \to e^{-[2i\Omega j - (j+1/2)\Gamma]t} a_{j,m_0}^{(j)} e^{(\pi/2)X} | j, j \rangle, \quad (49)$$

where

$$a_{j,m_0}^{(j)} = (-1)^{m_0-j} 2^{j+1/2} \sqrt{\frac{(m_0+j)!}{(2j)!(m_0-j)!}}$$

An expression for $\psi(t)$ can also be obtained by expanding $e^{-(i/\hbar)(H_0+H_1)t} | jm_0 \rangle$ in the $| jm \rangle$ representation. Results identical with Equation 45 are obtained. It is important to note that this procedure does not depend upon the normalization discussion of the previous section. Finally note that $\langle \psi(t), \psi(t) \rangle = 1$.

Conclusions

The quantization of the damped harmonic oscillator carried out above can be readily extended to three or more dimensions. More generally it would be appropriate for the quantum theory of small vibrations including friction.

References

1. SWIATECKI, W. J. & S. BJØRNHOLM. 1972. Phys. Rep. **4C:** 327.
2. BONCHE, P., KOONIN S. & J. NEGELE. 1976. Phys. Rev. C **13:** 1226.
3. MORSE, P. M. & H. FESHBACH. 1953. Methods of Theoretical Physics.:298. McGraw-Hill Book Co. New York, N.Y.
4. The quantization has also been treated by the Feynman path integral method starting from the in Lagrangian Equation 2. NEMES, M. C. & A. F. R. DE TOLDEO PIZA. Private communication.
5. DE SHALIT, A. & H. FESHBACH. 1974. Theoretical Nuclear Physics. Appendix. John Wiley & Sons, New York, N.Y.

A DIAMOND AS BIG AS THE RITZ*

Elliott Flowers, Alak Ray, Malvin Ruderman, and Edward Spiegel

Department of Physics and Astronomy
Columbia University
New York, New York 10027

> It's not very big, for a mountain. But . . . it's
> solid diamond. One diamond, one cubic mile without
> a flaw. Aren't you listening?
>
> F. Scott Fitzgerald
> "A Diamond as Big as the Ritz"
> 1922

> Who ordered that?
>
> I. I. Rabi
> 19??

Introduction

Though matter in the solid state is a familiar feature of our environment, it is in fact relatively scarce in the cosmos. Thus, some 90% of the observable mass of the universe is in the form of ordinary stars, which are primarily gaseous. Most of the rest of the observable matter is in the interstellar medium, and almost all of that is in the form of gas.

Not only does solid matter seem scarce in other parts of the universe, but when it is present it is not easy to detect, except when it is in the form of fine grains whose scattering properties make their presence in the interstellar medium readily apparent. Yet theoretical evidence does point to the existence of solids elsewhere, and not just in the other solar systems whose existence is a virtual certainty to many cosmogonists. For example, there is little doubt that solid matter exists within some stars. The enormous pressures at the center of white dwarfs produce phase transitions to the solid state when the temperatures fall below about 10^7 K. Similarly, the outer layers of a neutron star form a solid crust under the influence of the even greater pressures occuring just a few meters below the surfaces of these bodies.[1]

Of course, such matter is buried within the stars and is not accessible to direct observation. But even if it were exposed it would be virtually undetectable unless it were self-luminous. Boulders, golf balls, ball bearings, could all abound in space and not be noticed in any way. But it does seem reasonable to ask whether they may exist, or at least to speculate on ways that they may be formed. This leads us to look for situations in which relatively dense matter, although initially hot, may

---

*This work was supported by the National Science Foundation under Grant NSF PHY-7505660.

be cooled sufficiently by adiabatic expansion without too much rarefaction. This is what seems to occur in the aftermath of supernova explosions, and several authors have remarked on the possibility of forming interstellar dust in this way.[2,3] It may also be possible to form even larger solid bodies in supernova explosions, as we shall speculate below.

In supernova (SN) explosions, matter initially near the stellar cores and hence at very high densities and temperatures is rapidly expanded and cooled. In cases where the local radioactive content of the matter is negligible, the density and temperature of the expanding matter evolve along an adiabat, once the initial blast waves have passed through. For such a situation, we may ask under what initial conditions will the matter be condensed when the pressure becomes negligible. Those will then be the conditions for ejecting solids (or possibly liquids) into space from the explosion. To study this possibility, we first examine in the next section the way in which very dense matter such as that expanding from an SN explosion cools. Afterwards, we shall return to a discussion of possible implications.

Adiabatic Expansion of Hot, Dense Matter

A massive star at the endpoint of nuclear burning has a central temperature $\sim 10^{10}$ K ~ 1 MeV, a central density $\sim 10^8$ gm/cm^3, and is composed principally of iron-peak elements. Under these conditions, the core of the star contracts and heats, the nuclei disintegrate, and an implosion of the core takes place. The dynamics of what follows is poorly understood, but it is generally supposed that matter outside the collapsing core is compressed by blast waves and begins a rapid expansion, perhaps with the help of momentum transfer from trapped neutrinos. How much of the mass avoids collapsing into the core is not known, nor is even the initial temperature and density of the material which expands outward reliably estimated. For the moment we shall leave these quantities unspecified. But typical starting values of density and temperature may be 10^8 gm/cm^3 and 10^9 K with uncertainties of one or two orders of magnitude. Under these conditions, the matter is already not far from its melting point. The question then is: can the subsequent rarefaction caused by the expansion cool the matter sufficiently to form an ordinary solid.[7]

The cooling of the expanding matter takes place approximately at constant entropy, with the principal contribution coming from the ion entropy and a correction being provided by the degenerate electrons. In these hot circumstances, the thermal properties of condensed matter is adequately described by the Einstein model of a solid. In this model each ion, which vibrates independently as if it were feeling a restraining force from stationary neighboring ions, has the same characteristic frequency ω_E. The entropy contributed by the ions is

$$S_i \simeq 3N_i k \left[\ln\left(\frac{kT}{\hbar \omega_E}\right) - 1 \right], \tag{1}$$

where N_i is the number of ions, k is Boltzmann's constant, and $\hbar$ is Planck's constant divided by 2π. The characteristic frequency is well approximated over most of the conditions of interest by the ion plasma frequency

$$\Omega_p = \left(\frac{4\pi e^2 Z^2 N_i}{M_i V}\right)^{1/2},$$

where V is the volume, and M_i is the ion mass.

Throughout the process studied here, the electrons are degenerate. The entropy of N_e degenerate nonrelativistic electrons is

$$S_e = \frac{\pi^2}{2}\left(\frac{T}{T_f}\right)N_e, \tag{2}$$

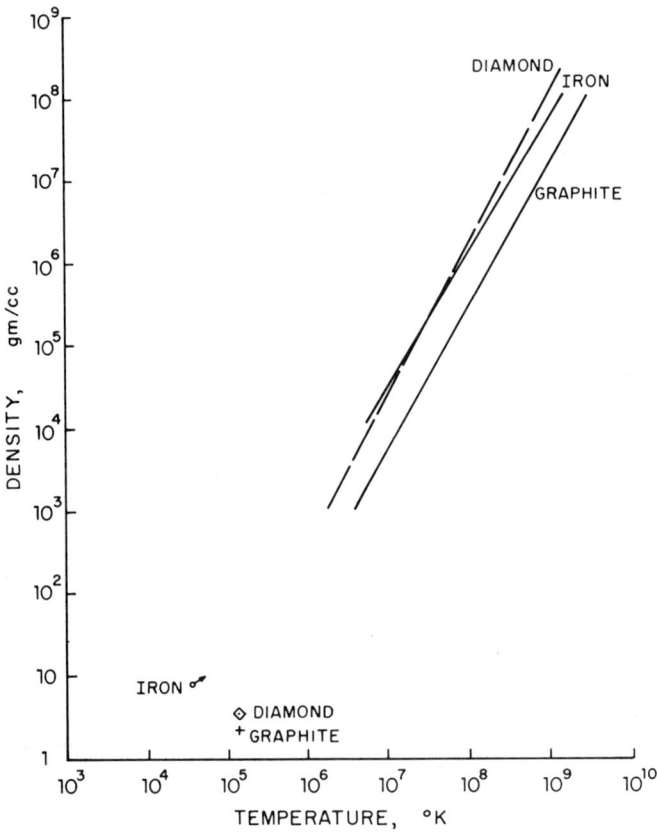

FIGURE 1. Isentropes of various substances that go through a temperature of $\frac{3}{2}\Delta/k$ at their normal density under standard atmospheric pressure. The initial rise of temperature on compressing iron under laboratory pressures, determined from Grueneisen's coefficient, is shown by the arrow. Also indicated are the temperatures corresponding to $\frac{3}{2}\Delta/k$ for diamond and graphite at their normal densities. The respective isentropes pass through these points and bound from below the initial conditions of density and temperatures for which adiabatic expansion can leave matter significantly condensed.

TABLE 1.
Properties of Some Elements*

Element	Heat of Vaporization, in eV/atom	Debye Temperature T_D, in °K	$\dfrac{\Delta}{0.7\,kT_D}$
C: graphite	7.4	420	291
C: diamond	7.4	2230	54
Fe	3.62	467	128
Si	3.08	640	80
Mg	1.3	400	53

*We use the approximation $\hbar\omega_E \sim 0.7\,kT_D$ in the text. The initial ion entropy depends only on the ratio $kT/\hbar\omega_E$, and the maximum possible final temperature which leaves most matter condensed is $\sim \Delta/k$; the last column in the table gives the argument of the logarithm of Equation 1 used in determining the isentropes of FIGURE 1.

where T_f is the Fermi temperature of the electrons. For ultrarelativistic electrons S_e is twice as great.

The thermal history of hot dense matter as it cools adiabatically is represented by $S_i + S_e$ = constant. The cooling curves in the ρ-T plane for some representative solids are illustrated in FIGURE 1. The method of drawing these curves was to choose endpoints corresponding to conditions under which the materials labeled are in condensed matter form and then to draw the cooling curves that pass through the endpoints. The required properties of the cited elements are given in in TABLE 1.

In order for there to be sufficient thermal energy to vaporize a liquid (or solid) as the pressure is adiabatically reduced, it is necessary that the initial thermal content per atom ($\sim \frac{3}{2}kT$) should exceed the heat of vaporization Δ. Thus for the solid to form at zero pressure, the material must achieve a temperature $T \lesssim \frac{3}{2} \Delta/k$ where $-\Delta$ is the binding energy of a particle in the solid. From these curves we see that if the temperature at the beginning of the expansion satisfies the approximate inequality

$$T_i \lesssim 10^8 (\rho_i/10^6)^{1/2},$$

where ρ_i is the initial density, solids may be expected to form.

The qualitative content of the cooling curve can readily be seen. Let Z_{eff} be the number of electrons per ion. At normal densities of condensed matter Z_{eff} is the number of valence electrons while at high densities $Z_{\text{eff}} \sim Z$. (The latter case corresponds to our initial condition with $\rho \sim 10^9\,\text{g/cm}^3$, $T \sim 10^8\,K$ and $T_f \sim 10\,\text{MeV}$, i.e., $T_f \sim 10^{11}\,K$.) The ratio of electron to ion entropy is

$$\frac{S_e}{S_i} \simeq Z_{\text{eff}}\left(\frac{T}{T_f}\right),$$

and we see that ion contribution to the entropy dominates the cooling law. Except for the approach to zero temperature, the adiabatic cooling curve is given by the prescription T/ω_E = const; hence, to a good approximation $T \propto \rho^{1/2}$. The small corrections for the electron thermal energy are easily made.

Discussion

In applying the cooling curves given above to material expanding away from a supernova explosion, there are two situations to be kept in mind. For stars in the mass range 3–8 $M_\odot$ the explosion probably occurs during the carbon flash. This takes place after the core has been transformed to carbon, when the carbon begins to burn with a central temperature $\sim 3 \times 10^8$ K and central densities in the range 10^8–10^{10} g/cm^3. In these conditions the core is supported by electron degeneracy pressure, which is insensitive to temperature. Therefore the energy released by carbon-burning heats the core without causing it to expand. The increased temperature raises the reaction rate and a thermal runaway ensues. The result may be a disruption of the star. If the burning proceeds to ^{56}Ni the consequent radioactive heating would make the adiabatic cooling curve inapplicable for the material that was in the core.

In heavier stars, the central density is lower and the heating during carbon burning may lift the degeneracy without disruption of the star. In that case the burning continues to the formation of iron when no more exothermic reactions are available. The core contracts and heats, and the iron in the center then disrupts in the reaction Fe $\rightarrow$ 13α + 4n, causing implosion of the core. The way in which an explosion of the rest of the star may be produced is problematical. A burst of $\nu\bar{\nu}$ is produced, which may eject the outer layers. Alternatively, a strong shock may be produced, which could drive the nonimploded material outward.

In both the mass ranges just mentioned, a shock wave is expected to run through the star, possibly raising the temperature to the point where nuclear reactions occur. The end products of such reactions provide one of the major unknowns of this discussion. If radioactive species with half-lives between minutes and months are produced, the cooling law discussed above no longer applies, as we have already noted. But if such species are not formed over the bulk of the star, the uncontaminated portions will cool nearly isentropically.

When radioactive heating is not important the qualitative cooling law given at the end of the previous section, $T \simeq T_i(\rho_i/\rho)^{1/2}$, may be used as guide, with T_i and ρ_i being the temperature and density at the start of the expansion. If the initial density is $\sim 10^8$ g/cm^3 we see that for $T_i \lesssim 10^8$ K the matter reaches temperatures $\sim 10^4$ K when it has expanded to densities ~ 1 g/cm^3. With the more precise results given in Figure 1, even higher initial temperatures will lead to solids; alternatively, with $T_i \sim 10^8$ K even a bit of radioactivity might be accommodated. Of course, we have left out of account possible turbulent heating and magnetic effects, but these do not seem to be large. It seems, therefore, that there are grounds for thinking that appreciable portions of some supernovae could produce solids.

As the expansion continues beyond the point just discussed, the shell will fragment into clumps, whose size is difficult to predict. However, we may note that, for at least those parts of the matter that were convecting at the outset, there should be turbulent density fluctuations in the material. These would impress a preferred fragmentation scale that may provide a useful preliminary estimate of the sizes of the initial solid chunks.

In the standard stellar convection theories, the preferred scale of motion is (unreliably) estimated to be the pressure scale height $kT/(\mu m_p g)$ where μ is the mean molecular weight and g is the acceleration of gravity. This leaves a wide latitude of choices depending on local initial conditions, but even at distances far from the imploding core, sizes of the order of tens of thousands of km are possible. Moreover, the convective velocities involved are generally quite small and the turbulent breakdown time ($\sim$ the turnover time) is long compared to the time required to expand to the point of solidification. Indeed this latter time is ~ 1 day at the expansion speeds $\gtrsim 10^8$ cm/sec seen in supernova. Hence the initial turbulent eddies will not decay hydrodynamically in the time available. It is therefore conceivable that the fragmentation initially occurs into chunks with masses like those of giant planets.

These estimates differ from those in the previous discussions of the formation of solids in supernova shells (Hoyle and Wickramasinghe,[3] Salpeter,[2] Lattimer[4]). But those discussions were aimed at forming dust such as is seen in interstellar matter and were concerned with the outermost layers of the supernova. They made the assumption that solids formed from nucleation and subsequent growth. But even if we use that approach to estimate the growth of initially small gains in the density regime discussed above, the chunks become literally mountainous in the time available.

When the chunks first form, they will be surrounded by vapor whose pressure is hard to estimate. It is clear though that further expansion will rapidly lower the vapor pressure and hence permit some evaporation of the solids. This evaporation occurs at a rapid rate until the solids cool to temperatures less than the binding energy of the particle in the solid. The cooling for the large bodies that form at first is slow enough so that considerable evaporation may occur. But note that a portion of the evaporated material may form atmospheres around the planetary solids, and dust particles could form in these.

One may envision other complications in the subsequent evolution of the expanding shell. If the initial lumps represent a heritage from earlier turbulent density fluctuations, they will have a range of sizes and densities. Probably the larger lumps will be loose amalgams of smaller lumps. Moreover, since the solids that form initially have densities comparable to the material from which they formed, they would be fairly closely packed. Then the large chunks would interact gravitationally and we may expect tidal disruption. Fragments will fly about at orbital velocities, colliding, disrupting, melting, and pulverizing. In the ensuing rocky inferno, a wide range of fragments would be produced, and we do not know how to compute the spectrum of sizes.

This assortment of solid objects would have an equally variegated range of chemical compositions, depending on the initial mass of the supernova. From the more massive supernovae, chunks of all sizes containing Si, Mg, C, and so forth may come from the various layers while from the less massive ones we may get an abundance of solid carbon. The latter would probably be mostly in the form of graphite, but, if at the outset of the expansion the incipient lattice structure were face-centered cubic, we may even get the diamond of the title. This, of course, was our goal when we set out to shop for Professor Rabi's birthday present: a unique gift for the occasion. Like all good shoppers we have been willing to go to great lengths to find the desired object. But what about delivery? Perhaps our present is best left as a diamond in the sky since even a small sample arriving at the earth with characteristic supernova velocity $\gtrsim 10^3$ km/sec would easily dissipate more energy than the great Siberian meteorite; a collision with one of the large chunks would be disastrous. Except for such unlikely and unwanted collisions, the larger chunks that may be expelled by a supernova will be very difficult to detect. Though they may be self-luminous initially, their light would be completely swamped in that phase by the supernova itself. When they have moved away from the supernova, they will have cooled somewhat and the luminosities would be low on account of their sizes, which are small by astronomical standards. Nor can they accumulate in the galactic disc: the escape velocity from our galaxy is ~ 300 km/sec and this is much less than the presumed explusion speed from the supernova. In a million years, the ejected objects will have traveled more than 3,000 light years.

At the other end of the size spectrum, we have particles which can scatter light and hence offer greater prospects for detection, but it is then not easy to identify their source. Here we have to distinguish two cases: particles that escape the galaxy and those which do not. The distinction, partly based on size, is made by the stopping power of the interstellar medium. An interstellar medium of density ρ_I will reduce the velocity of a particle of size a and density ρ_p by a factor ~ 3 in a distance $(\rho_p/\rho_I)a$. With $\rho_p/\rho_I \sim 10^{25}$ we find that the stopping distance is 3000 light years for particles of the order of a few micrometers. Thus, most of the particles that are trapped in the galaxy would become part of the interstellar medium and would look like fairly typical interstellar dust. But a range of particle sizes exist, which do get out of the galaxy and which can scatter light effectively. Such particles may be more identifiable because of their unusual location. In fact some galaxies do have halos containing dust, a celebrated example being the irregular galaxy M82, but we are not in a position to ascertain whether such dust may emanate from supernova explosions. The possibility that dust halos are provided by the several million supernova expected in the time for the dust to reach the halo may not be great, but it may be no less unlikely than other proposed origins for this dust.

We should like to mention one other possible place to look for dust from a supernova. In the Orion nebula, an active site of current star formation, there is an extensive gas cloud with high number density, as high as 10^{10} cm^{-3} according to some observers.[5] Recently, Rowan-Robinson[6] has invoked the existence of particles in the size range 10–100 μm to explain the far-infrared observations of this object. Such particles are much larger than the 0.1 μm of typical interstellar grains and their origin poses something of a puzzle. Though Rowan-Robinson's suggestion is disputed (Werner), theoreticians have not hesitated to look for explanations for such large particles, thus far in the context of normal processes of dust formation. Another possible explanation seems natural in the present discussion: a supernova may have gone off in this cloud and ejected particles of the size postulated by Rowan-Robinson. Such particles could be prevented from escaping by a cloud of density $\gtrsim 10^5$ cm^{-3}.

The prospect of a supernova explosion in such a cloud has already been contemplated by students of dense molecular clouds (e.g., C. Lada, Columbia Astronomical Colloquium). Stars are formed in these dense clouds and a sufficiently massive star could well evolve to the supernova stage in the cloud's lifetime. If the cloud is dense enough it could withstand the effects and these are currently under study.[7] It may be that such events are not all that rare, in which case some of the solids ejected from supernova would be retained in the galaxy. Professor Rabi's birthday diamonds could exist nearby. But if they do not, he will surely get over his disappointment when he learns what we have ordered for his 120th birthday.

Summary

Stable matter of very high density ρ whose initial temperature is less than about $2 \times 10^8(\rho/10^6)^{1/2}$ K is seen to become mainly liquid or solid on adiabatic expansion to densities ~ 1 g/cm^3. We have considered the speculation that significant amounts of such condensed matter might be produced in some supernova explosions where, after nuclear burning and shock heating, a very hot dense core expands adiabatically. A wide range of sizes of the expelled solid objects may be possible, of which all but the smallest of these would normally escape from the galaxy. The cut-off size for the particles prevented from escaping by interstellar gas is $\sim$ micrometers unless the supernova occurs in a dense cloud. In that case the critical size is much larger and may be comparable to larger-than-usual particles whose occurrence in a dense gas cloud in Orion has recently been proposed.

References

1. RUDERMAN, M. A. 1970. J. Pure Appl. Chem. **22:** 429.
2. SALPETER, E. E. 1974. Rev. Mod. Phys. **46:** 433.
3. HOYLE, F. & N. C. WICKRAMASINGLE. 1970. Nature **226:** 62.
4. LATTIMER, J. 1977. Bull. Amer. Astron. Soc. **9:** 2. LATTIMER, J. M., D. N. SCHRAMM & L. GROSSMAN. Preprint.
5. LEITCH-DEVLIN, M. A., T. J. MILLAR & D. A. WILLIAMS. 1976. Astrophys. Space Sci. **45:** 467.
6. ROWAN-ROBINSON, M. 1975. Monthly Notice R. Astron. Soc. **172:** 109. (See however comments by WYNN-WILLIAMS 1976. The Observatory—A Review of Astronomy). Vol. 96 (1012) 75.
7. SOLOMON, M. & M. THEYS. Private communication.

THE FINE STRUCTURE CONSTANT α

Vernon W. Hughes

Gibbs Laboratory
Physics Department
Yale University
New Haven, Connecticut 06520

I very much appreciate the opportunity to contribute an article for a volume in honor of Professor Rabi. Since so much of Professor Rabi's research involved the electromagnetic interaction both as a study of its own foundations and as a tool to learn about nuclei and elementary particles, I hope it is appropriate to review the history and status of our knowledge of the famous fine structure constant α, which characterizes the strength of the electromagnetic interaction.

Early History

The fine structure constant α can be defined by the relation:

$$\alpha = \frac{e^2}{\hbar c} \tag{1}$$

in which e is the magnitude of the charge on an electron, $\hbar$ is Planck's constant h divided by 2π, and c is the velocity of light. The quantity α is dimensionless and is approximately equal to $\frac{1}{137}$. Clearly α involves constants that characterize the discreteness of electric charge (e), quantum theory (h), and relativity theory (c).

Apparently the first suggestions[1,2] that this constant might be interesting were implied by Planck and Einstein in the period from 1905 to 1910. They noted that the elementary quantum of energy h has the same dimensions as e^2/c and approximately the same order of magnitude. They suggested that in a comprehensive theory there might be a relationship between the quantum structure of radiation and the elementary charge.

The fine structure constant was first introduced explicitly and related usefully to atomic theory by Sommerfeld.[3-6] He extended the old Bohr quantum theory of a hydrogen-like atom to include relativistic effects by solving the problem of relativistic Kepler motion together with the old quantum conditions. Equation 2 shows the resulting Sommerfeld formula for the energy levels, in which k is the azimuthal quantum number ($k = 1, 2, \ldots$), n_r is the radial quantum number ($n_r = 0, 1, 2, \ldots$), and n is the total quantum number. The quantity α ($\alpha = e^2/\hbar c$) occurs in the formula and was called the fine structure constant because it appears in the term of order $\alpha^2 Ry$ in the expansion of W which accounts for atomic fine structure. The Dirac theory gives the same formula (Equation 3) for the energy levels, but the interpretation of the quantum numbers is different. The quantum number k is given by: $k = j + \frac{1}{2}, (j = \frac{1}{2}, \frac{3}{2}, \ldots)$,

$$E = W + mc^2$$

$$W = mc^2\left\{\left[1 + \frac{\alpha^2 Z^2}{(n - k + \sqrt{k^2 - \alpha^2 Z^2})^2}\right]^{-1/2} - 1\right\}$$

(Sommerfeld Formula) (2)

$$\oint_0^{2\pi} P_\phi \, d\phi = kh; \quad \oint P_r \, dr = n_r h$$

$$n = n_r + k$$

$$W \simeq -Ry\frac{Z^2}{n^2} - \alpha^2 Ry\frac{Z^4}{n^3}\left(\frac{1}{k} - \frac{3}{4n}\right)$$

$$W = mc^2\left\{\left[1 + \frac{\alpha^2 Z^2}{(n - k + \sqrt{k^2 - \alpha^2 Z^2})^2}\right]^{-1/2} - 1\right\} \quad \text{(Dirac Formula)} \quad (3)$$

$$k = j + \frac{1}{2},$$

in which j is the total electronic angular momentum quantum number, including electron spin, which was not present in the old quantum theory.

We consider the comparison of Sommerfeld's fine structure theory with experiment. At the time of Sommerfeld's theory (1915), the constant α was known[7,8] to about 3 parts in 10^3 from independent measurements of c, e, and h/e, and the Rydberg constant Ry was known with much greater accuracy, so that the fine structure intervals could be calculated with an accuracy of better than 1%. Optical measurements[9] of fine structure intervals, principally for H and He$^+$, were of comparable accuracy and were in excellent agreement with the theoretical predictions from Sommerfeld's formula both as to the number and the spacing of the fine structure components. Later, after the development of quantum mechanics and the Dirac theory of the electron, the theory of fine structure was assumed to be correct, and α was determined from optical measurements of the fine structure of hydrogenic atoms.

Since World War II, the value of α has been based primarily on high precision radiofrequency and microwave spectroscopy measurements of atoms or solids. The original method of radiofrequency spectroscopy was, of course, Professor Rabi's molecular beam magnetic resonance method,[10] and much of the subsequent work in this field of spectroscopy is closely related in viewpoint and in spirit to the molecular beam magnetic resonance method.

Principal Measurements of α

The six principal modern measurements that determine the present-day value of α are the ac Josephson effect, the fine structure of hydrogen, the fine structure of helium, the hyperfine structure of muonium, the hyperfine structure of hydrogen, and the g-value of the electron.

The most precise value of α is obtained at present from the measurements of e/h through the ac Josephson effect and of the gyromagnetic ratio of the proton, γ_p, using the identity

$$\alpha^{-1} = \left(\frac{c}{4R_\infty \gamma_p} \frac{\mu_p}{\mu_B^e} \frac{2e}{h} \right)^{1/2} \quad (4)$$

in which c = velocity of light, R_∞ = Rydberg constant, μ_p = proton magnetic moment, μ_B^e = electron Bohr magneton, e = electronic charge, and h = Planck's constant. The ac Josephson relation applies to a junction consisting of two superconductors weakly coupled by an insulator and reads[11-15]

$$h\nu = 2eV, \quad (5)$$

where V is the dc potential difference across the junction and ν is the frequency of the alternating supercurrent. This relation is based on the viewpoint that the superconducting state is a highly correlated phase-coherent quantum state of macroscopic scale and that the supercurrent is due to the tunnelling of bound electron pairs. Theoretically, Equation 5 is believed to be general, exact, and independent of the detailed properties of the materials, geometry, or environment of the system. The experimental measurement of e/h involves application of a microwave frequency to the junction in order to zero beat with the ac supercurrent and observation of the resulting structure in the I-V dc curve. Measurements with various junctions of different materials and geometries have shown that the relative values e/h agree within the experimental error of about 1 part in 10^8. The value obtained for e/h is

$$2e/h = 4.835\ 934\ 20 \times 10^{14}\ \text{Hz}\ V_{NBS}^{-1}\ (0.030\ \text{ppm}) \quad (6)$$

in which the error is mainly associated with voltage measurements.

Recently a high precision measurement[16] of the gyromagnetic ratio of the proton in H_2O, γ_p', has been made by a nuclear induction experiment using a weak magnetic field of $12G$ produced by a solenoid. Electrical methods were employed to measure the effective dimensions of the solenoid to ppm level accuracy. The result quoted is

$$\gamma'_{pNBS} = 2.675\ 131\ 4(11) \times 10^8\ \text{rad s}^{-1}\ T_{NBS}^{-1}\ (0.42\ \text{ppm}). \quad (7)$$

The magnetic shielding correction is well known for H_2O so γ_p for the free proton can be given with essentially the same accuracy. We find by using in Equation 4 the values of e/h and γ_p from Equations 6 and 7 together with values of the other fundamental constants[15]

$$c = 2.997\ 924\ 58(1.2) \times 10^{10}\ \text{cm/sec}\ (0.004\ \text{ppm})$$
$$R_\infty = 1.097\ 373\ 143(10) \times 10^5\ \text{cm}^{-1}\ (0.01\ \text{ppm}) \quad (8)$$
$$\mu_p/\mu_B^e = 1.521\ 032\ 209(16) \times 10^{-3}\ (0.01\ \text{ppm})$$

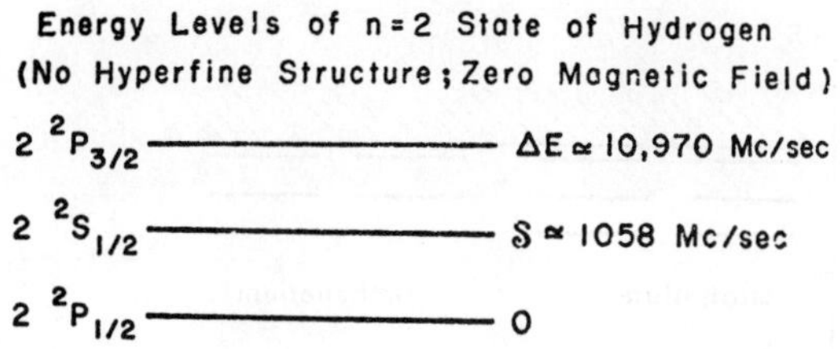

```
    Energy Levels of n=2 State of Hydrogen
    (No Hyperfine Structure; Zero Magnetic Field)

2 $^2P_{3/2}$ ─────────────────── $\Delta E \approx 10{,}970$ Mc/sec

2 $^2S_{1/2}$ ─────────────────── $S \approx 1058$ Mc/sec

2 $^2P_{1/2}$ ─────────────────── 0
```

FIGURE 1. Fine structure of hydrogen.

gives

$$\alpha^{-1} = 137.035\ 987(29)\ (0.21\ \text{ppm}). \tag{9}$$

One of the earliest precision determinations of α came from the classic experiments of Lamb and his colleagues on the fine structure in the $n = 2$ state of hydrogen (FIGURE 1). The theoretical formula for the $2^2P_{3/2}$–$2^2P_{1/2}$ fine structure interval for a hydrogen-like atom is given by

$$\Delta E(2^2P_{3/2} - 2^2P_{1/2})$$
$$= \frac{R_\infty(Z\alpha)^2}{16} \left\{ \left[1 + \frac{5}{8}(Z\alpha)^2\right] \left(1 + \frac{m}{M}\right)^{-1} - \left(\frac{m}{M}\right)^2 \cdot \left(1 + \frac{m}{M}\right)^{-3} \right. \tag{10}$$
$$\left. + 2a_e \left(1 + \frac{m}{M}\right)^{-2} + \frac{\alpha}{\pi}(Z\alpha)^2 \ln Z\alpha \right\},$$

in which m is the electron mass, M is the nuclear mass, and a_e is the anomalous g-value of the electron. The most accurate direct measurement of this interval in hydrogen has been made by an optical-pumping level-crossing method[15,17] giving

$$\Delta E(H, 2^2P_{3/2} - 2^2P_{1/2}) = 10\ 969.127(87)\ \text{MHz}\ (7.9\ \text{ppm}) \tag{11}$$

Using Equations 10 and 11 we find[15]

$$\alpha^{-1} = 137.035\ 44(52)\ (3.9\ \text{ppm}) \tag{12}$$

Other indirect determinations of this interval have been made using measured values of both the $(2^2P_{3/2}-2^2S_{1/2})$ and the $(2^2S_{1/2}-2^2P_{1/2})$ intervals.[15]

Study of the fine structure of helium in the 2^3P state has led to a precision determination of α.[18] The relevant energy levels for helium are shown in FIGURE 2. The 2^3S_1 state is 19.8 eV above the ground $^1S_0$ state and is metastable. The 2^3P state lies about 1 eV or 9231 cm^{-1} above the 2^3S_1 state. The fine structure is inverted with the $J = 0$ level being highest; the interval between the $J = 0$ and $J = 1$ levels is about 30 GHz and that between the $J = 1$ and $J = 2$ levels is about 2.3 GHz. The principal reason that a study of helium fine structure may have an advantage relative to hydrogen fine structure for determining α is that the lifetime

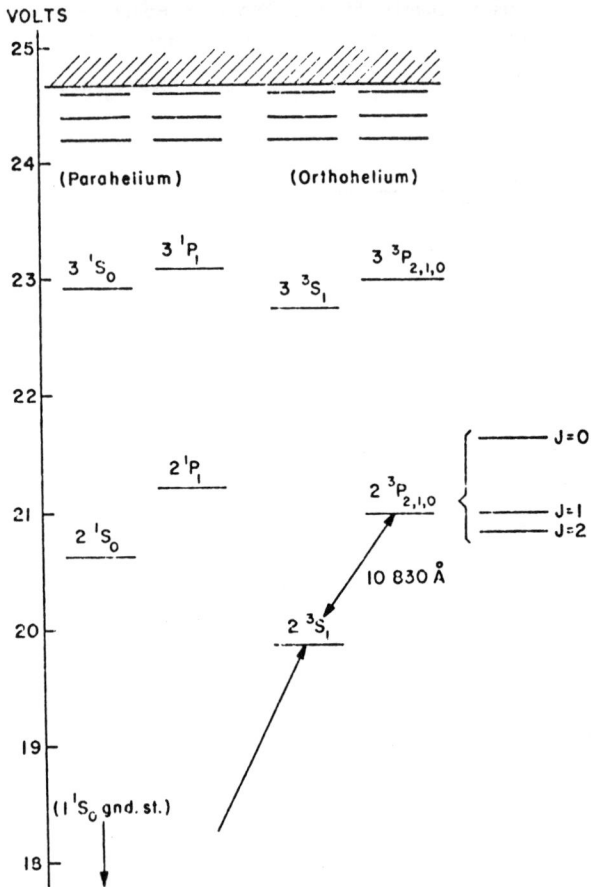

FIGURE 2. Helium energy levels.

of the 2^3P state of helium is about 100 times longer (10^{-7} sec) than that of the 2P state of hydrogen, and hence the ratio of the fine structure interval to the natural line width is about 100 times larger for He in the 2^3P state than for H in the 2P state. The helium fine structure interval $J = 0$ to $J = 1$ in the 2^3P state has been measured to a precision of 1.2 ppm, which is about an order of magnitude better than that achieved for the hydrogen fine structure interval $2^2P_{3/2}-2^2P_{1/2}$.

The experimental method, indicated in FIGURE 3, is the atomic beam magnetic resonance method adapted to optically excited states, which was first proposed by Professor Rabi and was first applied by Perl et al.[19] to measure the hyperfine structure of Na in the 3P state. In the He measurements done at Yale[20-22] He atoms in the ground 1^1S_0 state are excited by electron bombardment to the metastable 2^3S_1 state, and then pass through a conventional atomic beam magnetic resonance apparatus with inhomogeneous A- and B-fields and a homogeneous

C-field. In the C-region resonance optical radiation induces the transition from the 2^3S to the 2^3P state and, in addition, a microwave magnetic field induces a transition between different 2^3P fine structure levels. The microwave-induced resonance line has a microwave power-broadened width of about 5 MHz. The measured values for the $J = 0$ to $J = 1$ and $J = 1$ to $J = 2$ intervals in the 2^3P state are given in TABLE 1.

The theory for He fine structure is of course much more difficult than that for H fine structure. The energy levels in helium can be expressed by a power series in α

$$E_J = E_0 + \alpha^2 <\mathcal{H}_2>_J + O(\alpha^3) + \alpha^4 <\mathcal{H}_2 \frac{1}{E_0 - \mathcal{H}_0} \mathcal{H}_2>_J$$
$$+ \alpha^4 <\mathcal{H}_4>_J + O\left(\alpha^2 \frac{m}{M_{He}}\right), \quad (13)$$

in which E_0 is the Schroedinger term, $\mathcal{H}_2$ is the Breit interaction which contributes in first and second order perturbation theory, $\mathcal{H}_4$ is a higher-order spin dependent operator derived from the covariant two-particle Bethe-Salpeter equation, the term $O(\alpha^3)$ arises from the electron anomalous magnetic moment, and the term $O(\alpha^2 m/M_{He})$ arises from the nuclear recoil. Elaborate theoretical calculations have been made using extensive Hylleraas-type wave functions. A survey of these calculations is given by Lewis[23] and the latest theoretical results are given by Lewis et al.[24] and by Lewis and Serafino.[25] The theoretical values of the He fine structure intervals are given in TABLE 1. The value of α^{-1} obtained from the $J = 0$ to $J = 1$ interval is

$$\alpha^{-1} = 137.036\ 08(13)\ (0.94\ \text{ppm}). \quad (14)$$

The error is determined about equally from the experimental and theoretical inaccuracies. Efforts are in progress to improve both the experimental and theoreti-

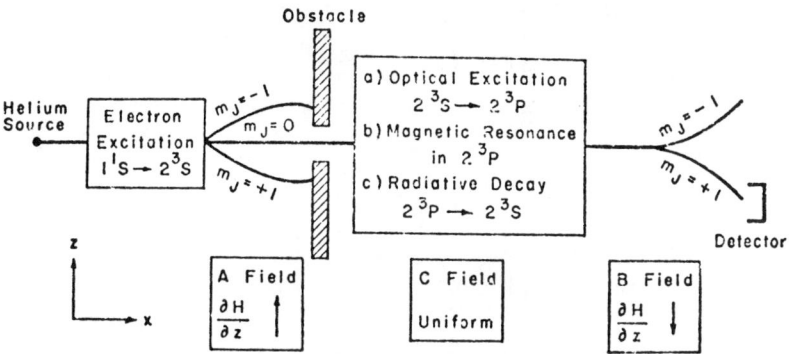

FIGURE 3. Schematic diagram of helium fine structure experiment.

TABLE 1
THEORETICAL CONTRIBUTIONS TO THE FINE STRUCTURE OF 2^3P HELIUM (IN MHz)*

Interval	$\alpha^4 mc^2$	$\alpha^5 mc^2$	$\left(\frac{m}{M}\right)\alpha^4 mc^2$	Second Order	$\alpha^6 mc^2$	ν_{theory}	ν_{exp}	$\nu_{\text{theory}} - \nu_{\text{expt}}$
ν_{01}	29 564.577 ±0.006 (0.21 ppm)	54.708	−10.707 ±0.000 44 (0.015 ppm)	11.657 ±0.042 (1.42 ppm)	−3.331 ±0.003 9 (0.13 ppm)	29 616.904 ±0.043 (1.44 ppm)	29 616.864 ±0.036 (1.2 ppm)	0.040 = 1.35 ppm
ν_{12}	2 317.203 ±0.001 8 (0.76 ppm)	−22.548	1.952 ±0.000 88 (0.39 ppm)	−6.866 ±0.081 (35 ppm)	1.542 ±0.006 8 (3.0 ppm)	2 291.283 ±0.081 (35 ppm)	2 291.196 ±0.005 (2.2 ppm)	0.087 = 37 ppm

*The values of α^{-1}, c, R_∞, and (m/M) are 137.035 987(29) (0.21 ppm), 2.997 924 58(12) × 10^{10} cm/sec (0.004 ppm), 109 737.314 3 cm^{-1} (0.009 ppm), and 1.370 934 × 10^{-4}, respectively. Thus $(1/2)\alpha^2 cR_\infty = 87.594\ 28$ GHz (0.42 ppm).

cal values of He fine structure intervals, so that a still more precise value of α will be obtained.

Muonium (μ^+e^-) is the simple hydrogen like atom consisting of a positive muon and an electron. Since the muon and the electron are structureless Dirac particles, the theory of the energy levels of muonium is a well-defined quantum electrodynamic problem. Hence study of its hyperfine structure interval provides an opportunity for a precise determination of α. FIGURE 4 shows the energy level diagram for the ground $n = 1$ state hfs levels in a magnetic field as given by the famous Breit-Rabi formula.[26] The microwave magnetic resonance method has been used to measure various transitions between these levels at both weak and strong magnetic field.[27] The most recent and precise measurement has been done

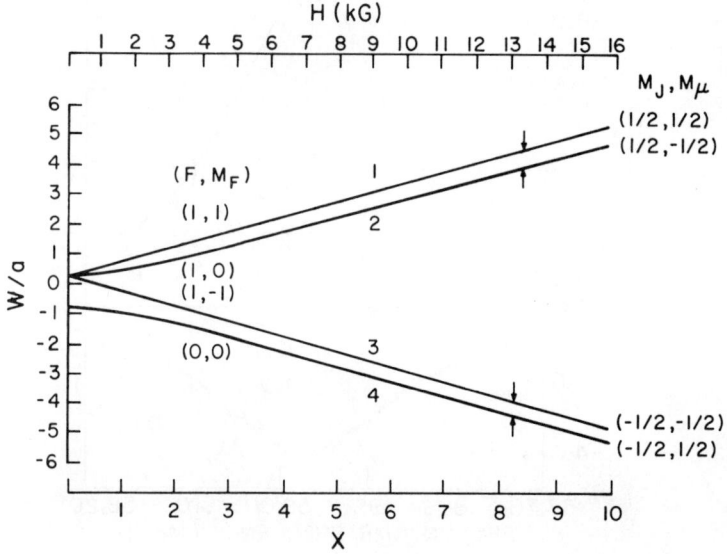

FIGURE 4. Energy levels for muonium in its $1^2S_{1/2}$ ground state in a magnetic field. The energy difference between $F = 1$ and $F = 0$ levels at $H = 0$ is the hfs interval $\Delta\nu$.

at a strong magnetic field of 13.6 kG, using the intense muon beam at the Los Alamos Meson Factory (LAMPF).[28] A resonance line is shown in FIGURE 5. The values of the hfs interval $\Delta\nu$ and of the ratio of muon to proton magnetic moments μ_μ/μ_p obtained in this experiment are:

$$\Delta\nu = 4\ 463\ 302.35(52)\ \text{kHz}\ (0.12\ \text{ppm}),$$
$$\mu_\mu/\mu_p = 3.183\ 340\ 3(44)\ (1.4\ \text{ppm}).$$
(15)

The theoretical formula for $\Delta\nu$ is given in TABLE 2.[29-31] This formula is based on conventional quantum electrodynamics and the assumption that the muon is a heavy structureless Dirac particle. The two-body bound state equation is solved in perturbation theory up to relative order α^3 in the radiative corrections and $\alpha^2 \ln\alpha$ m_e/m_μ in the relativistic recoil terms. The estimated theoretical error in the evaluation of the α^3 term is 0.6 ppm. The value obtained for α is

$$\alpha^{-1} = 137.036\ 02(9)\ (0.7\ \text{ppm}) \qquad (16)$$

where the error is due principally to the 1.2 ppm uncertainty in μ_μ/μ_p and to the estimated theoretical error of about 0.6 ppm. It seems likely that improvements in both experiment and theory will soon reduce the error in α^{-1} from muonium by a factor of about 3.

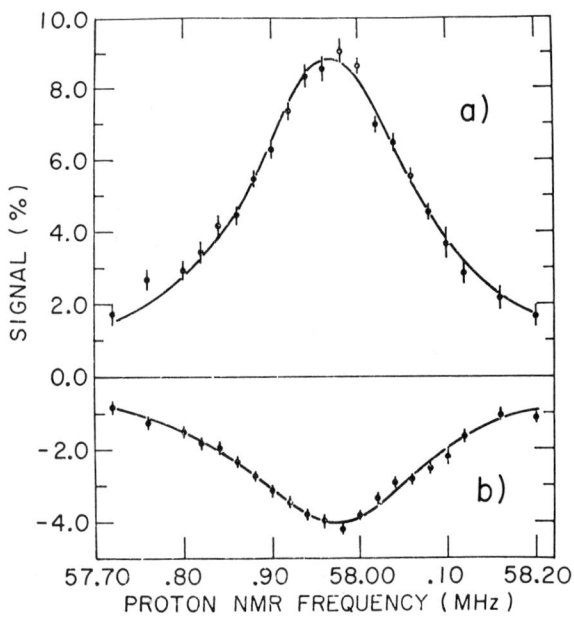

FIGURE 5. A resonance line for transition ν_{12} in a 1.7-atm Kr target, for (a) backward and (b) forward positron decay. The solid line is a least-squares fit of a Lorentzian curve to the data. The linewidth is 55 G and arises from the muon decay and power broadening. The data shown were obtained in 11 h.

The hyperfine structure interval of hydrogen in its gound state, $\Delta\nu$, can also be used to determine a value for α. The experimental value of $\Delta\nu$ is one of the most precisely known quantities in physics, and indeed serves as a frequency standard. The value of $\Delta\nu$ is given[15] in TABLE 3 to an accuracy of 1 part in $10.^{12}$ The theoretical formula is also given in TABLE 3 and is very similar to that of muonium. The principal difference occurs in the correction term δ_p, which includes not only the relativistic recoil terms present in δ'_μ for muonium but also contributions associated with proton structure, given by form factors of the proton for elastic electron-proton scattering as well as the spin dependent form factors char-

TABLE 2
Theoretical Formula for Muonium $\Delta\nu$ and Comparison with Experiment

$$\Delta\nu_{\text{theor}} = [\tfrac{16}{3}\alpha^2 cR_\infty (\mu_\mu/\mu_B{}^e)][1 + (m_e/m_\mu)]^{-3}[1 + \tfrac{3}{2}\alpha^2 + a_e + \epsilon_1 + \epsilon_2 + \epsilon_3 - \delta'_\mu]$$

$$a_e = \frac{\alpha}{2\pi} - 0.32848\frac{\alpha^2}{\pi^2} + (1.181 \pm 0.015)\frac{\alpha^3}{\pi^3}$$

$$\epsilon_1 = \alpha^2(\ln 2 - 5/2)$$

$$\epsilon_2 = -\frac{8\alpha^3}{3\pi}\ln\alpha\left(\ln\alpha - \ln 4 + \frac{281}{480}\right)$$

$$\epsilon_3 = \frac{\alpha^3}{\pi}(18.4 \pm 5)$$

$$\delta'_\mu = \frac{m_e}{m_\mu}\left\{\frac{3\alpha}{\pi}\left[1 - \left(\frac{m_e}{m_\mu}\right)^2\right]^{-1}\ln\frac{m_\mu}{m_e} + 2\alpha^2\ln\alpha\left[1 + \left(\frac{m_e}{m_\mu}\right)\right]^{-2}\right\}$$

$\alpha^{-1} = 137.035\ 987(20)\ (0.21\ \text{ppm})$

$R_\infty = 1.097\ 373\ 143(10) \times 10^5\ \text{cm}^{-1}\ (0.01\ \text{ppm})$

$c = 2.997\ 924\ 58(1.2) \times 10^{10}\ \text{cm/sec}\ (0.004\ \text{ppm})$

$\mu_\mu/\mu_B{}^e = (\mu_\mu/\mu_p)(\mu_p/\mu_B{}^e)$

$\mu_p/\mu_B{}^e = 1.521\ 032\ 209(16) \times 10^{-3}\ (0.01\ \text{ppm})$

$a_e = 0.001\ 159\ 656\ 7(35)\ (3\ \text{ppm})$

$m_\mu/m_e = 206.768\ 51(25)(1.2\ \text{ppm})$

$\mu_\mu/\mu_p = 3.183\ 341\ 7(39)(1.2\ \text{ppm}) \begin{cases} 3.183\ 340\ 3(44)\ [1.4\ \text{ppm}]\mu^+e^- \\ 3.183\ 346\ 7(82)\ [2.6\ \text{ppm}]\ \mu^+\ \text{in}\ H_2O \end{cases}$

$\Delta\nu_{\text{theor}} = \alpha^2[8.381\ 577\ 78 \pm 1.4\ \text{ppm}]$

$\Delta\nu_{\text{exp}} = 4\ 463\ 302.35(52)\ \text{kHz}\ (0.12\ \text{ppm})$

$\alpha^{-1} = 137.036\ 02(9)\ (0.7\ \text{ppm})$

TABLE 3
Hyperfine Structure Interval for Hydrogen*

$$\Delta\nu_{\text{theor}} = \left(\frac{16}{3}\alpha^2 cR_\infty \frac{\mu_p}{\mu_B{}^e}\right)\left(1 + \frac{m_e}{m_p}\right)^{-3}\left(1 + \frac{3}{2}\alpha^2 + a_e + \epsilon_1 + \epsilon_2 + \epsilon_3 + \delta_p\right)$$

$$\alpha_e = \frac{\alpha}{2\pi} - 0.32848\frac{\alpha^2}{\pi^2} + (1.181 \pm 0.015)\frac{\alpha^3}{\pi^3};\ \epsilon_1 = \alpha^2\left(\ln 2 - \frac{5}{2}\right)$$

$$\epsilon_2 = -\frac{8\alpha^3}{3\pi}\ln\alpha\left(\ln\alpha - \ln 4 + \frac{281}{480}\right);\ \epsilon_3 = \frac{\alpha^3}{\pi}(18.4 \pm 5)$$

δ_p = Proton recoil and proton structure term

$\delta_p = \delta_p(\text{rigid}) + \delta_p(\text{polarizability})$

$\delta_p = -34.6(9) \times 10^{-6} + \delta_p(\text{pol})\ [\ |\delta_p(\text{pol})| < 4\ \text{ppm}]$

$\Delta\nu_{\text{expt}} = 1420\ 405\ 751.7667(10)\ \text{Hz}$

*From Cohen and Taylor[15] and Brodsky and Drell.[52]

acterizing deep inelastic electron-proton scattering. The latter are associated with the proton polarizability correction in the term δ_p. The uncertainty in δ_p, principally that associated with the polarizability correction, is estimated to be about 4 ppm. Hence the uncertainty in α from H hfs as given in TABLE 3 is about 1.6 ppm.

The anomalous g-value of the electron, $a_e = (g_e - 2)/2$ provides another source for a precise value of α. The current theoretical value of a_e is[27,32-34]

$$a_e = \frac{\alpha}{2\pi} - 0.328\ 479 \left(\frac{\alpha^2}{\pi^2}\right) + (1.181 \pm 0.015)\left(\frac{\alpha^3}{\pi^3}\right), \quad (17)$$

where the numerical uncertainty in the sixth-order contribution is given. Efforts are beginning to calculate the eighth-order $(\alpha/\pi)^4$ radiative correction. Hadronic corrections will be at most of order $(\alpha/\pi)^4$ in magnitude.

Two modern measurements provide the experimental value of a_e. The first, which has until recently for many years given the most precise value, is the well-known g-2 experiment in which the difference frequency ω_a between the spin precession frequency and the cyclotron frequency in a magnetic field is measured.

$$\omega_a = \left(\frac{e}{mc}\right) a_e B. \quad (18)$$

The value obtained for a_e was[35,36]

$$a_e = (1\ 159\ 656.7 \pm 3.5) \times 10^{-6}. \quad (19)$$

Recently a second method based on observation of the spin-cyclotron beat frequency and the cyclotron frequency of isolated 1 meV electrons in an electric-magnetic field trap has given the still higher precision value[37]

$$a_e = (1\ 159\ 652.41 \pm 0.20) \times 10^{-6}. \quad (20)$$

Use of Equations 17 and 20 gives

$$\alpha^{-1} = 137.035\ 979(60)(0.43\ \text{ppm}) \quad (21)$$

in which the error is about equally due to theoretical and experimental uncertainties.

TABLE 4 and FIGURE 6 summarize the various determinations of α. They all

TABLE 4
VALUES* OF THE FINE STRUCTURE CONSTANT

Method	α^{-1} Value
Josephson e/h and γ_p	137.035 987(29) (0.21 ppm)
Hydrogen Fine Structure, $2^2P_{3/2}-2^2P_{1/2}$	137.035 44(52) (3.9 ppm)
Helium Fine Structure, $2^3P_0-2^3P_1$	137.036 08(13) (0.94 ppm)
Muonium Hyperfine Structure	137.036 02(9) (0.7 ppm)
Hydrogen Hyperfine Structure	137.035 97(22) (1.6 ppm)
Electron g-Value	137.035 979(60) (0.43 ppm)

*From Mohr.[53]

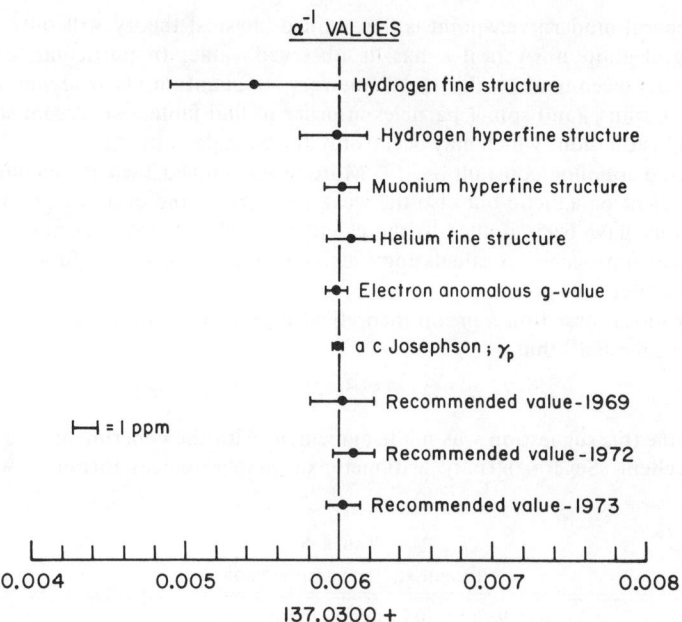

FIGURE 6. Values of α from different methods.

agree reasonably well within their errors. The most precise value is that based on the measurements of e/h by the ac Josephson effect and of γ_p.

The agreement of the values of α from these different sources can be regarded as confirming our quantitative understanding of the physics involved in these different systems.

A precise value for α is required in the overall determination of the fundamental atomic constants[15] and is essential for a comparison of theory and experiment for many basic quantities such as the Lamb shift, positronium fine structure, the muon g-value, and muonic atom energy levels. Our faith in QED is based in substantial part on these successful comparisons.

Theory of the Value of α

The fine structure constant α is a dimensionless pure number, which of course characterizes the strength of the electromagnetic interaction. It is reasonable to expect that the numerical value of α can be calculated in a comprehensive theory of the elementary particles, or perhaps of the universe.

Attempts to provide a theory for α date back at least to Eddington's famous proposal[38] of 1936, which relates α^{-1} to the number of dimensions of a certain wavefunction. In Heisenberg's unified field theory of elementary particles, which is characterized as a nonlinear spinor theory, quantum electrodynamics is incorporated and a value for α is calculated.[39] If magnetic monopoles exist[40] of strength g, one might expect the relationship $\alpha = e/(2g)$.

A general modern viewpoint is that unified physical theory will only provide meaningful, finite answers if α has its observed value. In particular, extensive studies have been made within the framework of quantum electrodynamics only for spin 0, spin $\frac{1}{2}$, and spin 1 particles in order to find finite, convergent solutions of the field equations which may occur only for a single value of α or of the unrenormalized coupling constant α_0.[41–44] More recent unified theories including not only the electromagnetic but also the weak and strong and even the gravitational interactions have been applied to the calculation of α.[45–48] As yet none of these theoretical approaches to calculating α can be regarded as successful or as a generally accepted viewpoint.

A proposal based on a group theoretical argument without obvious physical content suggested[49] that

$$\alpha^{-1} = 2^{19/4} 3^{-7/4} 5^{1/4} \pi^{11/4} = 137.036\ 082.$$

At the time this suggestion was made agreement with the experimental value of α was excellent. Several strictly arithmetic or numerological formulae were de-

TABLE 5
NUMERICAL EXPRESSIONS FOR α^{-1}

$\alpha^{-1} = 137.035\ 987(29)$ (0.21 ppm) experimental		
$\alpha^{-1} = 2^{19/4} 3^{-7/4} 5^{1/4} \pi^{11/4}$	$= 137.036\ 082$	Wyler
$\alpha^{-1} = 2^{-19/4} 3^{10/3} 5^{17/4} \pi^{-2}$	$= 137.035\ 938$	Roskies
$\alpha^{-1} = 2^{-13/4} 3^{17/4} 5^{2/3} \pi^{5/4}$	$= 137.036\ 163$	
$\alpha^{-1} = 2^{2/3} 3^{7/3} 5^{11/3} \pi^{-7/2}$	$= 137.036\ 120$	
$\alpha^{-1} = 2^{5/3} 3^{-8/3} 5^{5/2} \pi^{7/3}$	$= 137.036\ 007$	
$\alpha^{-1} = 2^{8/3} 3^{3/4} 5^{-1/2} \pi^{8/3}$	$= 137.036\ 289$	
$\alpha^{-1} = (5/2)^{1/2} (2)^{2/3} e^4$	$= 137.035\ 97$	Giaever
$\alpha^{-1}_{\text{Wyler}} - \alpha^{-1}_{\text{exp}} = 0.000\ 095 \pm 0.000\ 029$		

veloped by computer searches[50] that also gave values of α in agreement with the experimental value and hence lessened the marvel of the agreement of the group theoretical value of α with the experimental value. The current best experimental value for α does not agree well with the group theoretical value. TABLE 5 summarizes this situation.

All in all we do not yet have a convincing theory for α, or even a generally accepted viewpoint for such a theory. We must apparently await further understanding of physics before a good theory of the value of α can be given.[51]

References

1. EINSTEIN, A. 1909. Phys. Z. **10**: 185.
2. KLEIN, M. J. 1966. Phys. Today **19**(November): 23.
3. SOMMERFELD, A. 1915. Munchener Ber. 425, 459; 1916. Munchener Ber. 131.

4. SOMMERFELD, A. 1916. Ann. Physik **51**: 125.
5. SOMMERFELD, A. 1934. Atomic Structure and Spectral Lines. Vol. 1. 3rd edit. Methuen and Co. Ltd. London.
6. RUBINOWICZ, A. 1933. Handbuch der Physik. Vol. 24/1:1 J. Springer. Berlin.
7. MILLIKAN, R. A. 1917. The Electron. University of Chicago Press. Chicago, Ill.
8. COHEN, E. R., K. M. CROWE & J. W. M. DUMOND. 1957. Fundamental Constants of Physics. Interscience. New York, N.Y.
9. PASCHEN, F. 1916. Ann. Physik **50**: 901.
10. RABI, I. I., et al. 1939. Phys. Rev. **55**: 526.
11. JOSEPHSON, B. D. 1962. Phys. Lett. **1**: 251.
12. JOSEPHSON, B. D. 1965. Adv. Phys. **14**: 419.
13. PARKER, W. H., et al. 1969. Phys. Rev. **177**: 639.
14. FINNEGAN, T. F., et al. 1971. Phys. Rev. B **4**: 1487.
15. COHEN, E. & B. N. TAYLOR. 1973. J. Phys. Chem. Ref. Data **2**: 663.
16. OLSEN, P. T. & E. E. WILLIAMS. 1976. Atomic Masses and Fundamental Constants 5. J. H. Sanders & A. H. Wapstra, Eds.: 538. Plenum Press. New York, N.Y.
17. BAIRD, J. C. et al. 1972. Phys. Rev. A **5**: 564.
18. HUGHES, V. W. 1970. Facets of Physics. D. A. Bromley & V. W. Hughes, Eds. Academic Press. New York, N.Y.
19. PERL, M. L., I. I. RABI & B. SENITSKY. 1955. Phys. Rev. **98**: 611.
20. PICHANICK, F. M. J., et al. 1968. Phys. Rev. **169**: 55.
21. LEWIS, S. A., F. M. J. PICHANICK, & V. W. HUGHES. 1970. Phys. Rev. A **2**: 86.
22. KPONOU, A., et al. 1971. Phys. Rev. Lett. **26**: 1613.
23. LEWIS, M. L. 1975. Atomic Physics 4. G. zu Putlitz, E. W. Weber, and A. Winnacker, Eds.: 105. Plenum Press, New York, N.Y.
24. LEWIS, M. L., P. H. SERAFINO & V. W. HUGHES. 1976. Phys. Lett. **58A**: 125.
25. LEWIS, M. L. & P. H. SERAFINO. 1977. Second-order contribution to the fine structure of helium from all intermediate states. Phys. Rev. A (December).
26. BREIT, G. & I. RABI. 1931. Phys. Rev. **38**: 2082.
27. HUGHES, V. W. & T. KINOSHITA. 1977. Muon Physics. Vol. I. Chapt. 2. Academic Press. New York, N.Y.
28. CASPERSON, D. E., et al. 1977. Phys. Rev. Lett. **38**: 956, 1504.
29. BRODSKY, S. J. & G. W. ERICKSON. 1966. Phys. Rev. **148**: 26.
30. LAUTRUP, B. E., et al. 1972. Phys. Rep. Phys. Lett. C (Netherlands) **3C**: 193.
31. LEPAGE, G. P. 1977. SLAC PUB-1900. Phys. Rev. A (September).
32. CVITANOVIC, P. & T. KINOSHITA. 1974. Phys. Rev. D **10**: 3991, 4007.
33. LEVINE, M. J. & R. ROSKIES. 1976. Phys. Rev. D **14**: 2191.
34. CALMET, S. et al. 1977. Rev. Mod. Phys. **49**: 21.
35. WESLEY, J. C. & A. RICH. 1971. Phys. Rev. A **4**: 1341.
36. RICH, A. & J. C. WESLEY. 1972. Rev. Mod. Phys. **44**: 250.
37. VAN DYCK, R. S., JR., P. B. SCHWINBERG & H. G. DEHMELT, 1977. Phys. Rev. Lett. **38**: 310.
38. EDDINGTON, A. Relativity Theory of Protons and Electrons. Cambridge University Press, Cambridge, England.
39. HEISENBERG, W. 1966. Introduction to the Unified Field Theory of Elementary Particles. Intersciences. New York, N.Y.; DÜRR, H. P., et al. 1965. Nuovo Cimento **38**: 1220.
40. DIRAC, P. A. M. 1948. Phys. Rev. **74**: 817.
41. GELL-MANN, M. & F. E. LOW. 1954. Phys. Rev. **95**: 1300.
42. SALAM, A. 1963. Phys. Rev. **130**: 1287; SALAM, A. & R. DELBOURGO. 1964. Phys. Rev. **135**: B1398.
43. JOHNSON, K., et al., 1964, 1967. Phys. Rev. **136**: B1111. **163**: 1699.
44. ADLER, S. L. 1972. Phys. Rev. D **5**: 3021.
45. ROY, S. M. & A. S. VENGURLEKAR. 1976. Nucl. Phys. **B114**: 449.
46. ARNOLD, R. C. 1977. Phys. Lett. **67B**: 91.
47. MOTZ. L. 1977. Nuovo Cimento **37A**: 13.

48. TERAZAWA, H., et al. 1977. Phys. Rev. D **15**: 1181.
49. WYLER, M. A. 1969. C. R. Acad. Sci. Paris A **269**: 743; 1971. C. R. Acad. Sci. Paris, **272**: 186.
50. ROSKIES, R. 1971. Phys. Today **24**(November), 9.
51. HUGHES, V. W. 1968. A Tribute to I. I. Rabi. Columbia University Symposium, 1967.
52. BRODSKY, S. J. & S. D. DRELL. 1970. Ann. Rev. Nucl. Sci. **20**: 147.
53. MOHR, P. J. 1977. Atomic Physics 5. R. Marrus, M. Prior & H. Shugart, Eds.: 37. Plenum Press, New York, N.Y.

VECTOR AND TENSOR GAUGE PARTICLES IN SL(6,c) THEORY

C. J. Isham

*Imperial College
London, England*

Abdus Salam

*International Centre for Theoretical Physics
Trieste, Italy*

*Department of Physics
Imperial College
London, England*

J. Strathdee

*International Centre for Theoretical Physics
Trieste, Italy*

In a number of papers, the authors have presented an SL(6,c) gauge-invariant theory[1-6] of strong gravity, which would describe a nonet of spin-2^+ mesons interacting with hadrons. Such a theory is likely to produce a partial confinement of quarks through the well-known confining influence of spin-2 (strong) gravitons. However, physically, it is important not only to have (partial) confinement; the strong forces must also exhibit saturation. That is, there should be a force that makes for a binding of the three quark (or $q\bar{q}$) systems but not for two quark (qq) or four quark ($qqqq$) systems. Colored spin-1 gluons are supposed to provide such a force. All in all, therefore, a correct strong-interaction theory should contain spin-2 as well as spin-1 colored particles, mediating strong interactions. It is the purpose of this note—dedicated to Prof. I. I. Rabi—to show that a simple modification of the SL(6,c) theory proposed earlier is capable of describing the interactions of both (spin-2 as well as spin-1) types of mesons. We salute Prof. Rabi, not only for being a great physicist of this century, but also for his integrity and incisiveness, so completely mixed with his great humanity.

The Lagrangian proposed in Reference 1 for the description of systems with local SL(6,c) symmetry was so constructed that the gauge system reduced simply to a massive nonet of 2^+ particles (at least in the free field approximation). Our purpose now is to suggest a modification of this Lagrangian, which will allow massive octets of 1^+ (and 1^-) particles as well. This can be achieved by a rearrangement of terms without involving any new field variable.

Our gauge system comprises three distinct types of field, B_μ, S and L_μ, which transform under local SL(6,c) according to

$$B_\mu \to \Omega B_\mu \Omega^{-1} + \frac{1}{i} \Omega \partial_\mu \Omega^{-1},$$

$$S \to \Omega S,$$

$$L_\mu \to \Omega L_\mu \Omega^{-1}. \tag{1}$$

Their tensorial components can be exhibited in the Dirac basis as follows:

$$B_\mu = \left(B_\mu{}^k + \frac{1}{2}B_{\mu[ab]}{}^k \sigma_{ab} + B_{\mu 5}{}^k \gamma_5\right)\frac{\lambda^k}{2},$$

$$S = \exp i\left(P^k + \frac{1}{2}P_{[ab]}{}^k \sigma_{ab} + P_5{}^k \gamma_5\right)\frac{\lambda^k}{2},$$

$$L_\mu = (L_{\mu a}{}^k \gamma_a + L_{\mu a 5}{}^k i\gamma_a \gamma_5)\lambda^k. \tag{2}$$

All tensorial components are real in this basis. Notice that S is thereby constrained to be an SL(6,c) matrix. Its components P belong to a nonlinear realization of the local symmetry.

We do not wish to regard all the fields that appear in expressions (2) as independent variables. For the sake of convenience, we shall impose a number of constraints among them, by means of which some of the components are to be eliminated. These constraints must, of course, be compatible with the local symmetry. Now, according to Equation 1, the combinations

$$b_\mu = S^{-1} B_\mu S + \frac{1}{i} S^{-1} \partial_\mu S,$$

$$l_\mu = S^{-1} L_\mu S, \tag{3}$$

are local *invariants*. We are therefore entitled to impose the constraints

$$l_{\mu a 5}{}^k = 0,$$

$$l_{\mu a}{}^k - l_{a\mu}{}^k = 0, \tag{4}$$

which serve to eliminate $144 + 54 = 198$ components. (It should be remarked that these constraints are covariant with respect to the *global* Poincaré and SU(3) symmetries, under which $l_{\mu a}{}^k$ and $l_{\mu a 5}{}^k$ transform as tensor and pseudotensor nonets, respectively.) Which components are removed by means of Equations 4 is to some extent a matter of choice. For example, the components $L_{\mu a 5}{}^k$ and $L_{\mu a}{}^k - L_{a\mu}{}^k$ could be expressed in terms of P and the remaining L variables. Alternatively, P and certain of the L values could be expressed in terms of the remaining L variables. In this case, the elimination of P would leave the Lagrangian as a (highly nonlinear) function of L_μ and B_μ where the components of L_μ are subject to $198 - 70 = 128$ *nonlinear* but *covariant* constraints.

In order for the above considerations to have any sense, it is important that the action should be minimized by a Poincaré and SU(3) invariant vacuum solution, in which both $\langle L_\mu \rangle$ and $\langle S \rangle$ are nonvanishing. We shall in fact require that the values

$$\langle B_\mu \rangle = 0, \quad \langle S \rangle = 1, \quad \langle L_\mu \rangle = \gamma_\mu, \tag{5}$$

be a solution of the classical equations of motion.

Comparing Equation 5 with Equation 3 we find

$$\langle l_{\mu a}{}^k \rangle = \eta_{\mu a} \delta^{k0}. \tag{6}$$

This suggests that the 4×4 matrix field $l_{\mu a}{}^0$ is invertible. We therefore define the "vierbein" fields,

$$f_\mu = f_{\mu a} \gamma_a \quad \text{and} \quad f_\mu^{-1} = f_{\mu a}^{-1} \gamma_a, \tag{7a}$$

with

$$f_{\mu a} = l_{\mu a}{}^0 = l_{a\mu}{}^0. \tag{7b}$$

This definition is by no means unique. Equally acceptable would be the symmetric square root of the expression

$$\sum_0^8 l_{m\alpha}{}^k l_{\nu a}{}^k,$$

which includes octet contributions.* The basic requirements are that $f_{\mu a}$ should transform as a Poincaré tensor and global SU(3) singlet and be invertible, i.e.,

$$f_{\mu a} f_{\nu a}^{-1} = \eta_{\mu \nu}, \quad f_{\mu a} f_{\mu b}^{-1} = \eta_{ab}. \tag{8}$$

Both f_μ and f^μ are local invariants. They can be transformed, with the help of S, into local tensors, viz.,

$$F_\mu = S f_\mu S^{-1} \quad \text{and} \quad F_\mu^{-1} = S f_\mu^{-1} S^{-1}. \tag{9}$$

These fields, which we use to denote rather intricate nonlinear combinations of the basic fields L_μ and P, will prove very useful in the construction of the Lagrangian. With the covariant quantities defined by

$$\nabla_\mu L_\nu = \partial_\mu L_\nu + i[B_\mu, L_\nu],$$

$$C_\mu = \frac{i}{4} F_\lambda S \nabla_\mu S^{-1} F_\lambda^{-1} \left[= \frac{1}{4} F_\lambda \left(B_\mu - \frac{1}{i} S \partial_\mu S^{-1} \right) F_\lambda^{-1} \right],$$

$$C_{\mu\nu} = \nabla_\mu C_\nu - \nabla_\nu C_\mu, \tag{10}$$

we write the proposed Lagrangian in the form

$$\mathcal{L} = \frac{1}{8} \text{Tr} \left[-\frac{1}{\kappa^2} (\nabla_\mu L_\nu \nabla_\nu L_\mu - \nabla_\mu L_\mu \nabla_\nu L_\nu) \right.$$

$$-\frac{1}{16\kappa'^2} F_\lambda C_{\mu\nu} F_\lambda^{-1} C_{\mu\nu}$$

$$+ \frac{M'^2}{8\kappa'^2} F_\mu C_\nu F_\mu C_\nu$$

$$\left. - \frac{3M^2}{2\kappa^2} L_\mu L_\mu + \frac{M^2}{8\kappa^2} L_\mu [L_\mu, L_\nu] L_\nu \right], \tag{11}$$

where the constraints (4) and (7b) among L_μ, C_μ, and F_μ are tacitly understood. The parameters κ and κ' are coupling constants, while M and M' are masses.

*Such definitions will probably turn out to be equivalent in the long run when the possibility of making nonlinear field transformations in the Lagrangian is taken into account.

Since the Lagrangian (11) is invariant with respect to local SL(6,c) transformations, we can impose 70 gauge conditions on the solutions without losing any information. In contrast to the covariant equations of constraint discussed above, the gauge conditions must, as usual, violate the SL(6,c) symmetry in order to fix a gauge. For example, one might require solutions to satisfy $\partial_\mu L_\mu = 0$. However, by far the most convenient conditions are simply

$$S = 1 \quad \text{or} \quad P = 0. \tag{12}$$

In this gauge we have $L_\mu = l_\mu$ and $F_\mu = f_\mu$, or

$$L_\mu = \gamma_\mu + \varphi_{(\mu a)}{}^k \gamma_a \lambda^k,$$

$$F_\mu = f_{\mu a} \gamma_a,$$

and

$$C_\mu = \frac{1}{4} F_\lambda B_\mu F_\lambda^{-1}. \tag{13a}$$

The great advantages which follow from the introduction of F_μ and F_μ^{-1} into the Lagrangian can now be made apparent. Because of the identities (8) it follows that

$$C_\nu = \frac{1}{4} F_\mu B_\nu F_\mu^{-1} = \frac{1}{4} f_{\mu a} f_{\mu b}^{-1} \gamma_a \left(B_\nu{}^k + \frac{1}{2} B_{\nu[cd]}{}^k \sigma_{cd} + B_{\nu 5}{}^k \gamma_5 \right) \frac{\lambda^k}{2} \gamma_b$$

$$= (B_\nu{}^k - B_{\nu 5}{}^k \gamma_5) \frac{\lambda^k}{2},$$

i.e., the tensor part of B_ν is extinguished. We have used here the basic identity

$$\gamma_a \sigma_{cd} \gamma_a = 0. \tag{13b}$$

The C-containing terms in Equation 11 therefore reduce to the form

$$\frac{1}{\kappa^2} \left(-\frac{1}{4} B_{\mu\nu}{}^k B_{\mu\nu}{}^k + \frac{1}{2} M'^2 B_\mu{}^k B_\mu{}^k - \frac{1}{4} B_{\mu\nu 5}{}^k B_{\mu\nu 5}{}^k + \frac{1}{2} M'^2 B_{\mu 5}{}^k B_{\mu 5}{}^k \right). \tag{14}$$

where

$$B_{\mu\nu}{}^k = \partial_\mu B_\nu{}^k - \partial_\nu B_\mu{}^k - 2 f^{klm} [B_\mu{}^l B_\nu{}^m + B_{\mu 5}{}^l B_{\nu 5}{}^m]$$

$$B_{\mu\nu 5}{}^k = \partial_\mu B_{\nu 5}{}^k - \partial_\nu B_{\mu 5}{}^k. \tag{15}$$

Note that

$$B_{\mu\nu}{}^k = B_{\mu\nu}{}^{\text{Yang-Mills}} - 2 f^{klm} B_{\mu 5}{}^l B_{\nu 5}{}^m,$$

where

$$B_{\mu\nu}{}^{\text{Yang-Mills}}$$

is the conventional Yang-Mills covariant field strength. Thus, in the Lagrangian we have constructed, the (1^-) fields $B_\mu{}^k$ indeed act like Yang-Mills fields, while the (1^+) fields $B_\mu{}^{k5}$ do not. The terms in Equation 11 that do not contain C are the same as in References 1–6. They cause the propagation of a massive 2^+ nonet.

Thus, our Langrangian (11) describes the propagation of a 2^+ nonet, a 1^- Yang-Mills octet together with a 1^+ octet. The coupling strengths associated with the (1^-) and (1^+) fields, though equal among themselves, need not equal the coupling strength for the 2^+ fields.

To conclude, the gauge interactions of quark matter are specified by the Lagrangian

$$\mathcal{L}_{\text{matter}} = \frac{i}{2}(\bar{\Psi}L_\mu \nabla_\mu \Psi - \nabla_\mu \bar{\Psi}L_\mu \Psi) - m\bar{\Psi}\Psi,$$

$$= \frac{i}{2}(\bar{\Psi}L_\mu \partial_\mu \Psi - \partial_\mu \bar{\Psi}L_\mu \Psi) - \frac{1}{2}\bar{\Psi}\{L_\mu, B_\mu\}\Psi - m\bar{\Psi}\Psi,$$

$$= L_{\mu a}{}^k \frac{i}{2}\bar{\Psi}\gamma_a \lambda^k \overleftrightarrow{\partial}_\mu \Psi - m\bar{\Psi}\Psi$$

$$+ L_{\mu a 5}{}^k \frac{i}{2}\bar{\Psi}i\gamma_a\gamma_5 \lambda^k \overleftrightarrow{\partial}_\mu \Psi$$

$$- [d^{klm}(L_{\mu d}{}^k B_\mu{}^l + \epsilon_{abcd}L_{\mu a 5}{}^k B_{\mu[bc]}{}^l)$$

$$+ f^{klm}(-L_{\mu a}{}^k B_{\mu[ad]}{}^l + L_{\mu d 5}{}^k B_{\mu 5}{}^l)]\bar{\Psi}\gamma_d \frac{\lambda^m}{2}\Psi$$

$$- [d^{klm}(L_{\mu d 5}{}^k B_\mu{}^l - \epsilon_{abcd}L_{\mu a}{}^k B_{\mu[bc]}{}^l)$$

$$+ \cdot f^{klm}(-L_{\mu a 5}{}^k B_{\mu[ad]}{}^l + L_{\mu d}{}^k B_{\mu 5}{}^l)]\bar{\Psi}i\gamma_d\gamma_5 \frac{\lambda^m}{2}\Psi.$$

Notice that half of these terms disappear in the gauge $S = 1$ where $L_{\mu a 5}{}^k$ vanishes. In linear approximation one is left with

$$\mathcal{L}_{\text{matter}} = \bar{\Psi}\left(\frac{i}{2}\gamma_\mu \overleftrightarrow{\partial}_\mu - m\right)\Psi + \varphi_{(\mu a)}{}^k \frac{i}{2}\bar{\Psi}\gamma_a \lambda^k \overleftrightarrow{\partial}_\mu \Psi$$

$$- B_\mu{}^k \bar{\Psi}\gamma_\mu \frac{\lambda^k}{2}\Psi + \epsilon_{\mu bcd}B_{\mu[bc]}{}^k \bar{\Psi}i\gamma_d\gamma_5 \frac{\lambda^k}{2}\Psi.$$

As expected, vector field $B_\mu{}^k$ couples in this approximation like a true Yang-Mills field, while the axial vector $B_{\mu 5}{}^k$ does not. In fact, the axial vector field has no coupling to matter in this approximation. If we wish to construct a Yang-Mills theory for both 1^- and 1^+ particles, we shall need to extend the considerations of this paper to the larger symmetry group SL(6,c) × SL(6,c) in the manner of Section IV of Reference 1.

References

1. ISHAM, C. J., A. SALAM & J. STRATHDEE. 1973. Phys. Rev. **D8**: 2600.
2. ISHAM, C. J., A. SALAM & J. STRATHDEE. 1974. Phys. Rev. **D9**: 1702.
3. ISHAM, C. J., A. SALAM & J. STRATHDEE. 1973.
4. SALAM, A. 1974. *In* Fundamental Interactions in Physics. :55. Plenum Publishing Corporation. New York, N.Y.
5. ISHAM, C. J., A. SALAM & J. STRATHDEE. 1972. Lett. Nuovo Cimento **5**: 969.
6. SALAM, A. & J. STRATHDEE. 1976. International Centre for Theoretical Physics. Trieste, Italy.

SOME HISTORY OF THE HYDROGEN FINE STRUCTURE EXPERIMENT

W. E. Lamb, Jr.

*Department of Physics
University of Arizona
Tuscon, Arizona 85721*

Introduction

One of the major contributions of I. I. Rabi to physics was his organization of the Columbia Radiation Laboratory. This dates from early in World War II and was staffed largely by former students of Rabi and by physicists who taught at colleges and universities in the New York City area. Perhaps the greatest wartime achievements of this laboratory were the development* of the Rising Sun magnetron at wavelengths of 1.25 cm and shorter, and a measurement† of the absorption of microwaves in water vapor.

In the postwar years two Nobel Prizes have so far been awarded for researches carried out at the Columbia Radiation Laboratory. Rabi did not make any direct contribution to these works, but each of them received incalculable benefits from the scientific and physical environment made possible by him at the laboratory.

The researches of the Columbia Radiation Laboratory were for many years described in its *Quarterly Progress Report.* After the war this was distributed to a large number of scientists known to be interested in microwave physics. The proposal of Zeiger and Gordon (and Townes) for the work on the maser appeared in the December 1951 *Quarterly Progress Report* long before the published account‡ of success in 1954. The 1951 proposal has now been published elsewhere.§ The first account of the proposed Lamb-Retherford measurements appeared in the October 1946 issue. Success of the experiment was obtained in April 1947 and publication¶ followed in August of that year.

Thirty-one years have now passed since the release of the October 1946 Report. It seems appropriate to make it available to a wider audience in this Festschrift to Rabi. The complete test of the Lamb-Retherford proposal, with correction of a few misprints, is given below. This document makes possible for the historian an interesting comparison between the anticipated work and the actual research as published in August 1947.

---

*An account of war-time research on magnetron oscillators can be found in the volume *Microwave Magnetrons,* edited by G. B. Collins (1948. McGraw-Hill Book Company. New York, N.Y.).

†BECKER, G. E. & S. AUTLER. 1946. Phys. Rev. **70:** 300. And LAMB, W. E., Jr. 1946. Phys. Rev. **70:** 308.

‡GORDON, J. P., H. J. ZEIGER & C. H. Townes. 1954. Phys. Rev. **95:** 282.

§See "Physical Concepts in the development of the maser and laser" by W. E. Lamb, Jr. 1973. *In* Impact of Basic Research on Technology. B. Kursunoglu & A. Perlmutter, Eds.: 59–111. Plenum Publishing Corp. New York, N.Y.

¶LAMB, W. E., JR. & R. C. RETHERFORD. 1947. Phys. Rev. **72:** 241.

Microwave Physics

Experiment to Determine The Fine Structure of the Hydrogen Atom
(Lamb, Retherford)

The hydrogen atom is the simplest one in existence, and the only one for which essentially exact theoretical calculations can be made on the basis of the fairly well-confirmed Coulomb law of interaction and the Dirac equation for an electron. Such refinements as the motion of the proton and the magnetic interaction with the spin of the proton are taken into account in a rather good approximate fashion. Nevertheless, the experimental situation at present[1] is such that the observed spectrum of the hydrogen atom does not provide a very critical test either of the theory or of the Coulomb law of interaction between two point charges.

A critical test would be obtained from a measurement of the fine structure of the $n = 2$ quantum state. According to theory,[2] in the absence of an external field, this consists of three energy levels, denoted by the spectroscopic symbols $2^2S_{1/2}$, $2^2P_{1/2}$, and $2^2P_{3/2}$. The Balmer line H_α has this group of states for its lower level, but the line is given an additional fine structure due to the splitting to the upper $n = 3$ levels. This and the very considerable Doppler broadening make it difficult to set very precise limits on the validity of the theoretical predictions. The hyperfine structure is entirely beyond the range of examination by such methods. The $n = 1$ state is single. However, the transition from $n = 2$ to $n = 1$ gives a line in the far ultraviolet for which no accurate observations have been made.

According to theory, the $2^2S_{1/2}$ and $2^2P_{1/2}$ levels rather accidentally coincide, while the $2^2P_{3/2}$ level is 0.365 cm^{-1} above in energy. This means that an atom in either of the first two states should have an absorption frequency of 2.74 cm. This is sufficiently near X-band to suggest an attempt to obtain the radiative transition using microwave tubes of the type developed during the war.

There are two German papers dated 1932 and 1935, dealing with attempts to detect this transition. At that time, only spark gap oscillators of exceedingly low power output were available, and one had to work with a continuous spectrum, using interferometer methods to select a reasonably monochromatic range of radiation. The radiation was passed through an absorption vessel containing a hydrogen discharge of the Wood's type, and the attenuation was determined as a function of wavelength. The first worker, Betz,[3] claimed to have observed an absorption in the range expected. Three years later, working in the same laboratory, Haase[4] repeated the experiment more carefully and failed to find any effect. In addition, he made quantitative estimates which showed that the rate of production of excited hydrogen atoms was so low that no detectable energy could have been absorbed by all of them. (Further calculations made here confirm that the absorption cross-section is so small that a much larger number of excited H atoms would have been needed to obtain any measurable attenuation of the radiation.) The experiment is particularly difficult because in the Wood's discharge the absorption by the electron gas exceeds by far any expected absorption by the excited hydrogen atoms.

According to the selection rules, radiative transitions from the state $2^2S_{1/2}$ to the ground state $1^2S_{1/2}$ are forbidden as the transition involves a change of azimuthal quantum number from zero to zero, and there is no way for the atom to supply the angular momentum of the emitted light quantum. In higher approximations, of course, a very small transition probability will be obtained. Relativistic corrections would make the lifetime about a month and double quantum emission would make the lifetime about 1/10 second. Thus the $2^2S_{1/2}$ level may properly be called metastable. This is an important matter, as the success of any absorption experiment depends on the possibility of the accumulation of sufficient excited hydrogen atoms in order to obtain a measurable effect of one sort or another.

A number of workers[5] have attempted to show up this metastability of the $2^2S_{1/2}$ level of hydrogen, by measuring the absorption of the light from one hydrogen discharge in another one. If the state is metastable to any extent, there will be an increase in its steady-state population, and the corresponding fine structure components of the H_α line will have a higher intensity than those arising from the nonmetastable $2^2P_{1/2}$ and $2^2P_{3/2}$ states. Bethe[6] has discussed the theory of these experiments at considerable length in his *Handbuch der Physik* article, and shows that owing to Stark effect present in the absorbing Wood's tube, the $2^2S_{1/2}$ state is effectively hardly metastable at all. The reason for the large effect of any electric field on the state is that owing to the degeneracy between the two $j = \frac{1}{2}$ levels, one has a linear Stark effect instead of the usual quadratic one. Hence, the wave function of the perturbed states is a mixture of $2^2S_{1/2}$ and $2^2P_{1/2}$, and the latter can combine very strongly with the ground state (lifetime 1.6×10^{-9} seconds). Hence, even a field of 10 volts per cm can reduce the lifetime from 1/10 seconds to about 10^{-8} seconds. The experimental intensity measurements appear to be compatible with this slight metastability.

In order to make precise measurements on the fine structure, it would appear necessary to increase the lifetime of the $2^2S_{1/2}$ state. This could be done by working in a field-free region. Astronomers[7] have discussed the existence of metastable hydrogen in interstellar space, but microwave absorption measurements in such a region are probably out of the question. Another possibility is that by the application of a strong magnetic field, the Zeeman splitting of the degenerate states may be enough so that for the electric fields encountered, the stark effect is quadratic, and hence not very destructive of the metastability. We have made some theoretical estimates which indicate that electric fields as high as 10 volts per centimeter would be permissible if a magnetic field of a few thousand gauss were present. There are two kinds of electric fields acting on the H atoms in a discharge tube. One is essentially a static field arising from the electrodes and the space charge distribution. The other is due to the adjacent ions and electrons. The latter are in rapid motion, and the theory for their effect is not very reliable. Bethe calculated their field by neglecting their motion, and simply taking the electrons as distributed at random, but at rest. Breit and Teller[8] discuss the error in this calculation but do not come to any very clear-cut improvement in the theory.

We have the opinion that even with the aid of a magnetic field, it would be very difficult to detect the absorption of microwaves in a Wood's tube. The reasons for this were mentioned earlier, the high absorption due to the electrons and the small absorption coefficient of the microwaves in the metastable hydrogen. An alternate method would be to attempt to induce transitions from the metastable state using microwaves, and thus to upset the balance between the $2^2S_{1/2}$ and the other $n = 2$ states. This could be detected by measurements of the intensity of the H_α components (under Zeeman effect conditions), but this method does not seem well suited to the facilities of this laboratory.

It is proposed therefore to attempt to form some kind of a stream of metastable hydrogen atoms which fall upon a detector sensitive to them. When RF is absorbed by the atoms, they will go almost instantaneously to the ground state with the emission of a 1200 Å quantum. A decrease in the response of the detector will be then obtained. Calculations show that even intensities of the order of a microwatt are sufficient to obtain an appreciable depletion of the stream of metastable atoms. On the other hand, as pointed out above, an exceedingly large number of metastable H atoms would have to be present to give an absorption of a microwatt of power. The proposed method is similar to the molecular beam methods of Rabi[9] and the rejected method is analogous to the resonance method of Purcell,[10] and others.

There are several possible ways to detect the metastable atoms: (1) When a metastable atom hits a metal surface, there is a certain rather high probability that an electron will be ejected from the surface. This was discovered by Webb,[11] and confirmed by later workers of his school,[12] and recently in some very beautiful work of Dorrestein.[13] (2) The surface ionization detector[14] used in molecular beams depends on having an atom whose ionization energy is less than the work function of the metal. In this case with the metal at an elevated temperature, nearly 100% of the atoms striking the detector are re-emitted as positive ions. The requisite energy inequality is well satisfied in the case of metastable hydrogen atoms. A complication difficult to evaluate is the excited nature of the atom, and the possibility of large surface attractive forces which would keep the atom near the surface until it attained the ground state. Buehl[15] has reported the detection of mercury metastable atoms by this method. (3) A possible method that avoids the use of metal surfaces is to induce the transition to the ground state by the application of an electric field, and to detect the ultraviolet photon either by a photomultiplier tube, or by a Geiger-counter. In our first attempts we will try methods (1) and (2).

Hydrogen is a molecular gas normally. Presumably if H_2 is bombarded by electrons of sufficient energy, atoms in the $2s$ state will be formed. Little is known about the probability for such a process, or the kinetic energy given to the excited atom. If the excited atom has too high a kinetic energy (a few volts), its velocity through the magnetic field gives an induced electric field which is large enough to undo the good which the magnetic field was supposed to do. The proposed experiment will be tried with molecular hydrogen, but it seems safer to provide for the use of atomic hydrogen. Atomic hydrogen may be obtained from a Wood's discharge tube[16] or from a hot tungsten[17] surface. We have not yet decided which method is preferable, but will try the hot tungsten surface first. The atomic hydrogen will then be bombarded with electrons of the correct energy which are moving in an electric field free region. Any Stark effect due to the electron collisions will be minimized by the application of a magnetic field. A certain number of the hydrogen atoms will be excited by the electron stream, and will travel down a tube to the detector. They can be exposed to microwaves or electric fields on their way to the detector.

A possibility, which can only be hinted at here, is the production of a polarized beam of excited hydrogen atoms. Theoretically it is possible to run the source and detector in such a magnetic field that only atoms with spin parallel to the magnetic field are produced and detected. Any hyperfine transition induced by the RF would then give rise to a very sharp resonance decrease in the detected beam.

We have made quantitative estimates of the effects that should be observed, and the experiment seems feasible, if a detector with sufficient sensitivity can be found. An apparatus has been designed, and the parts are in process of construction in the shop.

References

1. BETHE, H. 1933. Handbuch der Physik. Vol. 24. Part **1**: 319. More recently, PASTERNACK, S. 1938. Phys. Rev. **54**: 1113.
2. BETHE, H. Handbuch der Physik. Vol. 24. Part **1**: 311.
3. BETZ, O. 1932. Ann. Physik **15**: 321.
4. HAASE, T. 1935. Ann. Physik **23**: 657.
5. References cited in Reference 2: 456.
6. Reference 2: 452.
7. STRUVE, O., K. WURM & L. HENYEY. 1939. Proc. Nat. Acad. Sci. U.S.A. **25**: 67.
8. BREIT, G. & E. TELLER. 1940. Astrophys. J. **91**: 215.
9. RABI, I. I., S. MILLMAN, P. KUSCH & J. R. ZACHARIAS. 1939. Phys. Rev. **55**: 526.
10. PURCELL, E., H. TORREY & R. POUND. 1946. Phys. Rev. **69**: 37.
11. WEBB, H. W. 1924. Phys. Rev. **24**: 113.
12. MESSENGER, H. A. 1926. Phys. Rev. **28**: 962.
13. COULLIETTE, H. J. 1928. Phys. Rev. **32**: 636.
14. SONKIN, S. 1933. Phys. Rev. **43**: 788.
15. DORRESTEIN, R. 1942. Physica **9**: 433, 447.
14. FRASER, R. 1937. Molecular Beams. Methuen & Co., Ltd. London.
15. BUEHL, A. 1933. Hev. Phys. Acta **6**: 231.
16. WOOD, R. W. 1921. Phil. Mag. **42**: 729.
17. ROBERTS, J. K. 1935, 1936. Proc. R. Soc. A **152**: 445; Proc. Cambridge Phil. Soc. **32**: 152.

RECOLLECTIONS OF A RABI STUDENT OF THE EARLY YEARS IN THE MOLECULAR BEAM LABORATORY

Sidney Millman

*American Institute of Physics
New York, New York 10017*

When I was invited to write a paper for the Rabi Festschrift I saw an opportunity to complement Jeremy Bernstein's prize-winning[1] *New Yorker* profiles on Rabi that appeared in two issues of that magazine in October 1975. The Bernstein article described in considerable detail Rabi's early life, his family background, his shift from a chemistry major at Cornell to graduate work in physics at Columbia, and his two postdoctoral years in Europe, leading to a faculty appointment in the Columbia Physics Department. Very little was said about Rabi's activities as a professor, about his intense interest in research, about his relations with his graduate students and his "postdocs" in the Molecular Beam Laboratory. How did the research program on nuclear spins and magnetic moments evolve into the Molecular Beam Magnetic Resonance Method, which Rabi invented and which led to his getting the Nobel Prize in Physics in 1944? Bernstein also devotes considerable space to Rabi's activities after World War II as the scientist-statesman. It is my intention to give the reader some feeling for the atmosphere in Rabi's Molecular Beam Laboratory, for Rabi as a research professor and for his relations with his students and postdocs. This will, of course, be from the point of view of one who as a student and postdoc in the Molecular Beam Laboratory interacted with him over a period of about 8 years. It will not be a systematic review of experimental techniques and theoretical developments for all the experiments carried out in Rabi's Molecular Beam Laboratory. The reader is referred to the excellent book on molecular beams by Professor Norman Ramsey, another former student of Rabi.[2]

I came to the Columbia Physics Department as a graduate student in the fall of 1931, originally for the purpose of satisfying the graduate requirement for teaching physics or mathematics in the New York City high schools. I continued on after the first year toward a Ph.D. since there were no teaching jobs available. In my second year of graduate studies, I got to know Professor Rabi from the courses I took with him in Statistical Mechanics and Quantum Mechanics. Aside from an occasional visiting professor, Rabi appeared to be the only professor in the Department to have sufficient mastery of the rapidly developing new physics to be able to teach those courses. He projected to his students an interest and understanding in depth of the quantum theoretical concepts. He also left the impression among the students that his interest was even greater in research than in the meticulous preparation of the lectures for the course. Nevertheless, by taking careful notes during the lecture and following it up by many hours of collateral reading in the Physics Reading Room and working out the assigned problems, a student could acquire a fairly good working knowledge of the quantum theories that helped to explain the research results of many experiments in atomic spectroscopy and other fields.

My closer association with Professor Rabi began one Saturday afternoon in May 1933, after I passed the Departmental qualifying examination, when Rabi called me into his office and asked me whether I would want to do my thesis research under his direction. He suggested the problem of determining the nuclear spin and magnetic moment of potassium. He pointed out that whereas other alkali atoms, such as cesium or rubidium, had fairly large, spectroscopically observed, hyperfine structure (hfs), the lower atomic number alkali atoms had much smaller hfs and that these had either not been observed or had such small values that one could not use the results for a definitive determination of either the nuclear spin or the magnetic moment. He suggested the molecular beam techniques, which were already being used by my colleague Bill Cohen. More specifically, Rabi suggested the application of the "zero moment" method that Cohen was going to use for cesium. This was all new to me and sounded quite exciting. I agreed right there and then to start on my research.

Rabi wisely assigned me for a three-months' apprenticeship with his research assistant (postdoc by modern terminology) Carl Frische, from whom I learned very valuable techniques for working with a vacuum system: glass blowing, the use of sealing wax for attaching glass diffusion pumps to a brass apparatus, the handling of the mercury McLeod gauges for monitoring the vacuum system, and so forth. This was very fortunate as it gave me a good start on my molecular beam experiments, since Rabi himself was not a particularly skillful laboratory man. He more than made up for it, however, by his brilliance in the selection of feasible physics experiments that are in the forefront of a developing field of research and that would attract widespread interest in the physics community.

The zero moment method for determining the nuclear spin and magnetic moment was based on a theory developed by Breit and Rabi in 1931,[3] giving the energy variation of an alkali atom in the normal $^2S_{1/2}$ state as a function of the magnetic field due to the changes in coupling that take place between the nuclear spin and the atomic angular momentum as the magnetic field is varied from the weak-field Zeeman region to the strong-field Paschen-Back region. The formula for the energy is given by the now quite well known expression:

$$W_m = \frac{-\Delta W}{2(2I+1)} \pm \frac{\Delta W}{2}\left(1 + \frac{4m}{2I+1}\gamma + \gamma^2\right)^{1/2}, \qquad (1)$$

with

$$\gamma = \frac{2\mu_0 H}{\Delta W}, \qquad \Delta W = hc\Delta\nu, \qquad (2)$$

where H is the applied magnetic field, I is the nuclear spin (in units of $h/2\pi$) and ΔW is the difference in energy (or $\Delta\nu$ if measured in wave numbers) at zero magnetic field between the quantum levels $F = I + \frac{1}{2}$ and $F = I - \frac{1}{2}$, m is the magnetic quantum number, μ_0 is the Bohr Magneton, h is Planck's constant, and c is the velocity of light. For example, for a nuclear spin of $\frac{3}{2}$ the F quantum numbers are 2 and 1, and the resulting magnetic quantum numbers m are 2, 1, 0, -1, -2 and 1, 0, -1, respectively.

For any magnetic quantum level m the effective magnetic moment of an atom in that state is

$$\mu_m = -\frac{\partial W_m}{\partial H} = \mp \frac{[2m/(2I+1)] + \gamma}{\{1 + [4m/(2I+1)]\gamma + \gamma^2\}^{1/2}} \mu_0. \qquad (3)$$

When a beam of alkali atoms is arranged to pass an inhomogeneous magnetic field, the force on the atom in a given direction will be the product of this effective magnetic moment and the field gradient in that direction, i.e.,

$$f_z = \mu_m \frac{\partial H}{\partial Z}. \qquad (4)$$

Thus, as the magnetic field is increased from zero there will be some values of the field for which the moment of the atom is zero, i.e.,

$$\gamma = \frac{-2m}{2I+1} \quad \text{or} \quad H = \frac{-m}{2I+1}\frac{hc\Delta\nu}{\mu_0},$$

and for such fields the atom in that quantum state will suffer no deflection, and a detector placed in the center of the atomic beam will detect a peak in the beam intensity. If one observes two or more peaks one can determine the nuclear spin by inspection, and the hfs value from the measurement of the magnetic field at which one such peak occurs. When only one peak is observed, one has to resolve the ambiguity between a nuclear spin of $(2/2)(h/2\pi)$ or $(3/2)(h/2\pi)$. This is illustrated by the two experimental curves.

FIGURE 1 is the curve obtained by Bill Cohen[4] for cesium. The equally spaced zero moment peaks, for $m = -1, -2, -3$, clearly identify the spin of ^{133}Cs to be $(7/2)(h/2\pi)$. For $I = (6/2)(h/2\pi)$ the peaks would correspond to m values of $-1/2$, $-3/5$, $-5/2$, and the relative magnetic field value locations would be at 1, 3, 5, respectively. FIGURE 2 is a similar curve as the one obtained by the author for potassium.[5] Note also the need for determining, by relative intensity measurements, whether the nuclear spin of ^{39}K is $(3/2)(h/2\pi)$ at $m = -1$ or $(2/2)(h/2\pi)$ at $m = -1/2$. Note also the effect of the less abundant isotope ^{41}K.

My molecular beam apparatus was going to be set up on the fifth floor of Pupin in the same room as Bill Cohen did his experiments. He started about a year earlier and naturally set up his apparatus at the side of the room closest to the sink, since running water was needed to cool the vacuum diffusion pumps and other parts of the apparatus. I had to build my apparatus in the diagonally opposite corner of the room. My first task was to go to the hardware store to procure the necessary pipes and stopcocks, do the thread cutting and assembly to provide the needed water cooling for my apparatus.

I was fortunate to have encountered very few of the problems one hears about in the assembly of vacuum tight systems and, thanks to the laboratory techniques I acquired from Frische and from Cohen, I was able to construct a working molecular beam apparatus in a few months. I should also mention that working with potassium atoms is relatively easy compared with hydrogen atoms. The detection of the atomic beam is fairly straightforward, even for a beginner. One uses a clean tungsten filament of about 0.003 cm in diameter, heating it to a dull red temperature by passing a small electric current. When potassium atoms impinge on the tungsten wire they are readily ionized, since the ionization potential of potassium is lower than the work function of tungsten. The ionized atoms evaporate and are collected by a surrounding electrode kept at a few volts negative with respect to the filament and amplified by a high resistance DC electrometer using an FP 54 vacuum tube. The current is in the range of about 10^{-15} to 10^{-11} amperes.

A few months later Rabi assigned another graduate student, Marvin Fox, to do his thesis research on the same apparatus. Fox was given the problem of determining the nuclear spin and magnetic moment of ^{7}Li. Together we made very rapid progress: good vacuum on the first run, a well-formed atomic beam on the second, and evidence for a zero moment peak on the third. For the third run we had inserted a magnet made of two water-cooled copper tubes embedded and insulated from a precisely machined duraluminum block. The magnet design was essentially the same as that used by Rabi, Kellogg, and Zacharias for their experiments with hydrogen atoms.[6,7] There was little doubt that there was a research thesis in the making for me. Obviously, many more runs had to be made in order to establish experimentally that the nuclear spin of ^{39}K is $(3/2)(h/2\pi)$ and not $(2/2)(h/2\pi)$. Moreover, there was some evidence in the early runs that we were seeing the effects due to the less abundant isotope ^{41}K ($\sim 6.9\%$), and I thought it would be nice to find out as much as one can about ^{41}K without extensive modification of the apparatus.

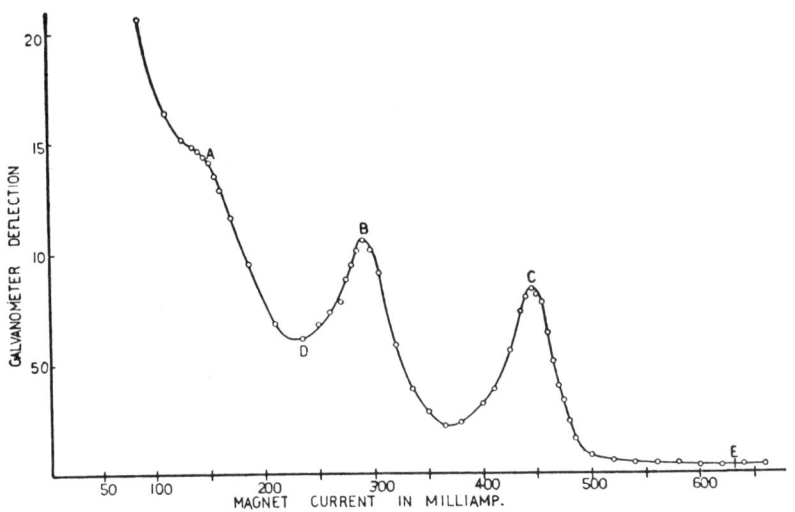

FIGURE 1. Zero moment peaks for ^{133}Cs. (From Cohen.[4] By permission of *The Physical Review*.)

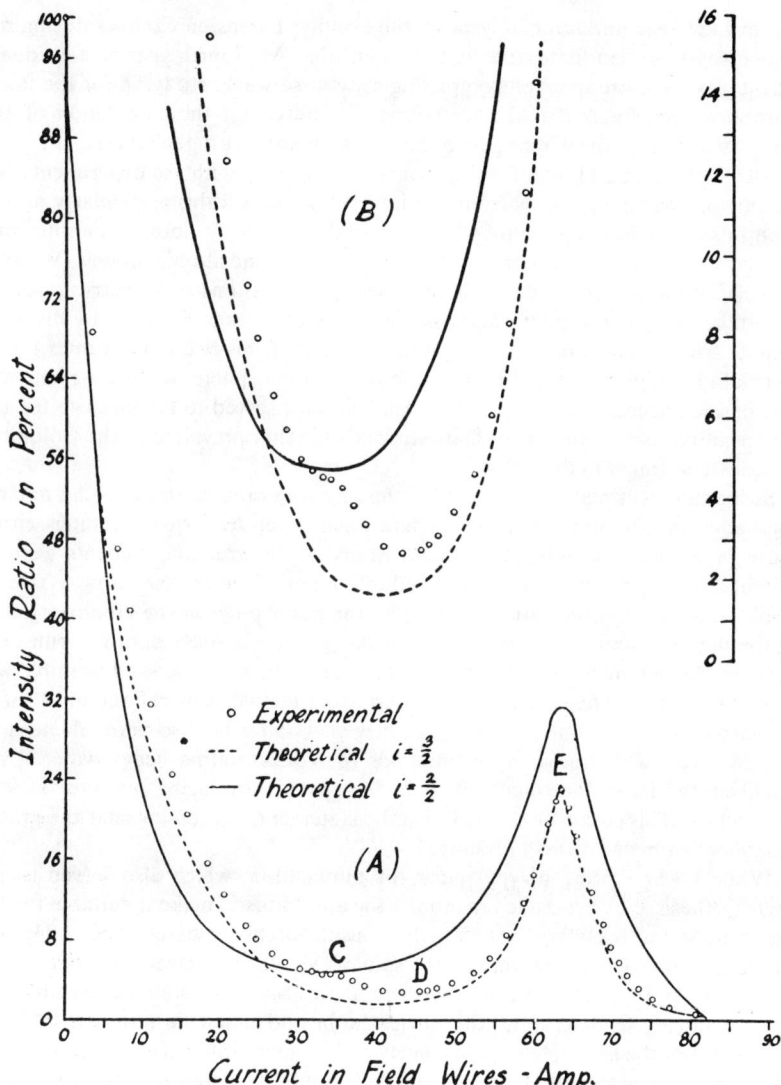

FIGURE 2. Zero moment peak for potassium. (From Millman.[5] By permission of *The Physical Review*.)

The rapid progress in my thesis research was pleasantly embarrassing, since by the end of the third year of my graduate studies I had taken only about 75% of the required course work for the Ph.D. The chief reason for deferring some of the courses was economic, since I was a paying graduate student, which was not that unusual at Columbia in the 1930s. There were only about six teaching assistantships in the Physics Department and only one or two fellowships. I did, however, pick up a few teaching hours a week in the evening Extension Courses during my last two years as a graduate student, at $2 per hour. My fourth year as a graduate student was therefore spent on completing my course work, studying for the fairly rigorous written finals that the Department required for the completion of the Ph.D., as well as on the winding up of my research and writing the thesis.

Although Fox and I didn't call on Rabi for any help with the experiments, we certainly profited enormously from our interaction with Rabi, particularly noting his enthusiasm when explaining what appeared to us fairly obtuse quantum mechanical theory in terms of fairly elementary concepts and atomic models. We also appreciated the informal atmosphere in which discussions were carried out. It didn't take very long to get to address him as "Rabi," as he is called by his most intimate friends. (No one ever calls Rabi by either of his two given names.) This informality, coupled with the highly stimulating atmosphere for delving into the most basic concepts of atomic and nuclear physics, served to compensate for the more negative aspects of the graduate student's life that prevailed at the Columbia Physics Department in the 1930s.

Surprising as it may sound in 1977, the physics graduate students did not get together every afternoon for coffee. There wasn't even tea Friday evenings either before or after the weekly colloquium. Many of the graduate students were at Columbia on a part-time basis and lived off campus. For me the subway ride to Brooklyn was often useful for reflecting on the happenings at the laboratory during the day, on some of the seemingly puzzling data obtained during a run, and occasionally coming up with the answers to the problems. One such example will be cited later. I shall never forget one phrase from an excellent colloquium Professor Harold Urey gave one Friday night a few years after he discovered deuterium. The talk dealt with fractionation methods for concentrating heavy water. Urey also listed the items that contributed to the cost of producing one liter of 99% heavy water. This cost estimate included the statement, "assuming that the graduate student's time was worth nothing."

While I was writing up my paper for publication, which also served as my Doctor's thesis, Fox proceeded to modify the apparatus to make it suitable for the determination of the hfs of ^{7}Li, which he used for his research thesis.[8] He was able also to put some upper limit on the spin of ^{41}K and to detect some effects due to ^{6}Li. Fox and I went on to use the same apparatus, with suitable modification in the magnets, to determine[9] the nuclear spin and magnetic moments of ^{85}Rb and ^{87}Rb. For the first time we had a fairly precise measurement of the ratio of hfs values for two isotopes of an alkali atom and could thus determine the ratio of the corresponding nuclear magnetic moments, since that does not depend on atomic screening and accurate knowledge of the atomic wave function at the nucleus.

Rabi was very eager to have the ^{6}Li spin and magnetic moment determined. Such a feat would exhibit the much greater power of the molecular beam zero moment method over standard spectroscopic techniques. It was also the only stable alkali nucleus with an even atomic weight, thus closely resembling deuterium. It was evident from Fox's work on ^{7}Li and the observed effect due to ^{6}Li that much greater resolution was needed, since the zero moment peak would occur at very weak magnetic fields and with weak fields one normally gets weak gradients and smaller deflecting forces. John Manley had, in the meantime, constructed a molecular beam apparatus with higher resolving power for an unambiguous determination of the nuclear spin of ^{41}K and a precise measurement of its hfs.[10] With suitable modification in the apparatus, Manley and I determined the nuclear spin of ^{6}Li and produced a fairly precise measurement of its hfs and therefore a fairly precise value for the ratio of the nuclear magnetic moments of the two isotopes of lithium.

Although Rabi followed the molecular beam work on the alkali atoms with keen interest, as I noted in the preceding pages, he was even more interested in the related molecular beam experiments on the gas apparatus that was constructed at about the same time as I started my research. The molecular beam experiments on hydrogen and deuterium involved the collaboration of two outstanding experimental physicists in Rabi's laboratory, Kellogg and Zacharias, and extended over a period of about 8 years. Jerome M. B. Kellogg started in Columbia as an instructor, was later promoted to assistant professor and, during the war, headed the Columbia Radiation Laboratory before going to the Los Alamos laboratory. Jerrold R. Zacharias started his postdoctoral research with Rabi while holding the position of Assistant Professor of Physics at Hunter College and teaching about 15 class hours a week. He exported molecular beams to MIT after World War II. He also became famous later for his coordinating work on PSSC High School Physics.

Experiments with hydrogen and deuterium were much more difficult than the corresponding molecular beam research with the alkali atoms. Although Rabi was well aware of that, he put a lot of effort into it because of the greater potential scientific dividends. He felt that the magnetic properties of the proton and deuteron were likely to play a much more fundamental role in atomic and nuclear physics. The main reason why working with hydrogen is so much harder than with potassium is in the detection of the atomic beam. The tungsten wire detector cannot be used because the ionization potential of hydrogen is much greater than the work function of any metal surface. Rabi, Kellogg, and Zacharias (RKZ) first used the method of atomic deposition of hydrogen on a layer of molybdenum oxide soot deposited on a glass plate. Later they used the Pirani gauge as a detector, which responds to a pressure change as the beam of hydrogen enters through a narrow slit in a small cavity placed in the path of the molecular beam. These slight pressure changes are detected by a change in the resistance of a wire inserted in that cavity. This method of detection is much slower than the hot tungsten wire method used in the detection of alkali atoms, and the signal to noise ratio is also much lower. In addition, the production of an atomic beam of hydrogen by the use of an appropriate discharge tube is not as simple as having a chunk of potassium inserted in a metal block covered by a properly designed pair of jaws to form a slit of a desired width.

Rabi used to spend more time in the laboratory where the hydrogen and deuterium experiments were performed than in our room. This was partly due to the long runs that were needed to accumulate the necessary data for any given experiment and, therefore, an appreciable fraction of these runs were carried out in the evenings. Since Rabi, as well as Kellogg, lived within a few blocks of the Pupin Physics Building, and Zacharias lived not far from Columbia either, they frequently returned to the laboratory after dinner. When watching a run in progress,

Rabi often whittled on a slab of wood and occasionally burst out in a rendition of a well known operatic aria, such as "Un bel di, vedremo" from Madame Butterfly.

Although the two teams of molecular beam researchers seemed to pursue different lines of research, there was considerable overlap in interest and sharing of experimental facilities, such as storage batteries for large DC currents needed for some of the magnets, or wave meters for the precision measurement of radio frequencies. In addition, Zacharias collaborated on both molecular beam systems. Of even greater significance is the fact that fundamental changes in methods of experimentation sometimes originated at the gas apparatus and carried over to the alkali apparatus and vice versa, as we shall soon see.

In February 1936 Rabi published the paper "On the Process of Space Quantization."[12] He examined the effect of a rapidly varying magnetic field in an oriented atom possessing nuclear spin and showed that by performing experiments similar to those done by Phipps and Stern[13] and by Frisch and Segre,[14] one can select atoms in specific atomic quantum states, send them through rapidly varying magnetic fields that will cause spin reorientations, and identify such changes by a properly placed refocussing magnetic field and beam detector. Rabi showed that one can in this way measure the sign of the nuclear moment, i.e., determine whether the magnetic moment points in the same direction as the nuclear angular momentum (spin) or in the opposite direction. This was first applied by RKZ[15] to determine the sign of the moment of the proton and of the deuteron. Henry Torrey[16] then built an alkali beam apparatus for determining the sign of the nuclear moments of ^{23}Na and ^{39}K and subsequently Zacharias and I[17] determined the signs of ^{7}Li, ^{85}Rb, ^{87}Rb, and ^{133}C$_s$. A schematic diagram of the apparatus used in the last mentioned work is shown in FIGURE 3. The four-wire transition field is similar to the one used by RKZ for their work on the proton and dueteron signs.

Rabi's paper on space quantization turned out to have a much greater impact on the molecular beam research that was to follow 2 years later than on the immediate results, since no great surprises were encountered by the experiments on the signs of the moments of the nuclei measured. However, in doing these experiments we were acquiring experience in spin flips, i.e., changes in quantum states, in transit, and with methods for refocussing and detecting parts of a molecular beam that have been deflected by the inhomogeneous magnetic field in the earlier stages of transit in the apparatus and have undergone such changes. These molecular beam techniques could easily be adapted to the molecular beam magnetic resonance method, which was inaugurated in the fall of 1937. The change in direction of our research activities was precipitated by the visit of Professor C. J. Gorter of the University of Leyden. He reported on his unsuccessful experiment performed a year or two earlier and published in 1936 in *Physica*.[18] He had attempted to detect nuclear transitions by using radiofrequency oscillating fields by measuring absorption of some of the energy in a radiofrequency circuit (similar to work done 10 years later by Purcell, Torrey and Pound, and by Bloch, Hansen and Packard). The discussion between Rabi and Gorter is referred to in the first publication of a nuclear moment resonance curve observed by the molecular beam magnetic resonance method by Rabi, Zacharias, Millman, and Kusch.[19] Gorter's own account of his visit to Columbia is described in his article published in the January 1967 issue of Physics Today,[20] based on a talk he gave at the occasion of his receiving the Fifth Fritz London Award.

Returning now to experiments with the zero moment method, we decided to extend it beyond the alkali atoms. Zacharias and I, after completing the work on the signs of the moments of the alkali nuclei referred to above, redesigned our apparatus to work on indium. The detection of an indium beam was going to be similar to that of alkali atoms, but it required much higher oven temperatures to produce sufficiently high vapor pressure (about 10^{-4} atmospheres) needed for an atomic beam. This did not present a major experimental challenge. We had to learn to work with molybdenum for an oven and slits. ^{115}In was also known to give rise to a large hfs constant in the ground state, which is $^2P_{1/2}$, and therefore much larger magnetic fields were required. This made the current-carrying two-wire field impractical. Accordingly, we designed a one-meter long electromagnet with four water-cooled copper windings. The cross section of this magnet is shown in FIGURE 4. The design of the cylindrical sections of the pole pieces forming the magnet gap, through which the atomic beam was to pass, was based on equipo-

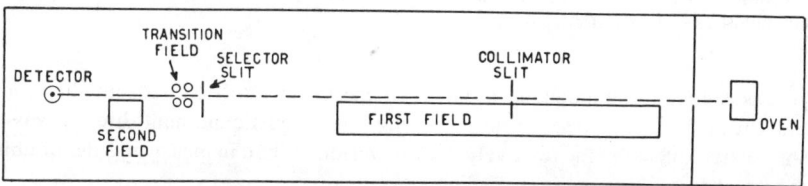

FIGURE 3. Schematic diagram of apparatus used for determining signs of nuclear moments. (From Millman & Zacharias.[17] By permission of *The Physical Review*.)

tential surfaces calculated from an appropriate two parallel wire current-carrying field as previously used in the experiments on the proton and on many alkali nuclei moments. The design of this magnet turned out to be a very useful experience for constructing an apparatus for the experiments with the molecular beam magnetic resonance method to follow. The only important change that was needed later was to make the gap smaller and make cylindrical surfaces with smaller radii of curvature.

The experiments on ^{115}In were carried out in a straightforward fashion. We measured its nuclear spin $[(9/2)(h/2\pi)]$ and the hfs value of the $^2P_{1/2}$ state[21] and even tried, with some limited success, to measure the magnetic moment of ^{115}In directly by resolving the doublet structure of some of the zero moment peaks. A doublet arises from the circumstance that the zero moment peak is contributed by atoms with the same m quantum number of two different F numbers, corresponding to $F = I + \frac{1}{2}$ and $F = I - \frac{1}{2}$. The effective moment of the atoms in these quantum states are not zero at the same value of the magnetic field and the differ-

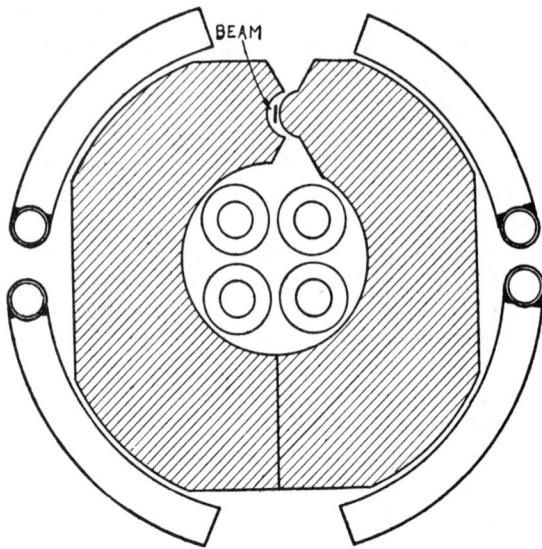

FIGURE 4. Cross section of the magnet used for indium. (From Millman et al.[21] By permission of *The Physical Review*.)

ence depends on the magnetic moment of the nucleus.* It turned out, however, that the advantage gained in thus measuring the magnetic moment directly was to some extent offset by the relatively poor precision we had in measuring the doublet spread.

Rabi's molecular beam postdoctoral research force was strengthened further in the summer of 1937 by the arrival of Polykarp Kusch. After obtaining a Ph.D. from the University of Illinois, he spent one postdoctoral year at the Physics Department of the University of Minnesota. Following the advice of Wheeler Loomis, Chairman of the Illinois Physics Department, Kusch accepted an instructorship in the Columbia Physics Department with the objective of doing research in Rabi's molecular beam laboratory. Rabi suggested that he join me in putting an all-out effort on the necessary modification of my apparatus for the study of nuclear magnetic resonances. From my point of view, and I believe from Kusch's too, it was a most felicitous suggestion. By that time I had completed two postdoctoral years in Rabi's laboratory, holding the Tyndall Fellowship ($972 per year) during the first year and the Barnard Fellowship ($1,325) during the second year. Rabi gave me a research assistantship, which was to last two more years. Kusch and I worked in the same room for the three years as true collaborators, neither of us feeling in any way more senior than the other. This happy association lasted until Kusch's term of three years as Instructor was up and he accepted a position at Westinghouse in Bloomfield, N. J.

*This does not follow from Equation 3, but a refinement of the theory of the behavior of an atom in an external magnetic field, as discussed in Reference 21, modifies somewhat the expression for the magnetic moment of the atom μ_m given in Equation 3.

Kusch, Zacharias, and I turned our immediate attention to the redesign and construction of the apparatus as required for molecular resonance. As I mentioned earlier, the design for the deflecting magnets was similar to the magnet used in the indium experiment except for the need for smaller gaps and smaller radii of curvature for the cylindrical surfaces for the deflection of molecules having no atomic moments but only nuclear and rotational moments. The two magnets were of unequal length (about 52 cm and 58 cm), the shorter one being placed closer to the oven having as much deflecting power as the larger magnet, which is at the end of the path of the molecular beam.

The second change required was the placing of a short, homogeneous, strong magnet between the two deflecting magnets, with a gap sufficiently large (~ 0.6 cm) to accommodate a radiofrequency oscillating field. This was accomplished by the insertion of a loop of wire in the form of a hairpin (with its axis parallel to the direction of the beam) which is connected outside the vacuum chamber to a radio-frequency current source. The rest of the apparatus was standard molecular beam gear. A schematic diagram of this apparatus is given in FIGURE 5.

The first nuclear magnetic resonance (NMR) curve obtained with the molecular beam magnetic resonance method was published in the February 15, 1938 issue of *The Physical Review*[19] and is reproduced here in FIGURE 6. The abscissa refers to the current in the homogeneous magnet, the C magnet of FIGURE 5. This magnet produces a constant magnetic field at right angles to the direction of the beam. The weak oscillating field is at right angles to the plane of the paper.

At resonance the precession frequency of the nuclear magnetic moment about the constant magnetic field is the same as the frequency of the oscillating field. This can bring about a change in the component of the magnetic moment parallel to the field. Since conditions in the A and B magnets, the deflecting and refocussing magnets, were supposed to have been adjusted to yield maximum beam intensity at the detector for no change in the magnetic properties of the molecules during transit, any change produces a drop in intensity.

The advantages of this method of measuring nuclear magnetic moments—more accurately, the ratios of magnetic moment to nuclear spin, or nuclear g values—over previous methods are numerous. First, it is a direct measure of nuclear moment as compared with calculating it from the measured hfs of an atomic state by some semiempirical and semitheoretical expression involving

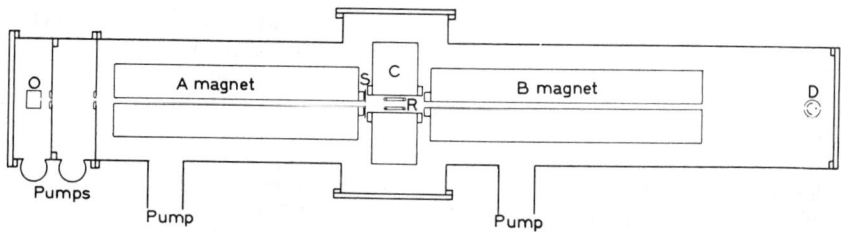

FIGURE 5. Schematic diagram of apparatus used in the first experiments with the molecular beam magnetic resonance method. (From Rabi *et al.*[23] By permission of *The Physical Review*.)

imprecise knowledge of the wave function at the nucleus, except for hydrogen. Second, in an NMR measurement the resonance frequency and the magnetic field can be measured with greater precision than the hfs values of atomic states that had been measured thus far. Finally, many more nuclei become accessible to NMR measurements by this method because of the wide range of magnetic field and oscillating frequencies available and because many molecules can be formed that contain alkali atoms, and this can greatly aid the beam detection. This becomes immediately apparent from noting the large number of publications on nuclear magnetic moments that resulted from experiments with the alkali beam ap-

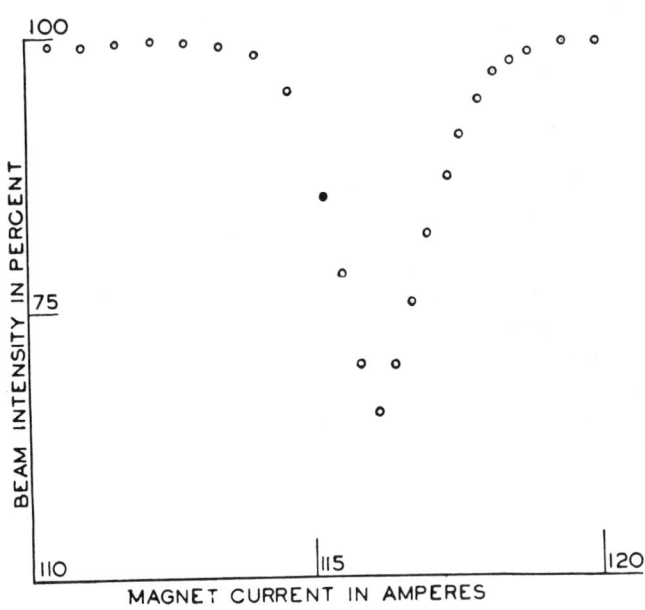

FIGURE 6. First nuclear magnetic resonance curve. (From Rabi et al.[19] By permission of The Physical Review.)

paratus in a relatively short time, as listed below:

^{6}Li, ^{7}Li, ^{19}F	March 1938[22] and March 1939[23]
^{14}N	December 1938[24]
^{9}Be	April 1939[25]
^{14}N, ^{23}Na, ^{39}K, ^{133}Cs	June 1939[26]
Boron Isotopes	July 1939[27]
^{27}Al	August 1939[28]
Isotopes of Rb and Cl	September 1939[29]
Precision Measurement of ^{1}H, ^{7}Li, ^{19}F, and ^{23}Na	July 1941[30]
^{13}C, ^{135}Ba, and ^{137}Ba	July 1941[31]

When the initial measurements on the modified alkali beam apparatus were completed, i.e., the work on ^{6}Li, ^{7}Li, and ^{19}F, Zacharias rejoined Kellogg to redesign the gas apparatus for measuring the nuclear properties of the proton and the deuteron by the molecular beam magnetic resonance method. They were joined by an outstanding graduate student of Rabi's, Norman Ramsey, referred to at the beginning of this article in connection with the book on molecular beams he published 18 years later. Like Zacharias, he also founded a new molecular beam laboratory after the war when he became Professor of Physics at Harvard.

Rabi began spending more time with Kusch and me as new measurements were being turned out at a rapid pace. He was particularly interested in some of the spurious resonances we observed at times. Rabi often advanced the most fantastic hypotheses for the possible origin of these effects. It was quite exciting. We could hear him talk to Ed Condon, who was at Princeton at that time, projecting potentially important discoveries if the effect in question were confirmed. At the same time Rabi encouraged us to be the devil's advocates to insure that the observed effects are tracked down and established as spurious. Needless to state, nothing was ever published unless it was proved beyond doubt that an observed effect is real and has a sound physical explanation.

We developed fairly reliable methods for testing the authenticity of a nuclear resonance. There was the obvious constancy of the ratio of the resonance frequency to the strength of the magnetic field, f/H. This is easily tested in any run. Another straightforward test, but not realized at any one given run, is the observation of a magnetic resonance of a given nucleus with two or more molecules containing the same atom, such as ^{7}LiCl and ^{7}LiF; the f/H value at resonance for ^{7}Li should be the same. Some spurious resonances might result from nonadiabatic transitions that may occur in the region between magnets; and these could be affected by changes in the strength of the fields in the gradient-producing magnets. One can also study the effects on spurious resonances produced by changes in the amplitude of the radiofrequency field. Some resonances exhibited unduly large widths and perhaps even some reproducible subsidiary minima due to interactions between the electric quadrupole moment of a nucleus and the electric field gradients associated with molecular rotational states. The latter effect was the subject of a theoretical paper published in January 1945 by Bernard T. Feld and Willis E. Lamb, Jr.[32]

There was one effect that was observed very early in our resonance curves and persisted in all our runs, namely, that the resonance curves were not quite symmetrical. At best, one side appeared to rise from the minimum more steeply than the other. At times one could see a well pronounced "shoulder" on one side. The explanation for this puzzling effect occurred to me on one of the subway rides to Brooklyn, after a run during the day when this effect was more pronounced than usual, and turned out to be quite simple if one views the resonance phenomenon from the point of view of the moving molecule, as Rabi frequently was wont to do. The asymmetry turns out to be caused by the asymmetry inherent in the construction of the hairpin-shaped wire producing the radiofrequency oscillating field, as we shall see.

The molecular beam magnetic resonance (MBMR) method for determining nuclear magnetic moments measures the Larmor precession frequency of a nucleus in a known strong magnetic field. The theory upon which the method is based, requires the application of a small perturbing field, at right angles to the fixed strong field, rotating with the same frequency and in the same sense as the Larmor precession of the nucleus possessing angular momentum and magnetic moment. If this requirement is fulfilled, the method is capable of measuring not only the magnitude of the nuclear moment but also its sign, since the sense of precession depends upon whether the magnetic moment is in the same direction as the mechanical moment (moment positive) or in the opposite direction (moment negative). However, when an oscillating field is substituted for the rotating field, as has been done in the MBMR experiments, one effectively introduces two component magnetic fields rotating in opposite directions, one of which rotates in the same sense as the precession and is, therefore, effective in producing the nuclear reorientations that make it possible to identify the Larmor frequency. The other rotating component does not interfere in any way with an experiment designed to measure the magnitude of the magnetic moment, but prevents the determination of the sign of the nuclear moment, since, if the perturbing field is one of pure oscillations along its entire length, there is no way of finding out which of the two components is effective in producing transitions.

A schematic perspective diagram of the hairpin-shaped copper tubes carrying the oscillating field, in the neighborhood of the molecular beam, is shown in FIGURE 7.[33] The strong homogeneous field is horizontal and is perpendicular to the beam axis. In the greater portion of the beam path in the oscillating field, from F to G, the lines of force are vertical, but at D the lines are more nearly horizontal. With respect to the moving molecule the field has turned in a counterclockwise sense. This is true when the fields point up in the region F to G and along the beam direction at D, or when the fields are reversed. The effect is in the same sense at the detector end of the hairpin. Thus, with respect to the moving molecule, the perturbing field is strictly oscillatory only in the region from F to G, while at the end regions the field is a superposition of oscillation and rotation. If we assume that

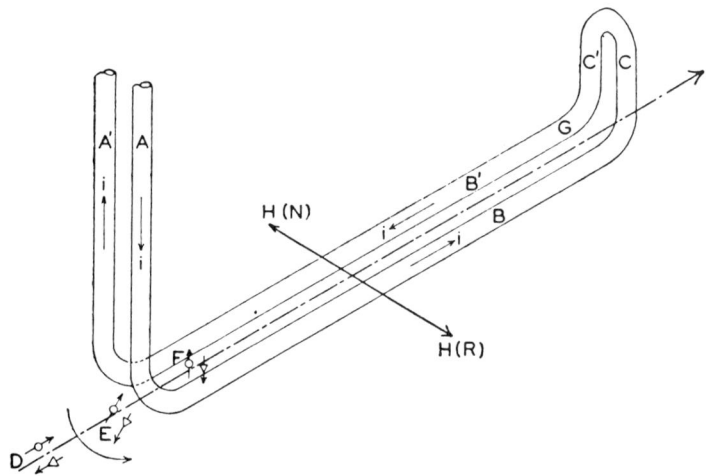

FIGURE 7. Schematic diagram of the hairpin-shaped oscillating field. (From Millman.[33] By permission of *The Physical Review*.)

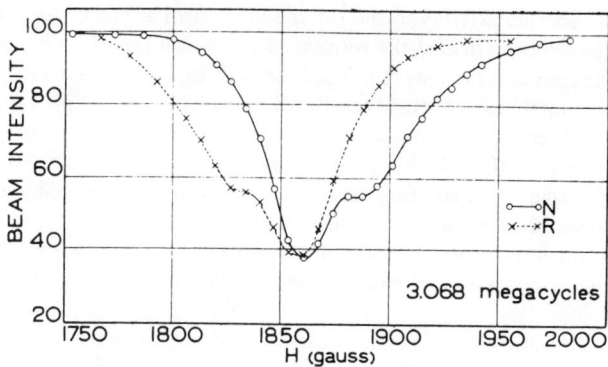

FIGURE 8. Two ^{7}Li nuclear resonance curves from Li$_2$ molecules. (From Millman.[33] By permission of *The Physical Review*.)

the turning of the field through an angle of 90° takes place in a distance about twice that between the axes of B and B^1, then for molecules having thermal velocities of the order of 10^5 cm/sec it is equivalent to a rotational frequency Δf of 3×10^4 cycles per second, which is of the order of 1% of the oscillating frequencies f used in these experiments. Since the oscillating field can be resolved into two component fields, one rotating in the same direction as the end rotation just described and the other in the opposite sense, the effective perturbing field at the end regions rotates with a frequency $f + \Delta f$ if the Larmor precession is in the same sense as this end rotation, and with a frequency $f - \Delta f$ if the precession is in the opposite sense. A resonance curve plotting intensity against the value of a strong homogeneous field H will show a principal minimum at the position H given by $f = \mu H/Ih$ and an additional minimum (which may or may not be resolved from the principal minimum) at a field value given either by $H + \Delta H = (f + \Delta f)Ih/\mu$ or by $H - \Delta H = (f - \Delta f)Ih/\mu$. Since the sense of the Larmor precession depends only on the direction of H and on the sign of the magnetic moment μ, we may determine the sign of μ by a mere glance at a resonance curve obtained for a known H direction, or better still by observing two curves corresponding to opposite field directions.

When I returned to the laboratory the following morning the ideas just discussed were put to test with no modification in the apparatus whatsoever. We simply departed from the usual procedure and recorded our resonance data first with the homogeneous field in one direction (which we normally used in calibrating the magnetic field) and then with the reversed direction. FIGURE 8 shows a pair of such resonance curves obtained for ^{7}Li with Li$_2$ molecules. The experimental conditions were identical for the two curves except that for one curve, H was in a direction opposite to that for the other. The additional minimum, in each case at about 30 gauss from the principal minimum, appears on the high field side in one curve and on the low field side in the other.

The asymmetry exhibited in either of the two curves in FIGURE 8 is the most pronounced ever observed. The curves were obtained under experimental conditions that are particularly favorable for showing to what extent the end effects of the oscillating field can distort the resonance curve. On the other hand, the asymmetries in the resonance curves published for ^{6}Li, ^{7}Li, and ^{19}F were very slight.

The following three important results followed from this simple experiment:

(1) We had receovered our ability to measure the signs of nuclear magnetic moments, which we thought we gave up by introducing an oscillating radiofrequency field instead of a rotating field, which would be very difficult to achieve while satisfying other desirable boundary conditions.

(2) We had improved the precision of measuring the nuclear moments by averaging the values obtained with the two directions of the homogeneous magnetic field.

(3) We had removed the persistent irritant of having a physical effect that is not understood.

While Kusch and I continued to measure magnetic moments of a large number of nuclei using the alkali beam apparatus, Kellogg, Ramsey, and Zacharias applied the MBMR method to the measurement of the magnetic moments of the proton and the deuteron[34] and established the existence of an electrical quadrupole moment in the deuteron[35] and measured its value with considerable precision. Ramsey used the same apparatus for the measurement of the rotational magnetic moments of the H_2, D_2, and HD molecules.[36]

With the increased precision obtainable with the MBMR method for measuring nuclear magnetic moments, we were in a position to begin making comparisons between the ratio of the moments of the isotopes of a given nucleus obtained by this method with that derived from the ratio of the hfs of the ground state of the atom corresponding to the two isotopes. Such a comparison could help establish whether the hfs of an atomic energy level can be accounted for entirely by the assumption that the interaction between the nucleus and the external electrons is purely electromagnetic. In the case of the Li isotopes it was already observed that the MBMR method yielded a more precise measure of the ratio of nuclear magnetic moments than that derived from the ratio of the corresponding hfs values. Kusch and I therefore decided to apply the MBMR techniques to the measurement of the ground state hfs of some of the alkali atoms. In our latest molecular beam apparatus we had most of the essential features required for such experiments.

One important difference in working with atomic states as compared with diatomic molecules is the requirement of much higher radiofrequencies—several hundred to 10,000 megahertz instead of a few megahertz. Fortunately, by the late 1930s the vacuum tube technology as well as radiofrequency circuit design had advanced far enough to make the use of radiofrequencies up to about 500 megahertz feasible. For atoms with much larger hfs values we could resort to the use of transitions between m states of a given F value and apply the Breit-Rabi expression in the weak field Zeeman region, the validity of which was established with very high accuracy, to calculate the hfs value for ^{133}Cs, which is about 20 times the frequency obtainable with the then available Western Electric 316A tube.

The strong inhomogeneous deflecting magnets used for deflecting molecules with nuclear moments were much too powerful for deflecting alkali atoms in the various magnetic states. The solution to that problem was to use much smaller currents in the four turns surrounding the iron core. The low fields were also desirable to insure that an observable change in the magnetic moment is produced when the m quantum number is changed by the oscillating radiofrequency field, which can then be detected in a manner similar to a nuclear resonance. The homogeneous C magnet also had to be used at much lower magnetic fields than it was originally designed for.

One of the many Zeeman patterns obtainable in these studies is shown in FIGURE 9.[37] It illustrates the remarkable resolving power obtainable by this method and the unusually low magnetic field ($\frac{1}{4}$ gauss) at which Zeeman pattern was obtained.

In the first publication on the Radiofrequency Spectra of Atoms we reported on the measurements of the hfs values of the ground states of ^{6}Li, ^{7}Li, ^{39}K and

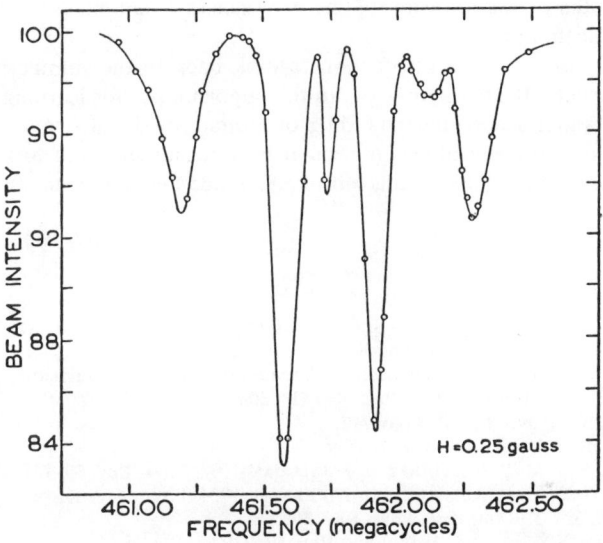

FIGURE 9. Zeeman pattern for ^{39}K by the new molecular beam techniques. (From Kusch et al.[37] By permission of *The Physical Review*.)

^{41}K with a precision of the order of 0.01%. The ratio of the hfs values for ^{7}Li and ^{6}Li agreed with the remeasured corresponding value of the ratio of the moments within the precision of the latter measurements, which was only about 0.04%.

Kusch and I went on to measure the Radiofrequency Spectra of ^{23}Na, ^{85}Rb, ^{87}Rb, and ^{133}Cs.[38] The results for the Rb isotopes were compared with those obtained from direct nuclear moment measurements. A discrepancy of 0.6% was found, but that was too close to the precision of the direct measurement of nuclear moments.

Zacharias constructed an alkali beam apparatus especially designed for the measurement of the radiofrequency spectrum of the low abundance radioactive isotope of potassium ^{40}K.[39] He was able to determine the spin and measure of the hfs value and nuclear moment and show that the moment is negative.

My research activities in the Molecular Beam Laboratory continued after I obtained a position as Instructor in the City College Physics Department in September 1939. I carried the usual teaching load of 15 class hours a week, and there was of course no additional remuneration involved for doing research full time during the summer months or during the winter and spring vacations. Kusch and I continued molecular beam research in association with some of Rabi's graduate students.[40] This work ended abruptly in March 1942 when I took a leave of absence from Queens College to join Kellogg and Kusch in starting a research program at the Columbia Radiation Laboratory, which Rabi had set up as a branch of the MIT Radiation Laboratory radar effort. Zacharias was already well established at the MIT Center. Although the research on magnetrons and microwave transmission didn't have much resemblance to molecular beams, it was not at all adversely affected by the problem solving experiences I had encountered in Rabi's Molecular Beam Laboratory.

It is a source of great satisfaction to look back to the stimulating years in Rabi's Molecular Beam Laboratory, to the opportunities for learning experiences in research and a deeper understanding of atomic physics, to have played some part in the development of new molecular beam techniques and to have made a small contribution to the scientific output of a fundamental branch of physics.

References

1. 1976 winner of the American Institute of Physics—United States Steel Foundation Science-Writing Award in Physics and Astronomy.
2. RAMSEY, N. 1956. Molecular Beams. Clarendon Press. Oxford, England.
3. BREIT, G. & I. I. RABI. 1931. Phys. Rev. **38**: 2082.
4. COHEN, W. 1934. Phys. Rev. **46**: 713.
5. MILLMAN, S. 1935. Phys. Rev. **47**: 739.
6. RABI, I. I., J. M. B. KELLOGG & J. ZACHARIAS. 1934. Phys. Rev. **56**: 157.
7. RABI, I. I., J. M. B. KELLOGG & J. ZACHARIAS. 1934. Phys. Rev. **56**: 163.
8. FOX, M. & I. I. RABI. 1936. Phys. Rev. **50**: 220.
9. MILLMAN, S. & M. FOX. 1936. Phys. Rev. **50**: 220.
10. MANLEY, J. H. 1936. Phys. Rev. **49**: 921.
11. MANLEY, J. H. & S. MILLMAN. 1937. Phys. Rev. **51**: 19.
12. RABI, I. 1936. Phys. Rev. **49**: 324.
13. PHIPPS, T. E. & O. STERN. 1931. Z. Phys. **73**: 183.
14. FRISCH, R. & E. SEGRE. 1933. Z. Phys. **80**: 610.
15. KELLOGG, J., I. RABI & J. ZACHARIAS. 1936. Phys. Rev. **50**: 472.
16. TORREY, H. 1937. Phys. Rev. **51**: 501.
17. MILLMAN, S. & J. ZACHARIAS. 1937. Phys. Rev. **51**: 1049.
18. GORTER, C. 1936. Physica **9**: 995.
19. RABI, I. I., J. R. ZACHARIAS, S. MILLMAN & P. KUSCH. 1938. Phys. Rev. **53**: 318.
20. GORTER, C. 1967. Physics Today **20**(1): 76.
21. MILLMAN, S., I. RABI & J. ZACHARIAS. 1938. Phys. Rev. **53**: 384.
22. RABI, I. I., S. MILLMAN, P. KUSCH & J. R. ZACHARIAS. 1938. Phys. Rev. **53**: 495.
23. RABI, I. I., S. MILLMAN, P. KUSCH & J. R. ZACHARIAS. 1939. Phys. Rev. **55**: 526.
24. MILLMAN, S., P. KUSCH & I. RABI. 1938. Phys. Rev. **54**: 968.
25. KUSCH, P., S. MILLMAN & I. RABI. 1939. Phys. Rev. **55**: 666.
26. KUSCH, P., S. MILLMAN & I. RABI. 1939. Phys. Rev. **55**: 1176.
27. MILLMAN, S., P. KUSCH & I. RABI. 1939. Phys. Rev. **56**: 165.
28. MILLMAN, S. & P. KUSCH. 1939. Phys. Rev. **56**: 303.

29. KUSCH, P. & S. MILLMAN. 1939. Phys. Rev. **56**: 527.
30. MILLMAN, S. & P. KUSCH. 1941. Phys. Rev. **60**: 91.
31. HAY, R. H. 1941. Phys. Rev. **60**: 75.
32. FELD, B. T. & W. E. LAMB. 1945. Phys. Rev. **67**: 15.
33. MILLMAN, S. 1939. Phys. Rev. **55**: 628.
34. KELLOGG, J. M. B., I. I. RABI, N. F. RAMSEY & J. R. ZACHARIAS. 1939. Phys. Rev. **56**: 728.
35. KELLOGG, J. M. B., I. I. RABI, N. F. RAMSEY & J. R. ZACHARIAS, 1940. Phys. Rev. **57**: 677.
36. RAMSEY, N. 1940. Phys. Rev. **58**: 226.
37. KUSCH, P., S. MILLMAN & I. RABI. 1940. Phys. Rev. **57**: 765.
38. MILLMAN, S. & P. KUSCH. 1940. Phys. Rev. **58**: 438.
39. ZACHARIAS, J. R. 1942. Phys. Rev. **61**: 270.
40. HARDY, T. C. & S. MILLMAN. 1942. Phys. Rev. **61**: 459.

ALGEBRAIC INCOMPATIBILITIES BETWEEN ARNOWITT-NATH GAUGES AND SUPERSYMMETRIZED GRAVITY*

Yuval Ne'eman†

Department of Physics & Astronomy
Tel Aviv University
Tel Aviv, Israel

California Institute of Technology
Pasadena, California 91125

Preface

Between 1934 and 1939 I. I. Rabi and his collaborators applied powerful new experimental methods in measuring the magnetic moment of the proton. O. Stern had noted in 1933 (with O. Frisch) that the current value seemed abnormally large and did not obey the value predicted by Dirac's equation (presumably because of a "renormalization" by the strong interactions). Stern and Estermann had also checked the value for the deuteron (which implied a negative contribution due to the neutron) and Rabi and collaborators remeasured the deuteron's magnetic moment in 1936 and 1939 together with that of the proton. Their values were $2.85 \rightarrow 2.785 \pm 0.02 \rightarrow$ (present value) 2.7928456 ± 0.0000011 for the proton, and $0.85 \rightarrow 0.855 \pm 0.006$ for the deuteron (implying for the neutron in the most naive estimate: $-2.0 \rightarrow -1.93 \rightarrow$ present value -1.913148 ± 0.000066).

In the development of the theory of hadron physics, SU(3) symmetry has led to relations between various magnetic moments. However, to find a relation between those of protons and neutrons, one has in addition to assume spin-independence, i.e., quarks as constituents, and a quasifree behavior on their part. This produces directly the ratio $\mu_p/\mu_n = -\frac{3}{2}$, which is correct within 1%–2% only! Someday, we theoreticians may nevertheless achieve the precision of the masters of experimental physics (as had indeed happened in Quantum Electrodynamics)—or is this an impossible dream?

More recently, we have tried to look for more fundamental gauge principles that could explain the various forces and connect them together. One such principle, the "Color" $SU(3)_c$ gauge, mediated by massless vector potentials making an $SU(3)_c$ octet, explains the quasifree behavior of the quarks ("asymptotic freedom") and might hopefully explain their confinement.

Some other discoveries in the field of symmetry may provide a unification of all interactions with gravitation. This subject is at present being investigated by many authors. The following comment analyzes several different approaches to the same program.

---

*This work was partially supported by the U.S. Energy Research and Development Administration under Contract E-(11-1)-68.
†Partially supported by the United States—Israel Binational Science Foundation.

Introduction

After some earlier general suggestions of a possible connection between supersymmetry and gravity,[1] two approaches have been used in attempts to construct a theory of supergravity. On the one hand, Arnowitt and Nath[2,3] and Freund[4,5] have used a geometrical realization in which space-time x^μ is enlarged with the addition of Grassmann dimensions $\theta^{\alpha i}$ (α, a Majorana spinor index, $i = 1 \ldots N$ an "internal" index) and a generalized (self-sourced) Einstein equation is assumed for that superspace,

$$R_{AB} = 0 \quad (A, B = \mu \text{ or } \alpha i), \tag{1}$$

thus hinting at the possibility of a completely unified theory.

In a later version, as a result of attempts[6,10,11] to check the compatibility of Equation 1 with global Wess-Zumino supersymmetry,[7-9] Equation 1 was replaced[10,11] by an alternative "supercosmological" term on the r.h.s.:

$$R_{AB} = \tilde{\Lambda} g_{AB}. \tag{2}$$

Such a term had indeed been envisaged originally too[2,3] but had been discarded for observational reasons (assuming then that the $\tilde{\Lambda}$ would also show up in the space-time Einstein equation). It was found that $\tilde{\Lambda} \neq 0$ limited the theory to $N = 2$ (= electric charge) and that $\tilde{\Lambda} = 0$ required parity nonconservation.

Another approach has consisted of imposing supersymmetry as a local gauge in space-time, and proceeding in the footsteps of Yang and Mills, Utiyama–Kibble,[12-14] and others, to derive a gauged theory. Freedman et al.[15] presented a self-consistent scheme of such a field theory with no matter present except for the "Yang-Mills" multiplet itself (V_μ^a and ω_μ^{ab} the nonindependent vierbein field and its connection,[16] representing gravity, and $\psi_{\mu\alpha}$ with $J = \tfrac{3}{2}$ gauging supersymmetry transformations). More recently a general theory in the presence of matter has been suggested.[17-19]

In this note, we shall use an algebraic approach to explain the differences and incompatibilities between the two schemes.

Relevant Graded Lie Algebras

As a result of the physics interest, the mathematical theory of Graded Lie Algebras (GLA) has been extensively developed[20,21] and a complete classification has now been achieved.[22] The two main classes are the $SL(m/n)$ [or $SU(m/n)$ for a unitary form] and $OSp(r/2s)$ (the "ortho-symplectic" GLA). The even (i.e., ordinary) subalgebras are in the two cases,

$$SL(m/n) \supset SL(m) \times SL(n) \text{ [or } SU(m) \times SU(n) \text{ for } SU(m/n)], \tag{3}$$

$$OSp(r/2s) \supset O(r) \times Sp(2s). \tag{4}$$

The $SL(m/n)$ can be represented by $(m + n) \times (m + n)$ matrices, with the even subalgebra in the diagonal squares, and with the remaining off-diagonal $2mn$ generators belonging to the odd (Fermionic) type and behaving as $m \times n$ and their conjugates.

The OSp($r/2s$) are defined over a graded vector space, with r "even" ($= x^\mu$ type) dimensions and $2s$ "odd" (Grassman, $\theta^{\alpha i}$ type) dimensions. The O(r) and Sp($2s$) are again in the diagonal squares for x^μ and $\theta^{\alpha i}$, respectively; the $2rs$ odd generators consist in only one half of the remaining $4rs$ off-diagonal matrix elements. Thus, when acting between a generalized contravariant coordinate ζ^A (x^μ, $\theta^{\alpha i}$) and the derivatives $\partial_A(\partial_\mu, \partial_{\alpha i})$, *only one linear combination* $\alpha \zeta^A \partial_B + \beta \zeta^B \partial_A$ is allowed.[24]

Haag et al.[23] have shown that only two algebras do not conflict with the physical axioms and with a constraint requiring the generators to be expressible as space-integrals of local densities constructed from fields. These are, using the new knowledge about the GLA:

(1) For a massless system: SU(2,2/N), with even subalgebra SU(2, 2) × U(N) (for N = 1 this is the original Wess-Zumino algebra). SU(2, 2) is the space-time conformal group.
(2) For a massive system OSp($N/4$)$_{R\to\infty}$, with a Wigner-Inönu contraction[21]; the Sp(4) even subalgebra in the spinor square is identified with a real form of O(5), which is then contracted, yielding the Poincaré group. The $4N$ spinor generators are automatically also contracted and for N = 1 become the S_α of the Salam-Strathdee system.

Freund[24] has pointed out that the homogeneous algebra of the tangent (flat) space at any point of the Arnowitt-Nath (AN) superspace is OSp(3, 1/$4N$), and that the conformal algebra in the tangent space is OSp(4, 2/$4N$), thus playing the same role in superspace that the Lorentz and conformal groups fulfill in space-time. These are the symmetries of the Riemannian manifold in any small neighborhood (and "inertial" reference frame). In the sense of Kibble-Utiyama, they should be considered as the global symmetries whose gauging produces the new dynamical equations. In the following, we shall prove the following algebraic results:

(1) that inhomogeneous OSp(3, 1/$4N$) does not contain the Golfand–Likhtman/Salam-Strathdee OSp($N/4$)$_{R\to\infty}$ (the massive case);
(2) that OSp(4, 2/$4N$) does not contain the Wess-Zumino SU(2, 2/N) (the massless case);
(3) that the realizable symmetries of the Ogievetsky algebra[25] of the Arnowitt-Nath superspace indeed contain OSp($N/4$)$_{R\to\infty}$ for N = 2 only, when $\Lambda \neq 0$;
(4) that they may contain SU(2, 2/N) when $\tilde{\Lambda}$ = 0 in a chiral form, thus imposing parity violation.

Massive Case

We prove that inhomogeneous OSp(3, 1/$4N$)(i.e., adjoining ∂_μ and $\partial_{\alpha i}$) does not contain OSp($n/4$)$_{R\to\infty}$. This is trivial because the anticommuting generators of OSp(3, 1/$4N$) (with $J = \frac{3}{2}$ and $J = \frac{1}{2}$) cannot be identified with the S_α of supersymmetry

$$\{S_\alpha, S_\beta\} = (\gamma_\mu C)_{\alpha\beta} p^\mu, \tag{5}$$

since their anticommutators are included in $0(3, 1) \times \mathrm{Sp}(4N)$, i.e., in the Lorentz group and an SU(6)-like symmetry Sp(4N) and *not in the translations*. Moreover, the spinor generators are linear combinations of $x\partial_\alpha$ and $\theta\partial_\mu$ in which it is not possible to separate the $\theta\partial\mu$ parts corresponding to the S_α.

Massless Case

The realization of the Lorentz and conformal groups on AN space require their generators to act simultaneously on x^μ and $\theta^{\alpha i}$. Thus the $J^{\mu\nu}$ have to appear as the sum of the $0(3, 1)$ in the x^μ sector of $\mathrm{OSp}(3, 1/4N)$ and of the $0(3, 1)$ "spin" subgroup of Sp(4N) on the $\theta^{\alpha i}$ sector,

$$J^{\mu\nu} \sim x^\mu \partial_\nu - x^\nu \partial_\mu + \tfrac{1}{2} \bar\theta^{\alpha i} \sigma^{\alpha\beta} \partial_{\beta i}. \tag{6}$$

The same should be true of the conformal group SU(2, 2). However, it is impossible to represent the special conformal generators K^μ linearly on the $\theta^{\alpha i}$ sector. Dimensionally, K^μ has the dimensions L of a length x^μ (or of the inverse of p^μ) and $\theta^{\alpha i}$ has $L^{1/2}$ (alternatively, S_α has $L^{-1/2}$ from Equation 5, which fixes $L^{1/2}$ for $\theta^{\alpha i}$). Note that this definition of dimensionality is also fixed by the commutation relations with the dilation generator D within the Wess–Zumino algebra and is required by supersymmetry. To represent K^μ in $\theta^{\alpha i}$ only (since we are dealing with the even, ordinary, Lie subalgebra) we require a $\theta\theta\theta\partial_\alpha$ operator. This is not present in the $\theta^{\alpha i}$ square of $\mathrm{OSp}(4, 2/4N)$, with nonlinear operators appearing in the x^μ system only. The even subalgebra on the $\theta^{\alpha i}$ sector is still Sp(4N), and

$$\mathrm{Sp}(4N) \not\supset \theta\theta\theta\partial_\alpha. \tag{7}$$

The Ogievetsky Algebra

The Ogievetsky algebra[25] of a space is a realization of the Einstein-like group of general-coordinate-transformations (to "non-inertial" reference frames). For space-time it was shown that since any such transformation can be regarded as the action of a generator

$$\prod_{\substack{\mu=0\ldots 3 \\ n=0\ldots\infty}} (x^\mu)^{n_\mu} \partial_\nu,$$

all Einstein transformations would be generated by the infinite closure of

$$\mathrm{SL}(4R) \cup \mathrm{SU}(2, 2), \tag{8}$$

where the $J = 2$ generators in SL(4R) $(x^\mu \partial_\nu + x^\nu \partial_\mu)$ and the K^μ in SU(2, 2) (four generators $xx\partial_\mu$) are shown to generate in their commutation relations, first the other twelve $xx\partial_\mu$, then $xxx\partial_\mu$, etc.).

In the AN space, this is realized[24] by

$$\mathrm{SL}(4/4N) \cup \mathrm{OSp}(4, 2/4N). \tag{9}$$

Note that quasi-Einsteinian transformations on the $\theta^{\alpha i}$ sector terminate at $n_{\alpha i} = 4N$ due to the Grassmann antisymmetrizing property, and that this part of the Ogievetsky algebra is the finite $W(4N)$ of the Kats classification.[22]

In SL(4/4N), the $x\partial_\alpha$ and $\theta\partial_\mu$ can be separated, since we do have double the number of linear spinor generators as compared to OSp(3, 1/4N). Bracketing the translations ∂_μ in the 0(4, 2) subalgebra of OSp(4, 2/4N) with the $x\partial_\alpha$ of SL(4/4N), we get ∂_α, so that we can construct the S_α from ∂_α and $\theta\partial_\mu$ (we leave out the i index, which plays no role here):

$$S_\alpha = \bar{\partial}_\alpha + \frac{i}{2} \gamma^\mu \theta \partial_\mu. \tag{10}$$

To represent the R_α of SU(2, 2/1),

$$\{R_\alpha, R_\beta\} = - (\gamma_\mu C)_{\alpha\beta} K^\mu \tag{11}$$

$$\{S_\alpha, R_\beta\} = - (\sigma_{\mu\nu} C)_{\alpha\beta} J^{\mu\nu} + i C_{\alpha\beta}(D - \gamma_5 E), \tag{12}$$

we require, aside from the $x\partial_\alpha$, components in $x\theta\partial_\alpha$, $\theta\theta\partial_\alpha$, and $\theta\theta\theta\partial_\mu$ (at least in two of the latter). Such elements are generated in the GLA bracketing of

$K_\mu(x)$: $xx\partial_\mu \in 0(4, 2) \subset OSp(4, 2/4N)$, with $\theta\theta_\mu \in SL(4/4N) \rightarrow x\theta\partial_\alpha$ (in R_α)

$x\theta\partial_\mu$, with $\theta\partial_\mu \rightarrow \theta\theta\partial_\mu$,

$\theta\theta\partial_\mu$, with $x\partial_\alpha \rightarrow \theta\theta\partial_\alpha$ in R_α,

$\theta\theta\partial_\alpha$, with $\theta\theta\partial_\beta \rightarrow \theta\theta\theta\partial_\alpha$ [allowing for $K_\mu(\theta)$],

$\theta\theta\theta\partial_\alpha$, with $\theta\partial_\mu \rightarrow \theta\theta\theta\partial_\mu$ in R_α,

With $J^{\mu\nu}, S_\alpha, R_\alpha, \partial_\mu$ we can get the whole SU(2, 2/1) of Wess-Zumino, and using the O(N) internal symmetry of Sp(4N) $\subset$ OSp(4, 2/4N), we can reproduce the full set of $2N$ generators $S_{\alpha i}$ and $R_{\alpha i}$ and generate SU(2, 2/N) by closure. Note that the sequence of brackets also displays the manner in which the full $W(4N)$ will be reached.

The appearance of a generator in the Ogievetsky algebra does not yet ensure its realization as an active symmetry of the Lagrangian (or the S matrix). It is in the nature of a choice of a particular gauge for the translations, just as we can have a function $\alpha(x) = c_\nu x^\nu$ for the electromagnetic gauge. No new conservation law is generated in that case, and the Noether theorem yields a null-current whose charge vanishes for physical states (satisfying the equations of motion). In the case of gravity, the SL(4R) in the Ogievetsky algebra will lead to quadrupoles and so forth. However, the antisymmetric part of SL(4R) does formally coincide with orbital Lorentz transformations; applying the operators $x^\mu \partial_\nu - x^\nu \partial_\mu$ as physical generators rather than as a translation gauge choice, one gets a symmetry of ungauged (flat) Minkowski space, and the appropriate physical conservation laws.

In our analysis we shall thus have to check on two different criteria: (1) Does a given generator appear formally in the Ogievetsky algebra? and (2) Can it be realized physically? The latter question itself will be determined by two factors: the properties of the AN orthosymplectic space which determines the field equations, and the properties of physical space-time ("the vacuum") where supersymmetry is being enforced.

The $\tilde{\Lambda} \neq 0$ Case

We now identify D, the physical dilation operator $[D \in SU(2, 2)]$:

$$D = x^\mu \partial_\mu + \lambda \theta^{\alpha i} \partial_{\alpha i}. \tag{13}$$

This is a linear operator on the orthosymplectic AN space and should exist in $GL(4/4N)$. There are two such candidates:

$$D^F = x^\mu \partial_\mu + \theta^{\alpha i} \partial_{\alpha i}, \quad \tilde{Tr}(D^F) = 4 - 4N \tag{14}$$

$$D^\circ = x^\mu \partial_\mu + \frac{1}{N} \theta^{\alpha i} \partial_{\alpha i}, \quad \tilde{Tr}(D^\circ) = 0, \quad D^\circ \in SL(4/4N) \tag{15}$$

$$\tilde{Tr}(D^F D^\circ) = 0 \tag{16}$$

Here $\tilde{Tr}$ is the graded trace,[12] i.e.,

$$\tilde{Tr} = Tr \text{ (in } x^\mu \text{ sector)} - Tr \text{ (in } \theta^{\alpha i} \text{ sector)}.$$

Equation 16 implies graded-orthogonality and fixes this basis in the D system, once D° is picked by the graded-tracelessness condition of $SL(4/4N)$.

The operator D^F is the (unique) dilation operator of the AN orthosymplectic space.[21] Being in the center of $GL(4/4N)$ it has a vanishing bracket with all elements of $OSp(4/4N)$ since this is a subalgebra of $SL(4/4N)$.

For $\tilde{\Lambda} \neq 0$, the super-cosmological term in Equation 2 imposes a breakdown of scale invariance in AN space. The quasi-Einstein transformation D^F representing scaling, although formally in the Ogievetsky algebra, does not realize a symmetry.

As a result, only D° is available, and we make the identification,

$$D = D^\circ. \tag{17}$$

Using Equation 5, which fixes for Wess-Zumino supersymmetry,

$$[D, \theta^{\alpha i}] = \tfrac{1}{2} \theta^{\alpha i}, \tag{18}$$

for

$$[D, x^\mu] = x^\mu, \tag{19}$$

we require $\lambda = \tfrac{1}{2}$ in Equation 13, which fixes $N = 2$. Note that D is not broken by the $\tilde{\Lambda}$ term, since it was shown[10,11] that taking in Equation 2, $A = \mu$, $B = \nu$, one finds that the supercosmological term cancels out of the equation. The result is a sourced Einstein equation for ordinary space-time with no cosmological term. (Further scale breaking terms are introduced in the detailed scheme of spontaneous breakdowns at a later stage.)

Equation 17 ensures that D° act as a representation on (x, θ) space of dilations in x^μ space, which explains its survival as an active symmetry operator in the Ogievetsky algebra. The $\tilde{\Lambda}$ term also breaks chiral invariance.

In the alternative approach of a local supersymmetry gauge à la Yang-Mills[15-19] there will be no such limitations of $N = 2$. Any N which fits global supersymmetry should respond to a local gauge treatment. Indeed the present models are for $N = 1$, which is forbidden in the AN gauge. Limitations may come from imposing a ceiling on the spin of fields in the graviton supermultiplet, but this may be unnecessary, in a spontaneous breakdown which would provide masses for $J > 2$, for instance. Note that the AN gauge requires the existence of fields of spin $J = 2(N + 1)$.

The overlap with supergravity results does exist for $N = 2$, as has recently been shown.[26] The existence of global supersymmetry for the AN vacuum solution can then be used to impose supersymmetry on Equation 2; the graviton of the AN system is then forced to appear in a representation of $N = 2$ supersymmetry.

Actually, one more algebraic feature of $SU(2, 2/N)$ and $OSp(N/4)_{R \to \infty}$ plays a role in the extraction of a supersymmetric (either global or local) space-time from the AN gauge. The transformations wrought on the $\theta^{\alpha i}$ are not length-preserving. The generator S_α supplies δx^μ with a variation from the $\theta^{\alpha i}$ subspace, but does not do the inverse. Even when the R_α are introduced for $SU(2, 2/N)$, their physical role is completely different (they do not appear in the physical stability subalgebra, as S_α does, etc.). Supersymmetry preserves the space-time interval (including contributions from the θ subspace), but does not preserve an interval in AN (x, θ) space. *Moreover, the physical role of the interval ds requires ds^2 to be geometrically dilation covariant, a requirement violated by the constant K.* As shown for one case, in Reference 6 the equations of motion of global supersymmetry require taking the limit $K \to 0$ in the "vacuum metric,"

$$g_{\mu\nu}{}^{(0)} = \eta_{\mu\nu} \qquad g_{\mu\alpha i}{}^{(0)} = -i\beta(\bar{\theta}\gamma_\mu)_{\alpha i},$$

$$g_{\alpha i \beta j}{}^{(0)} = K C_{\alpha\beta}\delta_{ij} + \beta^2(\bar{\theta}\gamma_\mu)_{\alpha i}(\bar{\theta}\gamma_\mu)_{\alpha j}, \qquad (20)$$

(the γ_μ are 8×8 and given by ordinary $\gamma_\mu \otimes 1$). The constant K is in the denominator of g^{MN} and enters the expression for R_{MN}. With no $C_{\alpha\beta}$ in the metric, the Riemannian properties of the orthosymplectic space are inhibited.

The $\tilde{\Lambda} = 0$ Case

For $\tilde{\Lambda} = 0$, linear combinations of D^F and D° are allowed. Arnowitt and Nath[10,11] have shown that global supersymmetry can thus be imposed on the vacuum with a spontaneous breakdown of symmetry. This is the realization of the above-mentioned construction of $SU(2, 2/N)$ in the Ogievetsky algebra. Arnowitt and Nath show that this solution is chiral and violates parity. Indeed, the irreducible representations of $SU(2, 2/N)$ are chiral, and with the vacuum we do not have the freedom of restoring parity conservation by chiral doubling. The emergence of $SU(2, 2/N)$ rather than its nonchiral subalgebra $OSp(N/4)_{R \to \infty}$ is imposed by the geometry[27] of the original orthosymplectic space, through Equation 1.

The case $N = 1$ for instance, with parity conservation, is thus completely forbidden in the AN gauge, whereas it has been realized for supergravity of the Yang-Mills type, with and without the presence of matter fields as sources.[15-19]

Conclusions

(1) The tangent-space symmetry of the AN gauge[2,3] cannot be used directly, since it violates the HLS theorem. It is incompatible with supersymmetry. The

orthosymplectic analogs of the energy-momentum pseudotensor [i.e., the expressions representing the coupling to $g_{MN}(x)$] contain the graviton term but do not contain the supersymmetry current.

(2) However, spontaneous symmetry breakdown[10,11] may produce a globally supersymmetric physical space thus obeying the HLS theorem, provided the necessary Goldstone and Higgs mesons are introduced; this should happen not only to the unrealized parts of the IOSp(4/4N) internal symmetry generators,[27] but also to all other unrealized symmetries and their gauge fields. (The dimensionalities are $8N^2 + 22N + 10$ for IOSp(4/4N) and $N^2 + 8N + 15$ for SU(2, 2/N), with very little overlap beyond the Poincaré group).

(3) The orthosymplectic origin of the AN equation imposes restrictions on the internal symmetry. These restrictions do not occur in the Yang-Mills type supergravity, though other type limitations may appear in the latter system under further investigation. The existence of restrictions on N in the AN gauge may however be considered either as an advantage (in the sense of explaining observed internal symmetries) or as a limitation on the further applicability of the theory. This is a question whose answer will be given by future results in the field. However, it is good to remember that we have as yet no *experimental* evidence for either global or local supersymmetry, even though the theories are consistent with the physical axioms.

(4) The "spontaneous breakdown" of gauged AN orthosymplectic symmetry which is found to obey supersymmetry is not of the usual type. In a Weinberg–Salam U(2) gauge, for instance, the Goldstone-Higgs multiplet picks out a direction breaking global U(2) itself, and leaving invariant a subgroup of the global group. In the AN case, the spontaneous breakdown is supersymmetry invariant, while supersymmetry is not a subgroup of the global symmetry. Only the Poincaré group and the internal symmetry form a subgroup of both the vacuum group and the global one. The breakdown is thus the equivalent of choosing a specific gauge. This is as if we were to start from a manifestly non-Lorentz invariant Lagrangian and then pick the one specific gauge that makes it manifestly Lorentz invariant.

(5) Global supersymmetry defines eight invariant differentials,

$$dz^M \begin{cases} dl^\mu = dx^\mu - i\beta\bar{\theta}\gamma^\mu d\theta \\ d\theta^\alpha \end{cases}, \tag{21}$$

yielding in a Riemannian treatment the interval

$$ds^2 = dl^2 + K d\bar{\theta} d\theta, \tag{22}$$

from which the metric Equation 20 was derived. However, Salam–Strathdee dynamics imply $K = 0$, as required by dilation covariance of the line element. It is easiest to visualize the implications of this limit through the analogy of Galilean transformations. These define the four invariant differentials (dx^i, dt) but they preserve dt^2 only. Indeed, writing

$$K = -(c_0)^{-2}, \; d\hat{s}^2 = c_0^2 ds^2, \tag{23}$$

we have

$$d\hat{s}^2 = c_0^2 dl^2 - d\bar{\theta} d\theta = c_0^2 dl'^2. \tag{24}$$

The latter transformation is an orthosymplectic one which does not exist in supersymmetry (and is not allowed by HLS), just as the Galilean group does not include Lorentz transformations. Writing

$$dl' = \frac{d\hat{s}}{c_0} = dl\left(1 - \frac{v^2}{c_0^2}\right)^{1/2}, v_\mu{}^\alpha = \frac{d\theta^\alpha}{dl^\mu}, v^2 = \bar{v}v, \qquad (25)$$

we observe that $K \to 0$ corresponds to the θ space "interaction velocity" $c_0 \to \infty$. Global and gauged supersymmetry thus correspond to a Galilean-like collapse of the AN metric and of the dynamics represented by Equation 2, even for allowed N and parity assignments.

(6) Notwithstanding all these incompatibilities, the AN gauge provides a heuristic simple model for a possible future geometrization of supergravity, and of the Unified Field Theory this may usher in.

Summary

We have proved that ordinary Wess-Zumino supersymmetry is not contained in the tangent inhomogeneous or conformal groups of the Arnowitt–Nath superspace. It is contained in the infinite ("noninertial") general coordinate transformations of that space, with restrictions on internal symmetries (as found by Arnowitt and Nath) and Galilean collapse of the metric. These limitations do not exist in the alternative approach of Locally Gauged Supersymmetry.

Acknowledgments

The author would like to thank the Aspen Center for Physics for their hospitality during part of the time when this study was performed.

We would also like to thank Prof. R. Arnowitt for extensive comments, correspondence and discussions, and for making available to us the results of References 26 and 27 prior to publication.

References

1. VOLKOV, D. V. & V. A. SOROKA. 1973. JETP Lett. **18**: 529.
2. NATH, P. & R. ARNOWITT. 1975. Phys. Lett. **56B**: 177.
3. ARNOWITT, R. & P. NATH. 1976. Gen. Rel. Grav. **7**: 89.
4. CHO, Y. M. & P. G. O. FREUND. 1975. Phys. Rev. **D12**: 1711.
5. FREUND, P. G. O. 1976. J. Math. Phys. **17**: 424.
6. WOO, G. 1975. Lett. Nuovo Cimento **13**: 546.
7. GOL'FAND, YU. A. & E. P. LIKHTMAN. 1971. JETP Lett. **13**: 452.
8. WESS, J. & B. ZUMINO. 1974. Nucl. Phys. **B70**: 39.
9. SALAM, A. & J. STRATHDEE. 1974. Nucl. Phys. **B76**: 477.
10. NATH, P. 1975. Proc. Conf. on Gauge Theories and Modern Field Theory, Boston, M.I.T. Press. Cambridge, Mass. 1975. 281.
11. ARNOWITT, R. & P. NATH. 1976. Phys. Rev. Lett. **36**: 1526
12. YANG, C. N. & R. L. MILLS. 1954. Phys. Rev. **96**: 191.
13. UTIYAMA, R. 1956. Phys. Rev. **101**: 1597.

14. KIBBLE, T. W. B. 1961. J. Math. Phys. **2**: 212.
15. FREEDMAN, D. Z., P. VAN NIEUWENHUIZEN & S. FERRARA. 1976. Phys. Rev. **D13**: 3214. Phys. Rev. **D14**: 912.
16. DESER, S. & B. ZUMINO. 1976. Phys. Lett. **62B**: 335.
17. FREEDMAN, D. Z. & J. H. SCHWARZ. 1977. Phys. Rev. **D15**: 1007.
18. FERRARA, S., J. SCHERK & P. NIEUWENHUIZEN. 1976. Phys. Rev. Lett. **37**: 1035.
19. GELL-MANN, M. & Y. NE'EMAN. unpublished.
20. CORWIN, L, Y. NE'EMAN & S. STERNBERG. 1975. Rev. Mod. Phys. **47**: 573.
21. FREUND, P. G. O. & I. KAPLANSKY. 1976. J. Math. Phys. **17**: 228.
22. KATS, V. G. 1975. Functional Analysis (U.S.S.R.) **9**: 91.
23. HAAG, R., J. T. LOPUSZANSKI & M. SOHNIUS. 1975. **B88**: 257.
24. FREUND, P. G. O. 1976. J. Math. Phys. **17**: 424.
25. OGIEVETSKY, V. 1973. Lett. al Nuovo Cimento **8**: 988.
26. NATH, P. & R. ARNOWITT. 1976. Phys. Lett. **65B**: 73.
27. ARNOWITT, R. & P. NATH. 1977. Phys. Rev. **D15**: 1033.

RADIOACTIVITY'S TWO EARLY PUZZLES

A. Pais

Department of Physics
Rockefeller University
New York, New York 10021

1. Introduction

The first observation of radioactive transformations was made by Antoine Henri Becquerel in Paris, on Sunday March 1, 1896, and communicated by him to the Académie des Sciences the next day. He did not name his new phenomenon radioactivity—that term was first coined by the Curies and Bémont in 1898. (It entered the English literature for the first time in the November 16, 1898 issue of *Nature*.) Becquerel was unaware that this was the first observation ever of a nuclear process because even the existence of the nucleus was not as yet known. He did not realize at once that the bulk of the radiation he had observed consisted of very light particles, electrons, because the existence of the electron was not as yet known. What he did realize, however, was that he had made a startling discovery.

In this article I shall discuss two puzzles facing physicists in the first quarter of the 20th century as a result of the discovery of radioactive phenomena.

The first one was: What is the source of the energy that continues to be released by radioactive materials? Already in the year of discovery, Becquerel himself had been quite surprised at the persistence of the energy produced by what he initially called "uranic rays." From 1898 on, physicists began to pose such questions as: Could it be that energy is not conserved in these processes? Could there be something amiss with the second law of thermodynamics in radioactive transformations? Does the source of energy reside outside the atom or inside?

The second puzzle was: What is the significance of the characteristic half-life for such transformations? (The first determination of a life time for radioactive decay dates from the year 1900.) If in a given radioactive transformation all parent atoms are identical and if the same is true for all daughter products, then why does one radioactive parent atom live longer than another and what decides when a specific parent atom disintegrates?

It should be stressed that these problems did not hold center stage throughout the period under discussion, a period so rich in other developments. Rather, the puzzles to be discussed in this paper were principally the concern of a fairly modest-sized but élite club of experimental radioactivists. In those days, theoretical physicists did not play any role of consequence in the development of this subject, both because they were not particularly needed for its descriptive aspects and because the deeper questions were too difficult for their time. It is true that distinguished theorists (especially those belonging to an older generation) would on occasion express views on these issues from which we gain revealing insights into the climate of thought of the times. But these comments were not to be of lasting significance—with one most notable exception, the contribution by Einstein.

In the second of his 1905 papers on relativity[1] Einstein stated that, "If a body gives off the energy L in the form of radiation, its mass diminishes by L/c^2 .... The mass of a body is a measure of its energy.... It is not impossible that with bodies whose energy content is variable to a high degree (e.g., with radium salts) the theory may be successfully put to the test."

With the help of Einstein's discovery of the mass-energy equivalence, some of the questions related to the origins of the radioactive energy release could have been answered, at least in principle (see below). However, as a matter of historical fact this did not come to pass in the period under discussion. There appear to be three reasons for this. (1) The precepts of relativity were assimilated rather slowly.* (2) The level of accuracy of mass measurements was not adequate during this period. Thus in his 1921 review of relativity theory, Pauli notes that "*perhaps* the theorem of the equivalence of mass and energy can be checked *at some future date* by observations on the stability of nuclei"[2]† (my italics). (3) The life time question had to remain entirely unresolved until the advent of quantum mechanics, when it became possible for the first time to understand the *mechanisms* of radioactive decay. Prior to the understanding of these mechanisms it was inevitable that the origin of radioactive energy had to remain hazy as well even though, after the fact, much can be explained about the energy release by the simple application of conservation laws, independently from quantum mechanical arguments.

Yet, well before the proofs were there, a correct consensus began to emerge with regard to the energy puzzle. If around 1910 those who had labored and thought seriously about this question had been polled, there is no doubt that a majority would have expressed the belief that energy is conserved and that the energy source resides in the atomic interior. Had they further been asked about the explanation of the life-time puzzle, however, then the wisest would have readily admitted that this was a question beyond their horizon.

In any event, these questions, first stated around the turn of the century, remained unresolved until the summer of 1928 when (see below, section 6) it was found that," ... it has hitherto been necessary to postulate some special arbitrary 'instability' of the nucleus ... but ... disintegration is a natural consequence of the laws of quantum mechanics without any special hypothesis...."

That was good enough for α- but not for β-radioactivity. In fact, in just about that same year, 1928, new paradoxes emerged regarding the energy loss in β-decay. And so it has continued. The developments, which started in the 1890s, have posed challenge after challenge until this day. Now, as then, our endeavors are based on a jumble consisting of some fine but incomplete dynamics, some good but incomplete ordering principles, some consensus with no basis in facts—the whole of which is presently called particle physics.

2. First Interlude: Envoi

I first met I. I. Rabi in September 1946 at a meeting of the American Physical Society in midtown Manhattan. After a few pleasantries, he fired this question at

---

*As I shall discuss elsewhere, in a forthcoming history of particle physics.
†The validity of the energy-mass-velocity relation required by special relativity was well verified for the electron by about 1915.

me: "Do you think the polarization of the vacuum can be measured?" It was my first week in the United States, and I recall my astonishment at being in a new land where experimentalists would know, let alone bother, about vacuum polarization.

Shortly thereafter we discussed the same question again during the Princeton bicentennial. I did not hear any more about the subject until the morning of June 2, 1947. The setting was the Rams Head Inn on Shelter Island, during a small conference held under the auspices of the National Academy of Sciences with the support of the Rockefeller Foundation. On that morning, Willis Lamb gave the first report on his line shift measurements in hydrogen, which revealed an effect well over an order of magnitude larger than the vacuum polarization effect. And Rabi told us of his work with Nafe and Nelson on an anomaly in the hyperfine structure of atomic hydrogen and deuterium....

Since that time we have met often and under many circumstances. In particular, I have enjoyed his and Helen's hospitality numerous times in their Riverside Drive apartment. There are many reasons why it gives me pleasure to contribute to this volume. I shall name only two. First, Rabi always was, and still is a champion of the new and unexplored. Secondly, he has always laid great emphasis on the role of physics as part of the humanistic tradition. I chose my present topic with both these qualities of his in mind.

3. THE FIRST ENERGY CRISIS

Between 1898 and the early 1930s it happened three times that the discoveries of new natural phenomena were so unsettling as to make prominent physicists waver in their faith in the universal validity of the law of conservation of energy. The first of these crises, referred to in the Introduction, concerned radioactivity. Thirty years later it was radioactivity again (more specifically β-decay) that caused temporary doubt in some quarters about energy conservation. In between, agonizing attempts to reconcile quantum effects with classical reasoning led likewise, and again briefly, to suggestions that energy conservation might not hold strictly. It is the first of these three instances which shall concern us here. As a prelude to this subject, let us look briefly at the status of the conservation law toward the end of the nineteenth century.

In 1775, the Paris Academy of Sciences (still the Académie Royale, at that time) formally announced a significant decision:[3] "The Academy has resolved, this year, to examine no longer any solutions to problems on the following subjects: The duplication of the cube, the trisection of the angle, the quadrature of the circle, or any machine claiming to be a perpetuum mobile." The resolution was expatiated upon in a motivation with many a curious turn of phrase; for us, this simple categorical statement is of interest: "The construction of a perpetual motion machine is absolutely impossible." Evidently the illustrious Academicians grew tired of finding the inevitable flaws in papers submitted on this subject.

We now call the machine excommunicated by the Academy a perpetuum mobile of the first kind. The growing insight that such a device, which spontaneously creates energy, cannot be made was one of the main contributing factors to the formulation, more than 50 years later, of the universal energy principle, a major achievement of 19th century physics. Insofar as purely mechanical systems are concerned, the law of conservation of energy has much older roots.[4,5] Several of science's most illustrious names are associated with these early developments in mechanics. But the principle in its broader sense emerged only when the need arose to express quantitatively the convertibility of diverse forms of energy (mechanical, electrical, magnetic, chemical, physiological, etc.) into each other. The period of discovery of the macroscopic energy law (the first law of thermodynamics) in its generality, that is, applied to any form or several forms of energy, is approximately 1830–1850.

No single year can be associated with this discovery because it was made not by any one person but by many, working most often without initial awareness of each other's activities. A list of pioneers on the subject[6] contains no less than twelve names: Sadi Carnot, Colding, Faraday, Grove, Helmholtz, Hirn, Holtzmann, Joule, Liebig, Mayer, Mohr, Séguin. Four of these (Carnot, Hirn, Holtzmann, Séguin) became involved because of their interest in the effectiveness of steam engines, while two others (Helmholtz, Mayer) were initially intrigued by physiological questions. Given this large a number of dramatis personae, priority disputes were inevitable: "most intense battles took place about the priority of [these] ideas, during which execrable personal accusations and repugnant national chauvinism came into the open."[7] These controversies will not be discussed here. The interested reader can find several detailed accounts elsewhere.[6-8]

The curious case of Sadi Carnot (1796–1832) should be mentioned, however. He is of course justly famous as the discoverer of the second law of thermodynamics (for reversible systems). In actual fact he also discovered the first law. In the early 1820s he stated in his diaries that wherever there is destruction of mechanical work (puissance motrice) there is generation of heat (production de chaleur) and concluded: "one can therefore pose the general thesis that mechanical work is an invariable quantity in nature, that properly speaking it is never produced nor destroyed."[7] In addition he gave an estimate (somewhat low, but not at all so bad) of the mechanical equivalent of heat. But he never published! Long after his death, in 1878, this material was handed over to the French Academy by his surviving younger brother.[9] As Max Planck put it: "He [S.C.] has unquestionably the merit of having given the first evaluation of the mechanical equivalent of heat."[10] As Ernst Mach put it: "Since for practical reasons one cannot name the law [of the equivalence of heat and mechanical work] for all the people who took part in its discovery and its justification, it is advisable to associate [the law] with the names of those who in both respects must be accorded the priority of publication."[11] For this reason Mach speaks of the Mayer-Joule principle for the case that only mechanical work and heat are considered, and of the energy conservation law when *all* forms of energy are included. It is understandable that the first law is often referred to as *le principe de Carnot* in the French literature. Since the same appellation is also used for the second law, the reader of such papers is advised to find out from the context what the issue is.

Mach also observed[11] that the strongest emphasis on the universality of the conservation of energy stems from Robert Mayer (1814–1878) and Hermann

Helmholtz (1821–1894). Already the title of Helmholtz's important essay on the subject‡: "Über die Erhaltung der Kraft: eine physikalische Abhandlung", ("On the conservation of force: a physical memoir"),[12] is of considerable interest. What is here called force is what we now call energy. Current terminology in this respect is itself of 19th century origin. The first one to use the term "energy" in its modern technical meaning was Thomas Young (1773–1829): "The term energy may be applied with great propriety to the product of the mass or weight of the body, into the square of the number expressing its velocity . . ."[13] This quantity Mv^2 is the *vis viva* of Leibnitz. A factor $\frac{1}{2}$ is still lacking before we arrive at our familiar kinetic energy. This factor seems to have been supplied first by Gaspard Gustav de Coriolis (1792–1843).[14]

Nor should one fail to notice Helmholtz's emphasis on his subject as a treatise in *physics*. As he strongly urges, we are not dealing here with an axiomatic statement or a philosophical tenet, nor with a tautology (all such views were expressed at one time or another) but with a physical hypothesis that needs verification in each instance.§ The key to doing this is to find the equivalent of each energy form (via direct or indirect processes) in terms of mechanical work. Moreover, such an equivalence has an unambiguous meaning only after it is realized that the change in energy of a system from an initial to a final state is independent of the way in which the transition between these states takes place.

These, briefly, are the lines along which the conservation of energy came to be clearly understood as a physical principle of universal validity, as the 19th century drew to a close. In two respects there have been fundamental developments of the subject in the 20th century. First, a unification has taken place of the two basic laws: conservation of energy and conservation of matter. Unlike the former law, the latter one, a product of the 18th century, is associated with the name of one single scientist: Antoine-Laurent Lavoisier (1743–1794), a man to whom a greatful nation expressed its debt by putting him under the guillotine, an event of which Laplace has said: "It took them only an instant to cut off that head, and a hundred years may not produce another one like it." Secondly, the energy law appears in thermodynamic context as a macroscopic principle and on a different footing than the conservation of momentum and angular momentum. The modern association between conservation laws and invariance principles emphasizes the microscopic foundations of all three laws, treats them very much on a common level, and frees the conditions for their validity to a larger extent than before from dynamical details.

I shall return elsewhere to both these subjects. For the present purposes it is enough to conclude this brief survey with a comment by Max Planck, found on the opening page of his 1887 prize essay[8]:" . . . if today a quite new natural phenomenon were to be discovered, one would be able to obtain at once from [the energy conservation principle] a law for this new effect, while otherwise there does

---

‡ Both this essay and the first paper by Robert Mayer on the subject share the distinction of having been rejected for publication by *Poggendorf's Annalen*, the later *Annalen der Physik*.

§ "The principle is presented and should be understood as a hypothesis within physics totally divorced from philosophical considerations."[12]

not exist any other axiom which could be extended with the same confidence to all processes in nature."

To the best of our present knowledge, Planck was right. Yet in years to come the paradoxes posed by several new discoveries were initially so grave as to cause a temporary lack of confidence in the energy principle. Let us now turn to the first of these events.¶

Becquerel's surprise at the persistence with which the uranic rays kept pouring out energy was already mentioned above. In 1910, Marie Curie reminisced as follows about those early days: "The constancy of the uranic radiation caused profound astonishment to those physicists who were the first to be interested in the discovery of H. Becquerel. This constancy appears in fact to be surprising; the radiation does not seem to vary spontaneously with time...."[17] In order to appreciate this statement fully, three facts should be borne in mind: (1) The radiation emitted by uranium when unseparated from its daughter products does indeed represent to a very high degree a steady state of affairs. (2) It took 2 years from Becquerel's initial discovery until the first parent–daughter separation was effected. (3) It took another 2 years until it was firmly established that radioactivity does diminish with time.**

Speculation on the origin of radioactive energy started with Marie Curie's very first paper on radioactivity, [the one in which she announced her discovery of the activity of thorium (1898)]. There, cautiously, she suggests the possibility that the energy might be due to an outside source: "One might imagine that all of space is constantly traversed by rays similar to Roentgen rays only much more penetrating and being able to be absorbed only by certain elements with large atomic weight, such as uranium and thorium."[18] Also Becquerel made an analogy with an externally induced process, phosphorescence: "It would not be contrary to what we know about phosphorescence to suppose that these substances [U and Th] have a relatively considerable energy reserve, which they can emit for years, as radiation, without noticeable weakening."[19] However, he also stated that this analogy had its limitations. Phosphorescent phenomena exhibit a finite life-time (as Becquerel and his father well knew) and they can be affected by external agents. Neither of these properties seemed to apply to radioactivity: "... however, it has not been possible to induce any appreciable variation in the intensity of this emission."[19]

In that same year, 1898, Marie Curie discovered polonium, for which the liberated energy per unit weight of separated material was even larger than for

---

¶It may be noted that in 1882 Helmholtz expressed uncertainty about the applicability of the thermodynamic principles to "the fine structures of the organic living tissues."[15] Likewise Louis-Georges Gouy (1854–1926), one of the pioneers in refined experiments on Brownian motion, wondered in 1888 whether "le principe de Carnot ... serait seulement exact pour les mécanismes grossiers ... et cesserait d'être applicable ... [pour] des dimensions comparable à 1 micron," (... "whether the principle of C. would be exact only for large scale mechanisms ... and would cease to be applicable ... [for] dimensions of the order of one micron."[16] However, these comments are in the nature of asides and were not raised as central issues. For what follows, it may be of interest to observe that Marie Curie was aware of these remarks by Helmholtz and by Gouy.

**See further section 5 of this paper.

uranium and thorium. Thus the question of the origin of this energy became an even more burning one and she returned to it, listing a number of possible answers.[20] Here we find the first mention that one might have to face a contradiction with the conservation of energy. Furthermore she emphasized that the assumption of an external source would be nothing but an evasion of energy nonconservation —unless the nature of the external source were determined: "Any exception to Carnot's principle [first law!] can be evaded by the intervention of an unknown energy which comes to us from space. To adopt such an explanation or to put in doubt the generality of the Carnot principle are in fact two points of view which to us amount to one and the same as long as the nature of the energy here invoked stays entirely *dans le domaine de l'arbitraire.*" She also pointed out that the interior of the atom could be the energy source: "The radiation [may be] an emission of matter accompanied by a loss of weight of the radioactive substances."[20]

Not only the first but also the second law of thermodynamics was sometimes questioned as a result of this energy puzzle. For example, in his 1898 inaugural address as President of the British Association, the brilliant and erratic Sir William Crookes speculated, somewhere in between dissertations on food shortages and psychical research, whether one can "mentally modify Maxwell's demons" in such a way that radioactive substances release energy drawn from the air surrounding the active material.[21]

These various speculations set in motion a set of experiments designed to locate a possible outside source of radioactive energy. In an attempt to see whether the sun could be the cause, the Curies looked for diurnal variations in the activity of uranium. They found no effect.[22] Among others who addressed the same question, particular mention should be made of the team of Elster and Geitel.

Julius Elster (1854–1920) and Hans Geitel (1855–1923) had been high school friends. They both became teachers at the Gymnasium†† in Wolfenbüttel near Braunschweig. When Elster married and had a house built, Geitel moved in with the young couple and together the two friends built a laboratory in the new home. Here they started their research (often financed from their own pockets) which was to make them internationally renowned.‡‡ They experimented on photoelectric effects, on spectroscopy, on the conduction of electricity through gases, and especially on atmospheric electricity. These last experiments led to their classic work on the radioactivity of the atmosphere, research about which Rutherford spoke with great respect. Simultaneously with Crookes they discovered the scintillations of zinc sulfide screens by α-rays.[23]

In later years the two men loved to relate[24] their experiences in Berlin where the Prussian Minister of education tried to convince them to accept a joint offer as University professors at a first rate institute. They listened modestly but did not react. The minister believed that "die kleinen Oberlehrer der Provinz" were probably too awed and suggested that they take a few hours to think it over. They did so, came back and said no thank you. They had decided that the transition to the

---

†† It is not evocative enough to translate Gymnasium simply as high school. Let us say it is an academic high school preparatory to going to university.

‡‡ There were others as well who did their most creative work while teaching in a high school, Weierstrass for example.

academic world would inhibit their independent research. They were grateful for the honor but preferred to stay in Wolfenbüttel.

The two were inseparable. I cannot resist mentioning an anecdote related by d'Andrade[25]: "In their time there was a man who much resembled Geitel in appearance. A stranger meeting him said 'Good morning, Herr Elster' to which he replied 'Firstly I am not Elster but Geitel, and secondly I am not Geitel.'" Almost their complete oeuvre consists of joint publications. "We shall doubtless search in vain for a similar instance of private scientific partnership throughout a lifelong friendship. Each ascribed to the other the credit for a discovery published jointly."[26]

Their work on the origins of radioactive processes is contained in two papers. In the first one[27] they begin with the observation that if Crookes were right and the radioactive energy is taken from the surrounding air,[21] then the activity should decrease when the source is placed in a vacuum. They find no such effect. Next they turn to the conjecture of the Curies that the energy may be supplied by an X-ray-like radiation that is all pervasive in the atmosphere and reasoned that, if this were so, there should be a decrease in activity if the source were placed deep underground. So they requested and obtained permission to do an experiment in the Clausthal mines in the Harz mountains, under 300 meters of rock. They found no effect. They admit that perhaps the rock layer may not be all that good an absorber. Nevertheless they conclude, as early as 1898, that "the hypothesis of the excitation of Becquerel rays by radiation pre-existing in space appears *im höchsten Grade unwahrscheinlich* (improbable to the highest degree)." In their second paper[28] they report on attempts to increase the radioactive emissions by exposing a source to cathode rays; or to sunlight. They find no effect and conclude "*man wird vielmehr aus dem Atome des betreffenden Elementes selber die Lichtquellen ableiten mussen.*§§

It is important to stress at this point that the fascinating puzzles discussed in this paper were never any hindrance to progress in their days. If anything, the contrary is true. The field of radioactivity was young at the time when these questions arose; the tasks were enormous. While these problems were given much thought by the Curies, that never inhibited them from continuing their superb research. They were a stimulus to men like Elster and Geitel, as we have just seen. Others chose to state them as unresolved issues and then to move on to other pursuits. Such was largely the attitude of the English school. Rutherford, for example, simply noted in his 1899 memoir that, "The cause and origin of the radiation continuously emitted by uranium and its salts still remain a mystery."[29] J. J. Thomson always took the attitude that the atom itself was the energy source ("... the [radioactive] changes we are considering are changes in the configuration of the atom ..."[30]).

I referred earlier to the article by Marie Curie in which she listed possible options for the explanation of the energy release. It was written before but published after the discovery of radium by her, Pierre Curie, and Bémont. This last development once again brought the issue to the fore. The radium radiation was even

---

§§"Rather, one will have to derive the source of light [sic] from the atom itself of the element concerned."

more intense than for polonium! The question of nonconservation of energy came up once again: *On réalise ainsi une source de lumière, a vrai dire très faible, mais qui fonctionne sans source d'énergie. Il y a là une contradiction tout au moins apparente avec le principe de Carnot.*"[31]¶¶

Nonconservation of energy was never a widely held interpretation of these effects. In 1902 the Curies again gave a list of possible interpretations, on which this possibility no longer appears.[32] Yet in that same year, a visitor to England recalled that he "had been dining seated between Lord Kelvin and Professor Becquerel,... Lord Kelvin had turned to him and said that the discovery of Becquerel radiations had placed the first question mark against the principle of conservation of energy which had been placed against it since the principle was enunciated."[33]

It should also be stressed that such options as nonconservation of energy or external sources were not proposed lightheartedly. The idea that the atom itself is the source was not so easily swallowed at that time since it meant giving up the concept of an atom as an immutable entity. By 1900 the debate over the reality of atoms was well past its peak; but at that time the question was not universally regarded as settled. The Curies were proponents of the existence of real atoms, as their writings make abundantly clear. But to accept the atom itself as the source of the energy could only mean one thing to them: transmutation. And they could not simply accept this since to them at that time it seemed in conflict with the principles of chemistry as then known—which indeed it was. In 1900 Marie Curie summed up the dilemma in the following way[34]:

"Uranium exhibits no appreciable change of state, no visible chemical transformation, it remains, or so it seems, identical with itself, the source of energy which it emits remains undetectable—and therein lies the profound interest of the phenomenon. There is perhaps a disagreement with the fundamental laws of science which until now have been considered as general.... The materialistic theory of radioactivity*** is very attractive. It does explain the phenomena of radioactivity. However, if we adopt this theory, we have to decide to admit that radioactive matter is not in an ordinary chemical state; according to it, the atoms do not constitute a stable state, since particles smaller than the atom are emitted. The atom, *indivisible from the chemical point of view* [my italics] are here divisible, and the sub-atoms are in motion.... The materialist theory of radioactivity leads us... quite far. If we refuse to admit its consequences, our embarassment will not lessen. If radioactive matter does not modify itself, then we find ourselves again in the presence of the question: from where comes the radioactive energy? And if the source of energy cannot be found we are in conflict with Carnot's principle, a principle fundamental to thermodynamics.... We are then forced to admit that Carnot's principle [the second law!] is not absolutely general [and] ... that the radioactive substances are able to transform heat from the ambiant environment into work. *Cette hypothèse porte une atteinte aussi grave aux idées admises en*

¶¶"Thus one realizes a source of light [sic], quite weak to be sure, but which functions without a source of energy. There is here a contradiction, or so it seems, with the principle of Carnot [the first law!]."

***According to which radioactive atoms expel subatomic particles.

physique que l'hypothèse de la transformation des éléments aux principe de la chimie, et on voit que la question n'est pas facile à résoudre.†††

The transformation theory of Rutherford and Soddy, proposed in 1902, provided the break with the past that was clearly needed in order to answer Marie Curie's question. In this "great theory of radioactivity which these young men sprung on the learned, timid, rather unbelieving, and, as yet, unquantized world of physics of 1902 and 1903,"[35] they unabashedly put forward the idea that some atomic species are subject to spontaneous transmutation.[36] Forty years later, a witness to the events characterized the mood of the times as follows[37]: "It must be difficult if not impossible for the young physicist or chemist of the present day to realize how extremely bold it was and how unacceptable to the atomists of the time ... this is a point which must be stressed, for the younger generation is more likely to be familiar with the ordered simplicity of the radioactive series as we know them than with the chaotic state which preceded the transformation theory."‡‡‡

The main tenet of the transformation theory §§§ is: Radioactive bodies contain unstable atoms of which a fixed fraction decay per unit time. The rest of the decayed atom is a new radio element which decays again, and so forth, till finally a stable element is reached.

As Rutherford himself emphasized some time later (see below) there is no explicit reference in this theory as to the energy mechanism. Nevertheless, the successes of the transformation theory led Rutherford to express the following opinion: "This [transformation] theory is found to account in a satisfactory way for all the known facts of radioactivity, and a mass of disconnected facts into one homogeneous whole. On this view, the continuous emission of energy from the active bodies is derived from the internal energy inherent in the atom, and does not in any way contradict the law of conservation of energy."[40]

And so the energy debate might have quieted down were it not that, in March 1903, new fuel was added to it by the discovery that radioactive energy release surpassed *in magnitude* anything that had been known until then from chemical reactions. In that year Pierre Curie and Laborde[41] measured the amount of energy released within a Bunsen's ice calorimeter by a known quantity of radium. They found that one gram of radium can heat $\sim \frac{4}{3}$ grams of water from the melting point to the boiling point in one hour. These results caused a tremendous stir. The authors themselves referred once again to a possible outside energy source: "This release of heat can also be explained by supposing that radium utilizes an exterior energy of unknown nature." In a discussion of the new discovery, Kelvin

---

†††This hypothesis undermines the accepted ideas in physics as seriously as the hypothesis of the transformation of the elements does in chemistry, and one sees that the question cannot easily be resolved.

‡‡‡The events surrounding the enunciation of this theory have been described in more detail by Badash.[38]

§§§Why did Rutherford and Soddy not use the term "transmutation" but rather the more neutral one, "transformation"? The following exchange took place while they were at work on the separation of thorium X,[39] Soddy: "Rutherford, this is transmutation...". Rutherford: "For Mike's sake, Soddy, don't call it transmutation. They'll have our heads off as alchemists."

spoke of THE mystery of radium (his capitals) and continued: "It seems to me, therefore, absolutely certain that if emission of heat can go on month after month ... energy must be supplied from without ... I venture to suggest that somehow etherial waves may supply energy to radium...."[42] In a lecture on "The Present Crisis of Mathematical Physics" given in 1904, Poincaré brought up again the energy conservation question; "... These principles on which we have built everything, are they about to crumble away in turn? ... When I speak thus, you no doubt think of radium that grand revolutionist of the present time.... At least, the principle of the conservation of energy still remained with us, and this seemed more solid. Shall I recall to you how it was in its turn thrown into discredit?... This [activity of radium] was itself a strain on the principles ... But these quantities of [radioactive] energy were too slight to be measured; at least that was the belief, and we were not much troubled. The scene changed when Curie bethought himself to put radium in a calorimeter; it was then seen that the quantity of heat created incessantly was very notable...."[43]

The next section of this paper will relate how these developments became general public knowledge. From the physics point of view, these results became even more remarkable when it was found from additional experiments that ~75% of this effect was due to a daughter product of radium, the radium emanation (radon, $^{222}_{86}$Rn) although the amount of emanation present was actually extremely small. In fact the energy released by radon[44,45] is more than a million times greater than the heat evolved by the same volume of hydrogen and oxygen when they explode to form water. In 1905 Soddy wrote of these discoveries: "It is probably the most far reaching and revolutionary fact that has yet transpired in the study of radioactive substances. This enormous evolution of energy which accompanies the production of helium from the radium emanation establishes beyond question the new and fundamental character of radioactive change."[46]

Soddy also pointed out that the magnitude of the energy production made it ever more difficult to imagine it to be due to an external source: "It has been suggested ... that all space is traversed by undiscovered radiations to which ordinary matter is completely transparent, but to which radioactive substances are opaque. On this view, the energy traversing a cubic centimetre of space must be at least 60,000 calories per hour [in order to explain the heating effects due to radium]. The total quantity in the universe must therefore be so great that the hypothesis involves far greater difficulties than the effects it is designed to explain."[46]

Still the external source idea would not quite die.

In 1906, Sagnac raised a new possibility.[47] He asked: could gravitational energy be the external source? Might it be that the Newtonian attraction is universal for nonradioactive bodies while yet the Newtonian constant could have a *valeur spéciale* for radium? This led him to do a torsion balance experiment in which he compared the oscillations of equal weights of barium and radium. He found no effect.¶¶¶

In 1911, Rutherford again referred to the energy issue but expressed himself

---

¶¶¶A related negative result was obtained by Thomson[48] who determined the swinging time of a pendulum to which a bob of RaBr was attached. His work was differently motivated however. Much later, Rutherford and Compton also obtained a negative result when looking for the influence of gravitation on radioactivity.[49]

more cautiously than he had done earlier.[40] He observed[50] that the transformation theory leaves open the question of the inside versus the outside source, since all results of the transformation theory remain true for either hypothesis.

As late as 1919, Jean Perrin came forth with a new "ultra X ray" mechanism for explaining radioactivity as an externally induced effect.[51] This time the new radiation was supposed to come not from outer space but "from under our feet, from the fiery center of the planet." The scheme is discussed in detail on its astronomical and cosmological implications.

The matter was still being discussed in the year in which quantum mechanics was born.[52] This is not to say that the external source idea was in any way a serious issue at that late time. It does bring home the fact, however, that only in the quantum mechanical era could the definitive proof of the existence of internal mechanisms for energy generation be given, which settled the question for the ages.

The negative outcome of these various searches for an external energy source was a positive contributing factor to a major insight, which emerged early in the 20th century: Physical and chemical actions do not affect radioactive phenomena. In 1903, Rutherford and Soddy[53] elevated this to a new principle, the "conservation of radioactivity": "Radioactivity, according to our present knowledge, must be regarded as a process which lies wholly outside the sphere of known controllable forces, and cannot be created, altered or destroyed."

There is a vast body of experiments that bear on this question. Temperature independence was established by many. Pierre Curie went to London, to repeat with Dewar his radium heat production experiment, at liquid air temperatures. Marie Curie later went to Leyden to do the same at liquid hydrogen temperatures, with Kamerlingh Onnes. Rutherford stuck 4 mg of radium bromide inside a steel-enclosed cordite bomb, exploded the device, and concluded that there was no change in radioactivity at temperatures $\sim 2500°C$. Independence of pressure, of concentration, of the presence of strong magnetic fields, of irradiations of various kinds were established. Detailed discussions of these results are beyond the compass of this paper. For more information the interested reader is referred to older text books.[54]

Yet, strictly speaking, there is no such thing as the conservation of radioactivity. This became clear after the Second World War. I shall come back to this in section 6. Even so, the "conservation of radioactivity" served an excellent purpose in its time. In particular it helped Bohr to locate the atomic nucleus as the seat of all radioactive phenomena, as will be discussed elsewhere. As this "principle" so well illustrates, it is often better at an early stage to know the truth than to know the whole truth.

Nowhere in this section has there been mention of relativity theory. Some reasons for this absence have already been given in the Introduction. To conclude this section, I should like to make a further comment on this point.

As far as I know, Einstein's work did not lead to theoretical studies on the role of relativity in nuclear phenomena until the year 1913, when the question was raised whether small deviations from Prout's rule of integer mass multiples could be associated with an equivalent mass of the interaction energy between nuclear constituents.**** Also the related (and still open) question why some but not all nuclei are radioactive came under scrutiny. As a good example of such early work a paper by Swinne entitled, "On an Application of the Principle of Relativity in Radiochemistry,"[55] should be mentioned. It contains a relativistic treatment (in the sense of kinematics) of the α-radioactive process, based on the tacit assumption that this process is a *decay* of the type $A \rightarrow B + \alpha$; that is, there is no external source,†††† no doubt in accord with the general consensus of that time. However, it would have been just as easy to apply the same ideas to the "external source model" according to which α-radioactivity should be a *reaction* of the type $X + A \rightarrow B + \alpha$. It is obvious that a simple discussion of this reaction would yield criteria for the need of the source X. Yet (to my knowledge) such a reasoning was not proposed at any time.

4. Second Interlude: Atomic Energy

Speculations on the possible good and the possible evil of the atom go back to the founding fathers of radioactivity.

Here is Becquerel, in an early lecture[56]: "Today the [radioactive] phenomena are of transcendent interest, but in them almost infinitesimal amounts of energy are utilized. Whether ultimately science will have so far advanced as to permit of the practical utilization of the abundant store of energy locked up in every atom of matter is a problem which only the future can answer. Remember, at the dawn of electricity this was looked on as a mere toy, suitable only to amuse children by attracting bits of paper with a stick of rubbed sealing wax."

Here is Pierre Curie, also in a lecture[57]: "... it can even be thought that Radium could become very dangerous in criminal hands, and here the question can be raised whether mankind benefits from the secrets of Nature...."

And here is Soddy in an early popular book[58]: "If we pause but for a moment to reflect what energy means for the present, we may gain some faint notion as to what the question of transmutation may mean for the future to a fuelless world, once more dependent on a hand-to-mouth method of subsistence. It may still be centuries before this occurs, but neither the application of the discoveries of science nor even their achievement is to be compared with the struggle in winning them."

It is also noteworthy that, even in these early stages, questions concerning radioactive energy, and their implications, seized the imagination of the general public. This was due mainly to reports that began to circulate in the press, soon after the discovery by Curie and Laborde,[41] about a mysterious new energy stored in radium. For example, in a preview of the forthcoming 1904 International Electrical Congress in St. Louis,‡‡‡‡ the *St. Louis Post Despatch* of October 4,

---

****The existence of radioactive isotopes was known by then, and J. J. Thomson had achieved the first isotope separation in neon in that same year.

††††The author states that the masses are not known precisely enough to draw any firm conclusions on the questions just mentioned.

‡‡‡‡This meeting was attended by J. J. Thomson, Rutherford, and Boltzmann.

1903 carried an item about the unusual properties of radium. The article contains sweeping statements to the effect that radioactivity could cause a holocaust.§§§§

In that same year, 1903, the term "atomic energy" entered the language for the first time, in a most appropriate sense. It was first used by Rutherford and Soddy[60] *not* just for the energy released by a radioactive element, but much more generally for the energy locked up in *any* atom: "All these considerations point to the conclusion that the energy latent in the atom must be enormous compared with that rendered free in ordinary chemical change. Now the radio elements differ in no way from the other elements in their chemical and physical behavior. On the one hand they resemble chemically their inactive prototypes in the periodic table very closely, and on the other they possess no common chemical characteristic which could be associated with their radioactivity. Hence there is no reason to assume that this enormous store of energy is possessed by the radio elements alone. It seems probable that *atomic energy* [my italics] in general is of a similar high order of magnitude, although the absence of change prevents its existence being manifested." This, truly, is the physics of the 20th century.

The fact that today "atomic energy" is an expression firmly anchored in our everyday language has nothing to do, however, with the above marvelous lines. Rather, the present common usage of the term derives in the first instance from a report released by the President of the United States on the evening of Saturday, August 11, 1945. The title of this report came eventually to be "Atomic Energy for Military Purposes. The Official Report on the Development of the Atomic Bomb under the Auspices of the United States Government, 1940-45." It is now generally known as the Smyth Report.

Here is how I became aware of the fact that "atomic energy" is an expression reinvented, as it were, in 1945. After I finished the draft of the present paper, sometime in November 1976, I went to see my friend Henry DeWolf Smyth. We had the following conversation, the essence of which I report here with his permission.

A.P.: When you were writing your report, were you aware that the term atomic energy dates back to the beginning of this century?

H.D.S.: No, I was not.

A.P.: I have a second question. In the days of Rutherford and Soddy, the nucleus was not yet discovered. Therefore, the expression "atomic energy" was, so to speak, the only natural one they *could* use. Now I have been puzzled, ever since your report came out: why in fact did you not speak of "nuclear energy," "nuclear bomb," etc.? Having just finished writing the paper of which I told you, this seems a good time for me to ask you about this.

H.D.S.: Your question comes at a very opportune moment, since I have just published an article on the history of the Smyth Report. You will be glad to know that in my original draft I did use the word "nuclear" instead of "atomic." After the writing

---

§§§§The article was headed: "Priceless mysterious radium will be exhibited in St. Louis. A gram of the most wonderful and mysterious metal to be shown in St. Louis in 1904." The text contained these lines: "Its power will be inconceivable. By means of the metal all the arsenals in the world would be destroyed. It could make war impossible by exhausting all the accumulated explosives in the world.... It is even possible that an instrument might be invented which at the touch of a key would blow up the whole earth and bring about the end of the world."[59]

was done there followed a period of consultation with [Major General Leslie R.] Groves. In turn, Groves must have discussed my draft with his advisers [James B.] Conant and [Richard C.] Tolman, and possibly with others as well. In a subsequent discussion, we decided that the word "nuclear" was either totally unfamiliar to the public or primarily had a biological flavor, whereas "atomic" has a definite association with chemistry and physics. Since it became clear that the report was aimed at a wider audience than nuclear physicists, we decided that "atomic" was less likely to frighten off readers than "nuclear." So I accepted the change after a somewhat painful suppression of my purist principles.

With these words he gave me a copy of his article.[61] I gratefully accepted this gift from a man my respect for whom I have already expressed elsewhere.[62]

5. WHY A HALF-LIFE?

The first determination of a half-life for a radioactive decay was made by Rutherford in 1900.[63] In a study of the properties of thorium emanation (Rn_{86}^{220}) he found that the intensity of the radiations given out by his sample fell off with time in a geometric progression. Thus he was the first to note that if $N(t)$ is the number of active atoms at time t, then the decrease of N with t is well described¶¶¶¶ by

$$\frac{dN}{dt} = -\lambda N, \text{ or } N(t) = N(0)e^{-\lambda t}. \tag{1}$$

He called λ the radioactive constant and "it has been shown that $e^{-\lambda t} = \frac{1}{2}$ when $t = 60$ seconds," a quite respectable half-life determination (the modern value is about 55 sec). It is of course no accident that this first discovery concerned an element of medium short life. Much longer half-lives (such as the one for radium, ~1600 years) were also well established within the next few years, with the help of the theory of radioactive equilibrium between parent and daughter substances[64]. Equation 1 and its generalization to sequential decays was the first of two contributions by Rutherford to theoretical physics, an activity which he did not always hold in the highest esteem (his second contribution was his theoretical discovery of the central nucleus from the results of scattering experiments).

Today, even though we may not always be able to compute λ theoretically for any given radioactive decay, the meaning of λ is certainly quite clear. However different the respective mechanisms for α-, β-, and γ-decay are, in each case λ is a quantum mechanical transition probability per unit time. Thus radioactivity represents one instance among very many of a situation in which physicists of earlier days were unwittingly dealing with quantum effects.

At the turn of the century there already existed a body of knowledge on unstable systems of atomic dimensions. For example, much work had been done at that time on luminescent phenomena. This had made the lifetime concept familiar. It is true that insuperable problems arose for those who attempted to find mechanisms for these and similar processes; consider for instance Boltzmann's struggles with molecular dissociation.[65] However, if these various unstable systems were not

¶¶¶¶In the absence of any replenishing mechanism.

amenable to theoretical treatment, they did not appear to pose any manifest paradoxes, principally since the causes for instability could at least be identified on a phenomenological level. It is in this last respect that radioactivity created problems unique for their time. It seemed (in fact, it was true) that radioactive decays were contrary to the classical concepts of cause and effect. During the first two decades of this century, physicists had no reason to suspect that these paradoxes were not by any means typical for radioactivity only.

Jeans has given a graphic description of the situation: "Interesting but difficult questions arise when we discuss which atoms will disintegrate first, and which will live longest without disintegration. [Suppose that] 500 million atoms are due to disintegrate in the next second. What, we may inquire, determines which particular atoms will fill the quota? .... it seemed to remove causality from a large part of our picture of the physical world. It we are told the position and the speed of motion of every one [of a set of radium atoms], we might expect that Laplace's supermathematician would be able to predict the future of every atom. And so he would if their motion conformed to the classical mechanics. But the new laws merely tell him that one of his atoms is destined to disintegrate today, another tomorrow and so on. No amount of calculation will tell him which atoms will do this...."[66]

Nevertheless, there were those who, early on, began to think of dynamical models that would incorporate radioactivity. One of these early model builders was J. J. Thomson. Some details of his work on models will be discussed elsewhere. Here it suffices to state that he attempted to describe radioactivity in terms of classical mechanical pictures. Thus we can well understand the objections which Lord Kelvin wrote to Thomson: "What would be the difference, between radium atoms in a piece of radium bromide, of the atoms which are nearly ripe for explosion, and those which have the prospect of several thousand years of stable diminishing motions before explosion?"[67] Rutherford also saw this weak point of the classical model: " ... all atoms formed at the same time should last for a definite interval. This, however, is contrary to the observed law of transformation, in which the atoms have a life embracing all values from zero to infinity."[50]

In despair, one might of course reply to the question: How is it possible that one species of identical atoms is made up out of particles some of which live longer than others? by saying that "the different atoms of a radioactive substance are not in every respect identical,"[68] a possibility mentioned by Thomson in an address in 1909. However, he never came back to this.

Those who wisely left aside the question: "why does a radioactive atom change?" and focussed on the more modest problem: "how does it change?" were able to make some further progress on a more descriptive level, however. They focussed on the essential content of Equation 1, which is probabilistic: the probability that a given unstable atom decays is the same for all atoms (in a sample of a given species) and is independent of its age, but does depend on the specific element under consideration. "If the destroying angel selected out of all those alive in the world a fixed proportion to die every minute, independently of their age, ... and chose purely at random ... then our expectation of life would be that of the radioactive atoms."[69] With the help of this probability Ansatz, it is of course possible to refine Equation 1 (which can hold strictly only in the limit of very large N, since it ignores the discreteness of N) for the case of finite samples, to predict the average number of events in any finite time interval (for given λ) and to study the fluctuations of that number (the "Schweidler fluctuations"*****) as well as fluctuations for other variables such as heat production, ionization, spatial distributions, and so forth.[54,71]

Beyond that, there was not much more that could be done. The lifetime paradox simply did not lend itself to the statement of new hypotheses subject to test. The problem was so difficult that it was hard even to get a wrong idea about it.

In a review of alternatives by Debierne in his thesis[72] an old acquaintance briefly returns: the exterior source of radioactive decay. He notes that exponential decays occur in several chemical processes, such as monomolecular irreversible reactions (dissociations, etc.). He notes that in such instances thermal disorder plays a role. This leads him to ask whether the radioactive decay processes could be due to some exterior action, which, however, cannot respond to temperature variations. He concludes that such a mechanism is hard to conceive. Pursuing the thermal disorder analogy, he speculates that each unstable atom contains an extremely complex system in which high velocity particles create a state of "internal disorder." Several others likewise tried to associate the decay properties with fluctuations in highly complex internal motions.[73]

Considerations of this kind were the subject of an address by Marie Curie to the second Solvay Conference in 1913. In the subsequent discussion, Rutherford expressed his interest in the ideas of Debierne and summarized his own view as follows: "The law of radioactive transformation, which is universal for all radioactive substances, seems only to be explicable as a consequence of accidental disturbances ("troubles fortuits") in the nucleus, in conformity with probability laws. But, in the present state of knowledge, it does not seem possible to form a clear idea as to the very constitution of the atomic nucleus, nor of the causes which lead to its disintegration."[74]

And so these problems remained unresolved until several years after the birth of quantum mechanics.

6. POSTSCRIPTA: MODERN TIMES

1928: α-Decay Explained

It was realized by George Gamow in Goettingen,[75] in August 1928, and independently by Ronald W. Gurney and Edward C. Condon in Princeton one month later,[76] that α-decay results as a consequence of quantum mechanical tunneling through a potential barrier. Moreover, all authors had a further significant advance to report: the first explanation of the Geiger-Nuttall relation, known phenomenologically since 1912,[77] which establishes a connection between the lifetime of an α-emitter and the range of the produced α-particles.

In the letter to *Nature* by Gurney and Condon we hear for the last time the

---

*****After Egon von Schweidler who was the first to draw attention to such fluctuation phenomena.[70]

echos of a confusing past: "It has hitherto been necessary to postulate some special arbitrary 'instability' of the nucleus; but in the following note it is pointed out that disintegration is a natural consequence of the laws of quantum mechanics without any special hypothesis.... Much has been written about the explosive violence with which the α-particle is hurled from its place in the nucleus. But from the process pictured above, one would rather say that the α-particle slips away almost unnoticed."

Nonconservation of Radioactivity

If the principle of conservation of radioactivity (mentioned toward the end of section 3) were strictly valid, then it would follow that the decay constant λ is independent of all chemical and physical changes, for any radioactive process.

The first suggestions that this cannot be universally true were published in 1947, independently by Segré[78] and by Daudel.[79] The process they chose to discuss, K-capture in ^{7}Be, was not known to the founding fathers, to be sure. They noted that the chemical environment should affect the electron-capture rate, especially in light nuclei, since chemical changes imply changes in the electron density at the position of the nucleus. During the next decade, effects $\sim 0.1\%$ were indeed established experimentally.[80] In 1951 it was noted that similar considerations also apply to decays involving internal conversion, and an effect $\sim 0.3\%$ was observed by comparing different chemical embeddings for a cleverly chosen technetium-99 isomer[81]; in 1965, a niobium-90 isomer turned out to be even better, yielding an effect one order of magnitude larger.[82]

These and other manifestations of "nonconservation of radioactivity" have become a lively subject of research in recent times.[83] This is of course also due to the discovery of a quite different influence of the environment on radioactive decay: the Mössbauer effect where a nuclear decay, even though it originates in a single nucleus, is properly described only by treating the decay as a collective quantum mechanical property of the entire crystal in which that nucleus resides.[84]

The Exponential Law of Radioactive Decay

The question how well Equation 1 describes the temporal behavior of radioactive substances is quite an old one. Its early version was: Is λ a constant independent of time? Through the years this was verified experimentally, for long periods of time by Rutherford in 1911,[85] and later for "very young" sources of radium emanation,[86] eventually down to 10^{-5} sec after their creation.[87] A more recent study of ^{56}Mn showed no deviations of the exponential law during the first 34 half-lives.[88]

Quantum mechanical arguments show nevertheless that Equation 1 is not mathematically rigorous. Deviations occur for times both very small and very large compared with λ^{-1}.[89-94] Asymptotically for large times the exponential behavior turns into a power behavior. Experimental situations in which such deviations play a noticeable role have not been found to date.

Acknowledgments

It is a pleasure to thank Paula Morgan, Sara Pais, Sam Treiman, and George Uhlenbeck for helpful comments.

References

1. EINSTEIN, A. 1905. Ann. der Phys. **18**: 639.
2. PAULI, W. "Relativitätstheorie", in Enzyklopädie der Math. Wiss. Vol. 5, Part 2, Teubner, Leipzig, 1921.
3. Histoire de l'Académie Royale des Sciences, Anneé 1775, pp. 61–66; Imprimerie Royale, Paris, 1778.
4. HIEBERT, E. N. 1962. Historical Roots of the Principle of Conservation of Energy, Univ. of Wisconsin, 1962.
5. ELKANA, Y. 1974. "The discovery of the conservation of energy", Harvard University Press, 1974.
6. KUHN, T. S. 1962. In Critical Problems in the History of Science. M. Clagett, Ed.: 321. University of Wisconsin Press. Madison, Wisc.
7. MACH, E. 1896. Die Principien der Wärmelehre. J. A. Barth, Leipzig. Earlier, Mach had devoted a separate essay to the history of the energy conservation law.
8. PLANCK, M. 1887. Das Princip der Erhaltung der Energie. B. G. Teubner. Leipzig.
9. CARNOT, H. 1878. Lettre, C. R. Paris **87**: 967.
10. Reference **8**: 16.
11. Reference **7**: 241.
12. Most easily accessible in KAHL, R. 1971. Selected Writings of Hermann von Helmholtz. Wesleyan University Press. 1971.
13. YOUNG, T. 1807. A course of lectures on natural philosophy and the mechanical acts. Vol. I: 75 (Lecture 8). Taylor and Walton. London.
14. CORIOLIS, G. G. 1829. Du calcul de l'effet des machines, ou Considérations sur l'emploi des moteurs et sur leur évaluation pour servir d'introduction a l'étude spéciale des machines. Carilian-Goeury. Paris.
15. HELMHOLTZ, H. 1883. Wissenschaftiche Abhandlungen. Vol. 2: 972. Barth. Leipzig.
16. GOUY, L. G. 1888. J Phys. (Paris) **7**: 561.
17. CURIE, M. 1910. Traité de Radioactivité. Vol. 1. Chap. 3. Gauthier-Villars. Paris.
18. CURIE, S. 1898. C. R. Paris **126**: 1101.
19. BECQUEREL, A. H. 1899. C. R. Paris **128**: 771.
20. CURIE, M. 1899. Rev. Gén. Sci. **10**: 41.
21. CROOKES, W. 1898. Nature **58**: 438. Also 1899. C. R. Paris **128**: 176.
22. Reference 17. Vol. 1: 129.
23. ELSTER, J. & H. GEITEL. 1903. Phys. Zeitschr. **4**: 439.
24. POHL, R. 1924. Naturw. **12**: 685.
25. ANDRADE, E. N. DA C. 1964. Rutherford and the Nature of the Atom: 40. Doubleday & Co. New York, N.Y.
26. LAWSON, R. W. 1924. Nature **113**: 432.
27. ELSTER, J. & H. GEITEL. 1898. Ann. Phys. **66**: 735.
28. ELSTER, J. & H. GEITEL. 1899. Ann. Phys. **69**: 83.
29. RUTHERFORD, E. 1899. Phil. Mag. **47**: 109.
30. THOMSON, J. J. 1903. Nature **67**: 601.
31. CURIE, P., M. CURIE & G. BÉMONT. 1898. C. R. Paris **127**: 1215.
32. CURIE, P. & M. CURIE. 1902. C. R. Paris **134**: 85.
33. HAMMER, W. J. 1903. Radium: 18. Van Nostrand Co. New York, N.Y.
34. CURIE, M. 1900. Rev. Sci. **14**: 65.
35. RUSSELL, A. S. 1951. Proc. Phys. Soc. London **64**: 217.
36. RUTHERFORD, E. & F. SODDY. 1902. Phil. Mag. **4**: 370, 569.
37. ROBINSON, H. R. 1943. Proc. Phys. Soc. London **55**: 161.

38. BADASH, L. 1966. Sci. Amer. **215**(2): 89.
39. HOWORTH, M. 1958. The Life Story of Frederick Soddy: 83. New World Publishers. London.
40. RUTHERFORD, E. 1904. Radioactivity: 2–4. Cambridge University Press. Cambridge, England.
41. CURIE, P. & A. LABORDE. 1903. C. R. Paris. **136**: 673.
42. Lord KELVIN. 1904. Phil. Mag. **7**: 220.
43. POINCARÉ, H. 1913. The Foundations of Science. Chap. 8. The Science Press. New York, N.Y.
44. RUTHERFORD, E. & H. T. BARNES. 1904. Phil. Mag. **1**: 202.
45. RAMSEY, W. & F. SODDY. 1904. Proc. R. Soc. **13**: 346.
46. SODDY, F. 1975. Radioactivity and Atomic Theory. Th. Trenn, Ed.: 84. John Wiley & Sons. New York, N.Y.
47. SAGNAC, G. 1906. J Phys. **15**: 455.
48. THOMSON, J. J. 1905. Trans. Int. Electrical Congress, Vol. 1: 234. J. B. Lyons Co. Albany, N.Y.
49. RUTHERFORD, E. & A. H. Compton. 1919. Nature **104**: 412.
50. RUTHERFORD, E. 1911. Radioactive Transformations: 267. Yale University Press. New Haven, Conn.
51. PERRIN, J. 1919. Ann Phys. **11**: 5. And 1920. Revue du mois (Febr. 10): 113.
52. BRINER, E. 1925. C. R. Paris **180**: 1586.
53. RUTHERFORD, E. & F. SODDY. 1903. Phil. Mag. **6**: 576.
54. MEYER, S. & E. SCHWEIDLER. 1927. Radioaktivität. Teubner Verl. Leipzig.
55. SWINNE, R. 1913. Phys. Zeit. **14**: 145.
56. CROOKES, W. 1910. Proc. R. Soc. **A83**: xx.
57. CURIE, P. 1967. In Nobel Lectures in Physics: 78. Elsevier Publishing Co. New York, N.Y.
58. SODDY, F. 1912. Matter and Energy. Henry Holt Co. New York, N.Y.
59. JAUNCEY, G. 1946. Amer J. Phys. **14**: 227.
60. RUTHERFORD, E. & F. SODDY. 1903. Phil. Mag. **6**: 576.
61. SMYTH, H. D. 1976. The Smyth Report. The Princeton University Library Chronicle **37**:173.
62. PAIS, A. 1969. In Oppenheimer. Charles Scribner's Sons. New York, N.Y.
63. RUTHERFORD, E. 1900. Phil. Mag. **49**: 1.
64. RUTHERFORD, E. J. CHADWICK & C. D. ELLIS. 1930. Radiations from Radioactive Substances. Cambridge University Press. Cambridge, England.
65. See the English translation: BOLTZMANN, L. 1964. Lectures on Gas Theory Part 2. Chap. 6. University of California Press.
66. JEANS, J. 1943. Physics and Philosophy: 149, 150. Cambridge University Press. Cambridge, England.
67. LORD RAYLEIGH. 1969. The Life of Sir J. J. Thomson: 141. Dawsons. London.
68. Reference 67: 142.
69. SODDY, F. 1920. The interpretation of radium: 114. Putnam's Sons. New York, N.Y.
70. SCHWEIDLER, E. v. 1905. Premier Congrés de Radiologie. Liège.
71. FÜRTH, R. 1920. Schwankungserscheinungen in der Physik. Chap. 6. Vieweg, Braunschweig.
72. DEBIERNE, A. 1914. Recherches sur les phénoménes de radioactivité. Gauthier-Villars. Paris.
73. LINDEMANN, F. A. 1915. Phil. Mag. **30**: 560. Also WOLFF, H. Th. 1920. Phys. Zeitschr. **21**: 175.
74. RUTHERFORD, E. 1921. In La structure de la matière: 67. Proc. 2nd Solvay Congress. Gauthier-Villars. Paris.
75. GAMOW, G. 1928. Z. Phys. **51**:204.
76. GURNEY, R. W. & E. U. CONDON. 1928. Nature **122**: 439. 1929. Phys. Rev. **33**: 127.
77. GEIGER, H. & J. M. NUTTALL. 1912. Phil. Mag. **23**: 439.
78. SEGRÉ, E. 1947. Phys. Rev. **71**: 247.
79. DAUDEL, R. 1947. Revue Sci. **85**: 162.

80. BOUCHEZ, R. *et al.* 1956. J. Phys. Rad. **17**: 363.
81. BAINBRIDGE, K. T., M. GOLDHABER & E. WILSON. 1951. Phys. Rev. **84**: 1260. 1953. Phys. Rev. **90**: 430.
82. COOPER, J. A., J. M. HOLLANDER & J. O. RASMUSSEN. 1965. Phys. Rev. Lett. **15**: 680.
83. See the review by EMERY, G. T. 1972. Ann. Rev. Nuclear Sci. **22**: 165.
84. FRAUENFELDER, H. 1962. The Mössbauer Effect. W. A. Benjamin. New York, N.Y.
85. RUTHERFORD, E. 1911. Wiener Ber. **120**: 303.
86. POOLE, H. H. 1914. Phil. Mag. **27**: 714.
87. JOLIOT, F. 1930. C. R. Paris **191**: 132.
88. WINTER, R. G. 1962. Phys. Rev. **126**: 1152.
89. HELLUND, E. 1953. Phys. Rev. **89**: 919.
90. HÖHLER, G. 1958. Z. Phys. **152**: 546.
91. LÉVY, M. 1959. N. Cimento **14**: 612.
92. MATTHEWS, P. & A. SALAM. 1959. Phys. Rev. **115**: 1079.
93. PETZOLD, J. 1959. Z. Phys. **155**: 422. 1959. Z. Phys. **157**: 122.
94. SCHWINGER, J. 1960. Ann. Phys. (N.Y.) **9**: 169.

NOTES ON THE STATISTICAL DISTRIBUTION OF SINGLE POPULATION LEVEL SPACINGS AND LEVEL WIDTHS

James Rainwater

Department of Physics
Columbia University
New York, New York 10027

The study of the resonance interaction of neutrons of $\lesssim 10^4$ eV with nuclei has been important in the development of nuclear physics. In the early 1930s, the nucleus was expected to behave like a region of strong attractive potential for incident neutrons, leading to "size resonances" in the interaction as the atomic mass number A and E_n was varied over large amounts. After Fermi discovered slow neutrons from the moderation of faster (>1 MeV) neutrons in hydrogen materials, it was soon discovered that medium A nuclei showed sharp resonances of ~0.1 eV width at a few eV neutron energy, with level spacings $\lesssim 100$ eV in many cases. This led Bohr and Kalckar to develop the "liquid drop model" of the nucleus. The low energy resonances are due to mainly $l = 0$ neutrons because of centrifugal repulsion for $E_n \lesssim 10$ keV for $l \lesssim 1$ or higher angular momentum neutrons. The resonances in the compound nucleus using $\lesssim 10$ keV neutron energy have total excitation energy above the compound nucleus ground state nearly equal to the binding energy (~6 to 8 MeV depending on A) of the extra neutron.

The theory of resonance reaction was developed by Breit and Wigner and later in more sophisticated form by Wigner and others. The early developments to 1936–7 were reviewed by Bethe and colleagues in an historically important set of *Reviews of Modern Physics* articles.[1-3] The experimental picture began to be filled in by time of flight neutron spectrometers in the 1940s using pulsed cyclotrons, and using mechanical "choppers" with fission reactors. The early experimental evidence was confused by poor experimental energy resolution for establishing level spacing systematics, and by the wide range of level strengths (neutron level partial widths Γ_n) for a given target nucleus. This led to uncertainty as to how many weak levels had been missed, or where more than one level was treated as one. The latter problem confuses both the analysis of the systematics of level spacings and of level Γ_n distributions. The data suggested a shortage of small level spacings compared with that expected for a random distribution of spacings, but it was not clear that it was not an artifact of insufficient experimental energy resolution. By about 1957, the experimental systems improvements had led to a belief by some that there was indeed a shortage of small level spacings for even atomic number $(I = 0)$ target nuclei of medium A. Wigner[4] was led (~1956) to propose a level repulsion formula

$$P_W(x) = (\pi x/2) \exp(-\pi x^2/4), \qquad (1)$$

where x is a nearest neighbor (single population) level spacing in units of the mean level spacing, $x = s/\langle s \rangle$. For small spacings the probability of a given s

increases in proportion to s. The exponential factor is the decrease in probability that a given x (or s) is a nearest neighbor spacing. This distribution was shown to be in (essential) agreement with experiment in the cases where best tests could be made. Wigner soon related this to the eigenvalue spacing distribution for (large) real symmetric square $N \times N$ matrices having randomly distributed matrix elements with the variance for the off-diagonal elements half that for the diagonal elements, both having zero mean. This is now called the "Gaussian Orthogonal Ensemble" (GOE). Wigner emphasized that the level density in this model formed a "semicircular distribution," peaked at the center and decreasing according to a semicircular law at the edges. The joint probability distribution is the "Wishart" distribution, which is proportional to the product of the level spacings $|E_k - E_j|$ for each level pair treated only once, times a Gaussian in $-\sum_j E_j^2$ which prevents a spacing "blowup."

Meanwhile, Porter and Thomas[5] suggested a theory for a single population (same A, J, parity) of levels for the neutron partial level width, Γ_n, distribution. Since the R-matrix theory of Wigner and Eisenbud[6] indicated that, for $l = 0$ neutrons, Γ_n should invlude an $E^{1/2}$ (or velocity) dependent factor, it was standard to use the "reduced neutron widths,"

$$\Gamma_n^0 = \Gamma_n(1\,\text{eV}/E_n)^{1/2}. \quad (2)$$

In the R-matrix formalism, Γ_{nj}^0 is proportional to γ_j^2, where γ_j is proportional to the amplitude of an eigenvalue set at the channel radius corresponding to the neutron entrance channel in the compound nucleus. The assumption was made that the γ_j had a random Gaussian distribution about zero mean. This implies that

$$P(y)\,dy = (2/\pi)^{1/2} e^{-y^2/2}\,dy, \quad (3)$$

where $y \equiv [\Gamma_n^0/\langle\Gamma_n^0\rangle]^{1/2}$. In terms of Γ_n^0, the distribution of the probability is proportional to $x^{-1/2} e^{-x/2}\,dx$, where $x \equiv \Gamma_n^0/\langle\Gamma_n^0\rangle$. This was called a χ^2 distribution of one degree of freedom. It gives a relatively large probability for small $x = \Gamma_n^0/\langle\Gamma_n^0\rangle$, with $\sim 8\%$ having $x < 0.01$ and $\sim 2.5\%$ having $x < 0.001$. The experimental distribution of single population levels tends to become incomplete above energies where very small Γ_n^0 values are too weak to be detected above the statistical noise fluctuations. Otherwise, the Porter-Thomas distribution applied well for medium to heavy A nuclei far from closed shells. Usually only levels of smallest Γ_n^0 are missed to significant energy, and the number of missed weak $l = 0$ levels is estimated by comparison with the Porter-Thomas function. The fit tends to be excellent over the remaining Γ_n^0 distribution. For many lower A nuclei and for nuclei near closed neutron shells, there seems to be an additional energy dependent "intermediate structure" modulation that causes "spurts" in the function $\sum_j \Gamma_{nj}^0$ which relates to the $l = 0$ strength function $S_0 \equiv \langle\Gamma_n^0\rangle/\langle s\rangle$, since S_0 is the slope of $\sum_j \Gamma_{nj}^0$ vs E over all $l = 0$ levels.

The theory that we consider here is best tested experimentally by using even-even nuclei near a peak in the $l = 0$ strength function and a minima of the p wave strength function. This tends to occur best in the rare earth region $150 < A < 190$. The optical model predicts size resonances in the $l = 0$, $l = 1$, and so forth, strength functions in certain regions of mass number A, with maxima for

$l = 0$ corresponding to minima for $l = 1$ and vice versa. Usually one observes some of the strongest $l = 1$ levels with Γ_n values overlapping those of the very weak $l = 0$ levels. To obtain "clean" tests for a pure single $l = 0$ population of levels (same A, J, parity), the data should include all of the $l = 0$ levels, but no $l = 1$ levels.

The optical model theory treats the nucleus as a region of complex attractive potential for the incident neutron for $r < R$, the nuclear radius. A boundary condition that $r\Psi = 0$ at $r = 0$ leads to maxima of S_0 vs A (or R) where $r\Psi = \phi$ goes through maxima (absolute value) at $r = R$ so $d\phi/dr = 0$. The imaginary part of the potential represents "absorption" to form the compound nucleus where the excitation energy is shared among many nucleons.

The original optical model of Feshbach, Porter, and Weisskopf[6] used a spherical nucleus model. Later models included the known large quadrupole shape distortions away from closed neutron and proton shell regions. This split the fourth maxima region for S_0, which occurs at $140 < A < 200$. Plots of recent theoretical curves of S_0, S_1, and other parameters vs A, and comparison with experiment are given in the introduction to Reference 7, which is a compendium of experimental results on neutron resonance parameters.

The work of Wigner on level repulsion for single populations of levels (same A, J, parity) created great interest in the associated mathematical problems for many theorists. Many papers resulted based in large part on the results of large scale Monte Carlo computer calculations of the properties of the energy eigenvalue sets for $N \times N$ square matrices. Emphasis was placed on next nearest neighbor spacing distributions, and so forth.

An advance in the analysis was made in a series of papers by Dyson alone,[8-10] and in collaboration with Mehta,[9,10] where a new mathematically equivalent circular ensemble was introduced. For the relevant circular orthogonal ensemble (COE), the eigenvalues are treated as points on a unit circle, and the joint probability is proportional to the product of the chord lengths between all possible pairs of levels. They were able to show that a crystalline-like long-range order for the spacings is implied. In particular, their statistic Δ_3, which we denote simply Δ, is the mean square deviation between the "ladder function" $N(E)$ of (observed) levels vs E and a best fit straight line:

$$\Delta \equiv \min_{A,B} \int_0^{\Delta E} (N - AE - B)^2 \, dE/\Delta E. \qquad (4)$$

In the absence of correlations, it is easy to show that Δ should increase linearly with the total number of levels n for $n \gg$ the spacing correlation length. On the basis of Monte Carlo calculations, by our Neutron Spectroscopy Group at the Columbia University Nevis Laboratories, we find that,[11]

$$<\Delta> \; = n/(55 - 210/n), \qquad (5)$$

using 25,000 sets of 108 adjacent uncorrelated Wigner (UW) spacings and 15,000 sets each for various smaller numbers of adjacent UW spacings. The Δ distribution was quite widely spread out on either side of $<\Delta>$. In contrast, Dyson and Mehta show[8-10] for their COE that

$$\langle \Delta \rangle = (1/\pi^2)(\ln n - 0.0687), \tag{6a}$$

$$\text{Var } \Delta = (0.11)^2. \tag{6b}$$

This remarkable formula implies that $n \sim 20{,}000$ is needed before $\langle \Delta \rangle = 1.00$, showing the degree of long range crystalline-like regularity predicted. The spread of Δ values for a UW distribution is such that there is more overlap with Equation 6a than is useful for a clean experimental rejection of the UW case for data agreeing with Equation 6a, even for $n \sim 50$. It is useful also to consider the correlation coefficient for adjacent nearest neighbor level spacings. We define

$$\rho(A,B) \equiv \frac{\langle (A - \langle A \rangle)(B - \langle B \rangle) \rangle}{[\langle (A - \langle A \rangle)^2 \rangle \langle (B - \langle B \rangle)^2 \rangle]^{1/2}}. \tag{7a}$$

For the orthogonal ensemble (OE) and large n, Mehta has shown that one expects to obtain a negative ρ for adjacent level spacings.

$$\rho(S_j, S_{j+1}) \approx -0.27. \tag{7b}$$

We find that the sum of Δ and this quantity, abbreviated as $[\Delta + \rho]$, gives a considerably better separation of the predicted values for the UW and OE distributions than either Δ or ρ alone.

For the above comparison, one needs to know not only the expected mean values for such parameters as Δ and ρ, but one must know the probability distributions for each n. Camarda[11] developed a computer program for generating and diagonalizing suitable $N \times N$ random Gaussian real square symmetric matrices, which would have eigenvalues obeying the Wishart distribution. This was carried out for 78 sets each for $N = 21, 31, 41, 50$, and 62 sets with $N = 81$. The eigenvalue distributions were found to be in satisfactory agreement with the Wigner semicircle law. A parameter transformation was made to unfold the semicircle form to constant density after two or more eigenvalues were omitted at each end to minimize end-effect errors. The resulting transformed eigenvalue distributions were then subjected to various statistical tests.

Values of $\langle \Delta \rangle$ and $\text{Var}(\Delta)$ were calculated and found, as expected, to agree with the DM result for the COE to within statistical uncertainties. We are not aware of any previous explicit demonstration of this equality using finite matrices.

Results for these random matrix calculations are given in Tables 1 and 2. The calculations were performed using the IBM type 360-44 computer at Nevis, for which it is uneconomical to treat large numbers of matrices of larger dimensions. During discussions with Professor Dyson of our experimental results and the comparisons with the above theoretical analysis, he suggested that we try to make use of his Brownian-motion model[8-10] to simulate the behavior of larger random square matrices.

The details of the Brownian motion calculations are given in Reference 11. The procedure was worked out by Dr. Harry Camarda, with suggestions by Professor Dyson. It is much faster than the use of square random matrix evaluations and gave results equivalent to the use of 900 matrices 120×120. It agreed with the Mehta-Dyson predictions.

The first comparison with experiment where the experimental results were of

TABLE 1*
PREDICTED BEHAVIOR OF $\langle\Delta\rangle$ AND $[\text{Var}(\Delta)]^{1/2}$ AS A FUNCTION OF THE NUMBER OF LEVELS FOR WIGNER'S RANDOM MATRIX MODEL AND DYSON'S CIRCULAR ENSEMBLE†

	Monte Carlo Random Matrix Calculations				Theoretical Predictions of the Circular Ensemble	
No. of Matrices Diagonalized	Matrix Dimension	Eigenvalues Used	$\bar{\Delta}$	$[\text{Var}(\Delta)]^{1/2}$	$\langle\Delta\rangle$	$[\text{Var}(\Delta)]^{1/2}$
78	21	13	0.257 ± 0.011	0.098	0.253	0.11
78	31	23	0.324 ± 0.012	0.106	0.311	0.11
78	41	33	0.364 ± 0.014	0.124	0.347	0.11
78	50	42	0.372 ± 0.011	0.098	0.372	0.11
62	81	77	0.424 ± 0.012	0.094	0.433	0.11
900‡	120	109	0.470 ± 0.003	0.093	0.468	0.11

*From Liou et al.[11] By permission of The Physical Review.
†It is seen that for these tests the predictions of the two models are indistinguishable.
‡Calculated using Dyson's Brownian-motion model.

TABLE 2*
AVERAGE VALUES OF $\rho(S_i, S_{i+1})$ CALCULATED USING THE AVERAGE
SAMPLE SPACING $\bar{S}$ AND THE TRUE AVERAGE SPACING $\langle S \rangle$

No. of Matrices Diagonalized	Matrix Dimension	Eigenvalues Used	$\langle \rho(S_i, S_{i+1}) \rangle_{\bar{S}}$	$\langle \rho(S_i, S_{i+1}) \rangle_{\langle S \rangle}$
78	21	15	-0.250 ± 0.026	-0.236 ± 0.026
78	31	25	-0.241 ± 0.021	-0.233 ± 0.021
78	41	35	-0.269 ± 0.018	-0.266 ± 0.018
78	50	44	-0.274 ± 0.016	-0.272 ± 0.016
62	81	77	-0.256 ± 0.012	-0.256 ± 0.012
900‡	120	109	-0.277 ± 0.003	-0.276 ± 0.003

*From Liou et al.[11] By permission of The Physical Review.
†The difference between the values of $\langle \rho(S_i, S_{i+1}) \rangle$ obtained using $\bar{S}$ and $\langle S \rangle$ (for the same number of levels) is much less than the difference $1/n$ found in the uncorrelated case.
‡Calculated using Dyson's Brownian-motion model.

sufficiently high quality to yield a "complete $l = 0$ level population" containing sufficient adjacent $l = 0$ levels with no $l = 1$ level contamination for a good test of the Dyson-Mehta theory occurred for our resonance measurements using ^{166}Er as target nucleus.[11] A plot of the staircase of observed resonances N vs E is shown in FIGURE 1. Above 4200 eV there is evidence of missed levels, but an excellent straight line fit is obtained for the first 109 levels. The plot of $\Sigma g \Gamma_n^0$ vs E

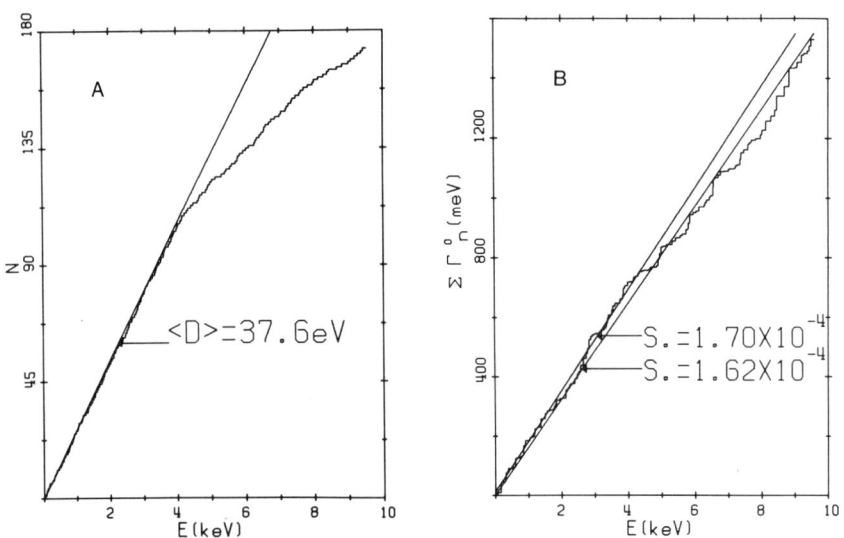

FIGURE 1. (A) Plot, N vs E, of number of observed levels vs energy for ^{166}Er. (B) Plot of $\Sigma \Gamma_n^0$ or $\Sigma g \Gamma_n^0$ vs energy for ^{166}Er. The slope gives the s-wave strength functions. (From Liou et al. By permission of The Physical Review.)

is shown in FIGURE 1b. It is less sensitive to missed levels, since they are mainly ones with very small Γ_n^0. The two straight lines through the origin correspond to two possible S_0 fit lines. The observed distribution of $(\Gamma_n^0)^{1/2}$ values and nearest neighbor spacings are compared with the Porter-Thomas and Wigner distributions to 4.2 keV and to 3.0 keV with satisfactory fits to within expected statistical fluctuations as shown in FIGURES 2a and b.

FIGURE 3a shows the least square straight line fit to the first 109 levels in ^{166}Er. The observed $\Delta = 0.455$ is in excellent agreement with the predicted value of (0.47 ± 0.11) by Mehta and Dyson but not to the value 2.05 predicted by Equation 5 for the UW case. The observed $\rho(s_j, S_{j+1}) = -0.22$ is in reasonable agree-

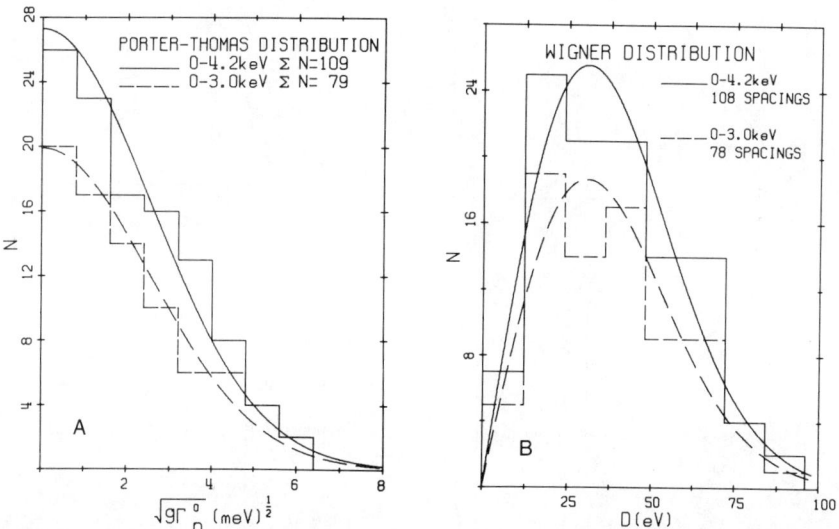

FIGURE 2. (A) Plot of the distribution of $(\Gamma_n^0)^{1/2}$ values for ^{166}Er to 4.2 and 3 keV. The Porter-Thomas curves are shown for comparison. (B) Plot of the adjacent spacing distribution for ^{166}Er to 4.2 and 3 keV and the comparison Wigner distribution curves. All experimental levels are included. (From Liou *et al*. By permission of *The Physical Review*.)

ment with the value -0.27 for the OE according to Mehta and Monte Carlo results of TABLE 2. The observed $(\Delta + \rho) = 0.235$ is such that the probability is 0.59 of obtaining a value this small according to the OE theory, but only 0.0004 for uncorrelated Wigner spacings. This is shown in FIGURE 3b. For the first 79 levels to 3 keV the observed $\Delta = 0.373$ vs 0.436 ± 0.11 for OE theory, and the observed $\rho(S_j, S_{j+1}) = -0.253$, giving $(\Delta + \rho) = 0.120$. The corresponding probabilities are 0.375 and 0.0014 for obtaining a value $\leq$ the observed value for the OE theory and UW theory, respectively as shown in FIGURE 3c. We have obtained similar but less definitive results for many other even-even target nuclei having $150 < A < 190$.

A competing approach has been carried out by groups in Mexico, U.S., and

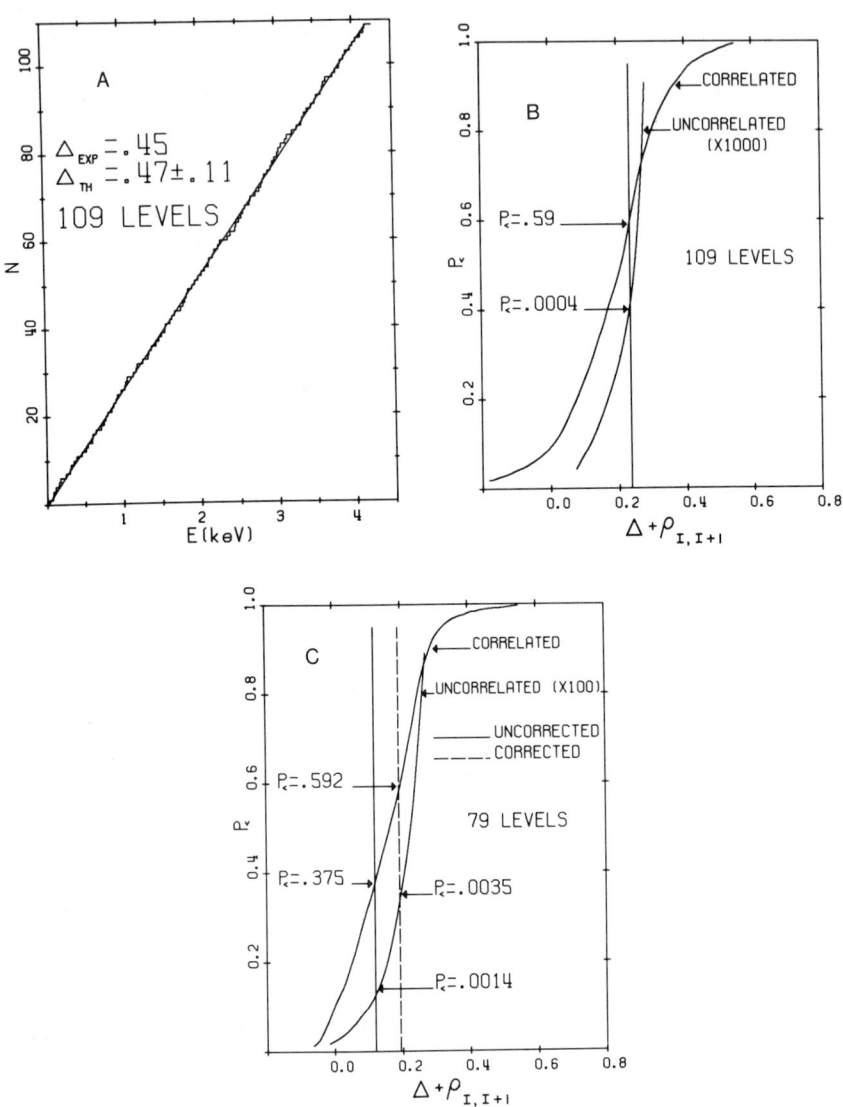

FIGURE 3. (A) Best straight-line fit to N vs E for all 109 observed resonances in ^{166}Er to 4200 eV. This gives $\Delta = 0.455$ for Dyson's parameter. (B) A plot of the probability of obtaining $\leq$ the indicated value of $\Delta + \rho(S_j, S_{j+1})$ for a set of 108 adjacent spacings, for uncorrelated Wigner adjacent spacings, and when correlations are included according to OE theory. The experimental result for ^{166}Er is indicated. (C) Same as (b) for the 78 adjacent spacings of ^{166}Er to 3 keV. (From Liou et al.[11] By permission of The Physical Review.)

Canada. They consider matrices of the type permitted by nuclear two body forces, and emphasize the statistic $\sigma(k)$ defined as standard deviation in $S(k)$, the spacing between levels having k levels between them, in units of $<S(0)>$, the mean nearest neighbor level spacing,

$$\sigma(k) \equiv S(k)/<S(0)>$$

The original Monte Carlo calculations using this model seem not to have been

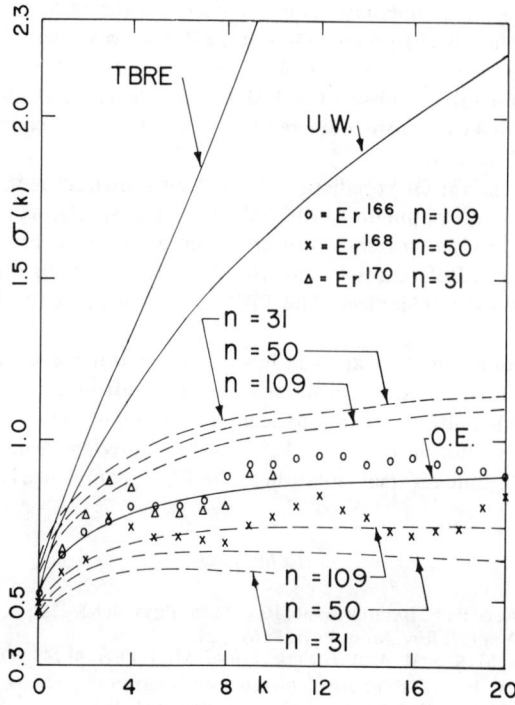

FIGURE 4. Comparison of experimental erbium data for $\sigma(k)$ vs k with Monte Carlo results. Here $\sigma(k)$ is the standard deviation of $S(k)/<S(0)>$, where $S(k)$ is the spacing of levels having k levels between them. The solid curves display the mean $\sigma(k)$ corresponding to the OE, UW, and Bohigas-Flores TBRE (References 12–14) cases. The dashed curves given the 10% and 90% limits for the OE case, for 31, 50, and 109 levels.

properly corrected for the energy dependence of the mean eigenvalue spacings and the predicted $\sigma(k)$ values increased with k even faster than for the UW theory. These earlier predictions were reported[12] at the 1971 Albany Conference on the Statistical Properties of Nuclei by Bohegas and Flores. FIGURE 4 shows the experimental $\sigma(k)$ vs k values for our ^{166}Er, ^{168}Er, and ^{170}Er data.[11] The experimental values are in excellent agreement with the predictions of the OE theory based on our Monte Carlo calculations. The earlier version of the two-body reaction ensemble curve is labeled TBRE. At the time of the Albany Conference, that curve tended to be in better agreement with the results of the poorer quality data then available. Subsequently, more careful examination[13-16] of the matter tends to reveal little if any difference between the TBRE and the OE predictions. The latest status was described in a paper[16] by P.A. Mello, J. Flores, T.A. Brody, J.B. French, and S.S.M. Wang at the July 1976 Conference on the Interactions of Neutrons with Nuclei at Lowell, Mass. In answer to a question, "If I understand you, it's not possible to distinguish between the two-body random ensemble and the Gaussian Orthogonal Ensemble, is that correct?" Dr. Mello replied, "Yes, I would say so. As far as fluctuations are concerned, the results are the same for the GOE as for the TBRE. You may find some differences if you do not eliminate, in the TBRE, fluctuations in scale, from matrix to matrix of the ensemble. If you take them out, then the results of the TBRE coincide as far as numerical calculations are concerned with those predicted by the GOE. But of course the density is quite different as you say."

The TBRE (and the OE) predictions have been shown to be in agreement with experiment for single populations of levels in atomic spectroscopy and for other data on nuclear levels after removal of gross energy dependence of level densities. This is discussed in References 15 and 16. Many references to other relevant papers are given in these references. The TBRE were considered to be more "realistic."

As a final point, Dr. H. Camarda has investigated the level spacing results to see if upper limits could be set on the presence of contributions of nontime reversal invariant contributions to the nuclear Hamiltonian. He concludes[17] that the experimental distribution of $\Gamma_n^o/<\Gamma_n^o>$ values provides a more sensitive test and implies $\lesssim 5\%$ time reversal violation at the 99.7% confidence level.

References

1. BETHE, H. A. & R. F. BACHER. 1936. Rev. Mod. Phys. **8**: 82–229.
2. BETHE, H. A. 1937. Rev. Mod. Phys. **9**: 69–224.
3. LIVINGSTON, M. S. & H. A. BETHE. 1937. Rev. Mod. Phys. **9**: 245–390.
4. WIGNER, E. P. 1957. Gatlinburg Conference on Neutron Physics: 59. ORNL 2309.
5. PORTER, C. E. & R. C. THOMAS. 1956. Phys. Rev. **104**: 483.
6. FESHBACH, H., C. E. PORTER & V. F. WEISSKOPF. 1954. Phys. Rev. **96**: 448.
7. MUGHABGHAB, S. F. & D. I. GARBER. 1973 Neutron Cross Sections. 3rd edit. Vol. 1. Brookhaven National Laboratory Report No. BNL-325 (National Technical Information Service. Springfield, Va.)
8. DYSON, F. J. 1962. J. Math. Phys. **3**: 140, 157, 166, 1199. Reprinted in Reference 10.
9. MEHTA, M. L. & F. J. DYSON. 1963. J. Math. Phys. **4**: 701, 713. Reprinted in Reference 10.
10. PORTER, C. E. 1965. Statistical Theories of Spectra Fluctuations. Academic. New York, N.Y. Has reprints of essentially all published and many unpublished papers on this subject up to early 1964.
11. LIOU, H. I., H. S. CAMARDA, S. WYNCHANK, M. SLAGOWITZ, G. HACKEN, F. RAHN & J. RAINWATER. 1972. Phys. Rev. **C5**: 974.
12. BOHEGAS, O. & J. FLORES. 1972. In Statistical Properties of Nuclei. J. Garg, Ed.: 195. Plenum Press. New York, N.Y.
13. BOHEGAS, O. & J. FLORES. 1971. Phys. Lett. **34B**: 261.

14. FRENCH, J. B. & S. S. M. WONG. 1970, 1971. Phys. Lett. **33B**: 449. **35B**: 5.
15. WONG, S. S. M. & J. B. FRENCH. 1972. Nucl. Phys. **A198**: 188.
16. MELLO, P. A., J. FLORES, T. A. BRODY, J. B. FRENCH & S. S. M. WONG. 1976. Spectrum fluctuations and the statistical shell model. Proc. Internat. Conf. on the Interaction of Neutrons with Nuclei, Lowell, Mass. (CONF-76-715). Available from NTIS. Springfield, Va. 22161.
17. Reference 16: 1454.

THE ELECTRIC AND MAGNETIC DIPOLE MOMENTS OF THE NEUTRON*

Norman F. Ramsey

Harvard University
Lyman Laboratory of Physics
Cambridge, Massachusetts 02138

Introduction

I have learned more physics and more about many other subjects from my Ph.D. Professor, I. I. Rabi, than from anyone else. The education began during my days as a graduate student and has continued through our collaboration and close association up to the present. As a result, it is a particular pleasure for me to contribute a scientific article in Rabi's honor.

Rabi is responsible in two different ways for the existence of the neutron studies reported in this paper. When I completed my molecular beam thesis research with Rabi, he encouraged my branching out into other areas of physics and gave added impetus to his recommendation by nominating me for a Carnegie Institution of Washington Fellowship in nuclear and particle physics at the then high energy of 3 MeV. Although Rabi undoubtedly expected me eventually to choose between these fields, I elected instead to participate in both of them. The present neutron studies result from the mutual interactions of the two fields. The subject studied is the neutron—one of the basic particles—but the magnetic resonance techniques used are modifications of Rabi's original invention and of the molecular beam methods I first learned as his student.

In 1950 Purcell and Ramsey[1] pointed out that the parity arguments then used to prove that particles and nuclei could not have electric dipole moments, must be based on an experimental rather than a theoretical basis. As a test of this assumption, Smith, Purcell, and Ramsey[2,3] used a neutron beam magnetic resonance apparatus to search for a neutron electric dipole moment and concluded that such a moment divided by the proton charge (μ_e/e) was experimentally less than 5×10^{-20} cm. Later, from the work of Lee and Yang[4] and Wu et al.,[5] it became apparent that the parity assumption was indeed invalid, but Landau[6] and others pointed out that the parity argument against an electric dipole moment could be replaced by one based on time reversal invariance. However, Ramsey[6] emphasized that time reversal invariance, like parity at an earlier time, was merely assumed and must rest on an experimental basis. In 1964 Christenson, Cronin, Fitch, and Turlay[8] discovered the CP-violating mode in the decay of the K_L° meson into two charged pions, which strongly suggested a violation of time reversal symmetry.

Since then a number of theoretical predictions[9] have been made for nucleon electric dipole moments on the basis of theories developed to account for the K_L°

---

*This work was partially supported by the U.S. Energy Research and Development Administration, The National Science Foundation, The French Commissariat a l'Energie Atomic and the Institut Laue-Langevin.

decay. Although the different predictions cover a wide range of values, some were as large as 10^{-19} cm for μ_e/e and most predicted 10^{-22} cm or larger. Since most of the range of predicted values was accessible to experimental search, several different experiments to measure the neutron electric dipole moment were started by Baird, Dress, Miller, and Ramsey,[9,10,13,14] by Nathan and Shull,[11] by Cohen, Lipworth, Silsbee, and Ramsey,[12] by Smith and Pendlebury,[15] and by Apostolescu, Ionescu, Bujor, Mecterts, and Petroscu.[16] The successive limits of the different experiments are given in TABLE 1. The greatest sensitivity at each time has been provided by the experiments of Dress, Miller, Ramsey, and Baird[9,10,13,14,15] and their most recently published experiment[13] provides the greatest sensitivity of any experiment so far published. This experiment was based on a neutron beam magnetic resonance study of 80 m/sec neutrons and provided a limit μ_e/e of 10^{-23} cm. It became apparent at the end of this Oak Ridge experiment that a further increase in precision would require a high flux of neutrons at velocities of 100 m/sec or less. For this reason, the apparatus was moved to Grenoble, France, to take advantage of the cryogenic moderator at the Institute Laue-Langevin (ILL) reactor.

TABLE 1
NEUTRON ELECTRIC DIPOLE MOMENT EXPERIMENTAL RESULTS

Value D (cm)	Laboratory (Year)	Reference
$<5 \times 10^{-20}$	ORNL (1951, 1957)	2
$(-2 \pm 3) \times 10^{-22}$	ORNL (1967)	9
$<3 \times 10^{-22}$	ORNL (1968)	14
$(+2.4 \pm 3.9) \times 10^{-22}$	MIT-BNL (1967)	11
$<1 \times 10^{-21}$	BNL (1969)	12
$<1 \times 10^{-21}$	Aldermaston (1968)	15
$<5 \times 10^{-23}$	ORNL (1969–1972)	10
$0.2 \pm 3.9 \times 10^{-22}$	Romania (1970)	16
$<1 \times 10^{-23}$	ORNL (1973)	13

The earlier apparatus with considerable modifications has now been operating at the Grenoble reactor and has produced a lower limit for the neutron electric dipole moment than any obtained previously. The experiments were done in collaboration with W. B. Dress and P. D. Miller of Oak Ridge National Laboratory in the United States, Paul Perrin of CENG (Centre Expérimentale Nucléaire de Grenoble) in Grenoble and Michael Pendlebury of Sussex University, England.

Method and Apparatus

The apparatus used in this experiment is essentially one to measure with high precision the precessional frequency of the neutron spin in a weak magnetic field with a neutron beam magnetic resonance apparatus similar to that used for measuring the magnetic moment of the neutron. A strong electrostatic field is then applied successively parallel and antiparallel to the magnetic field H. If the neutron had an electric dipole moment, the torque due to this dipole moment in the electric field would make the precessional frequency of the neutron spin some-

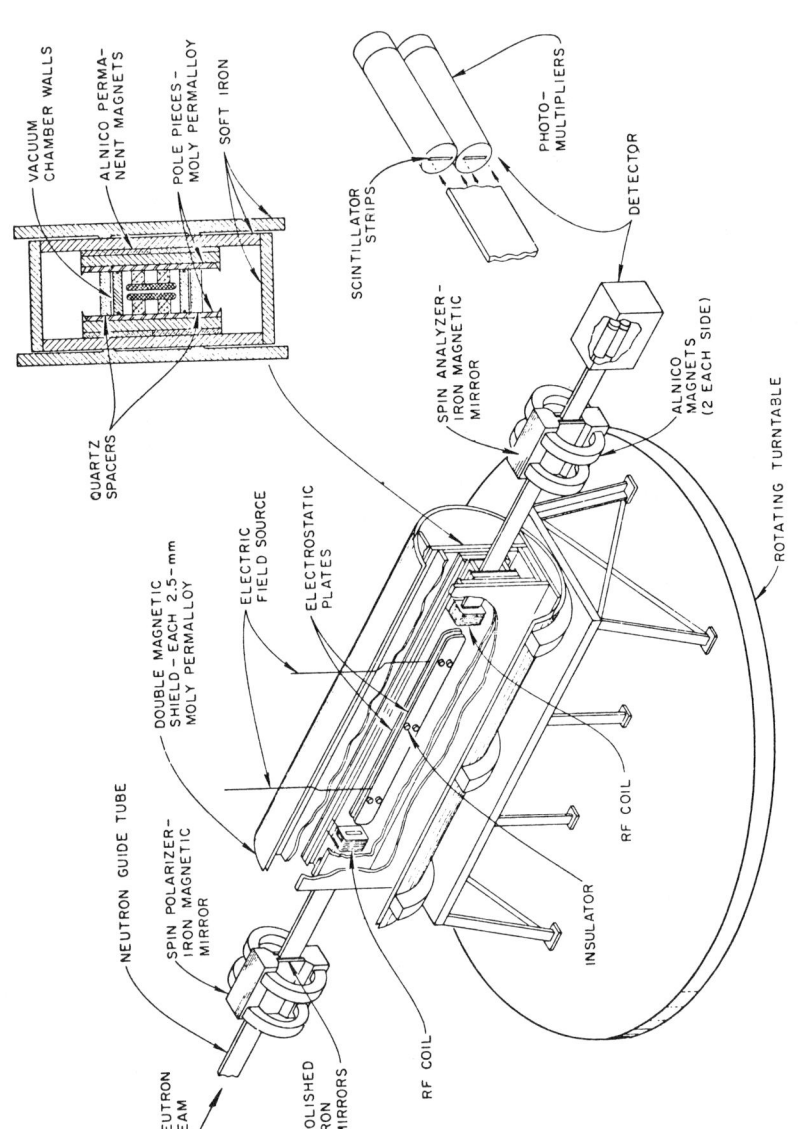

FIGURE 1. Experimental arrangement of the magnetic resonance spectrometer.

what greater, with the electric field in one direction and somewhat less in the opposite. By setting an experimental limit on the change in the precessional frequency, a limit is thereby set on the electric dipole moment of the neutron. The main requirements in the experiment are to achieve a very high sensitivity and to eliminate spurious effects that might either lead to a false apparent electric dipole moment or might obscure an actual moment.

A schematic view of the apparatus is shown in FIGURE 1. The neutron beam comes from the cryogenic moderator at the ILL reactor. The neutrons are conducted from the moderator through a neutron conducting tube of rectangular cross-sections on whose surface they are totally reflected at glancing angles of 2 degrees or less. The use of such neutron-conducting pipes, which becomes possible with sufficiently slow neutrons, markedly enhances the intensity by overcoming the normal diminution of beam intensity with the inverse square of the distance from the moderator. This gain of intensity is badly needed to compensate in part for the even greater loss of intensity by the selection of extremely slow neutrons.

As shown in FIGURE 1, the neutron beam goes through a portion of the pipe in which the walls consist of magnetized iron. Depending upon the orientation of the neutron spin, there is either a positive or negative magnetic interaction between the neutrons and the magnetic induction of the walls in the magnetized region. The combination of this positive or negative mean magnetic interaction with the coherent forward scattering amplitude of the neutrons by the wall material leads to total reflection at the walls for neutrons of one spin orientation while the neutrons with opposite spins are not reflected by that portion of the pipe and instead penetrate through the walls and are lost. Consequently following the spin-polarizing magnetic mirror, the neutrons are mostly polarized. The analyzing device to determine if there has been a change in the neutron spin orientation is a second spin-analyzing magnetic mirror. If the neutron spin remains unaltered between the first and the second of these magnetic field regions, most of the neutrons will be transmitted by the second region. If, on the other hand, the neutrons have been reoriented by approximately 180 degrees between the two iron mirror sections, the neutrons whose orientations have changed will not be totally reflected in the second magnetic mirror with a consequent reduction in beam intensity. Therefore, if the oscillatory fields are all in phase, the minimum of detected beam intensity occurs at the precessional frequency of the neutron. On the other hand, as shown by the author,[17] if the oscillatory magnetic field is provided in two separate segments with a 90 degree phase shift between them, the shape of the resonance curve is that of a dispersion curve with the steepest portion of the slope at the spin precession frequency as shown in FIGURE 2. If the frequency of the oscillator is set so that the detected neutron intensity is at the position of the steepest slope, the presence of a neutron electric dipole moment can be detected by successively reversing a strong electrostatic field. If there is an electric dipole moment, the torque due to the electric field will increase the precessional frequency of the neutron for one orientation of the field and decrease it for the opposite. At a fixed frequency of the oscillator, this change in the precessional frequency of the neutron spin will then be detectable with high sensitivity as a change in the neutron beam intensity.

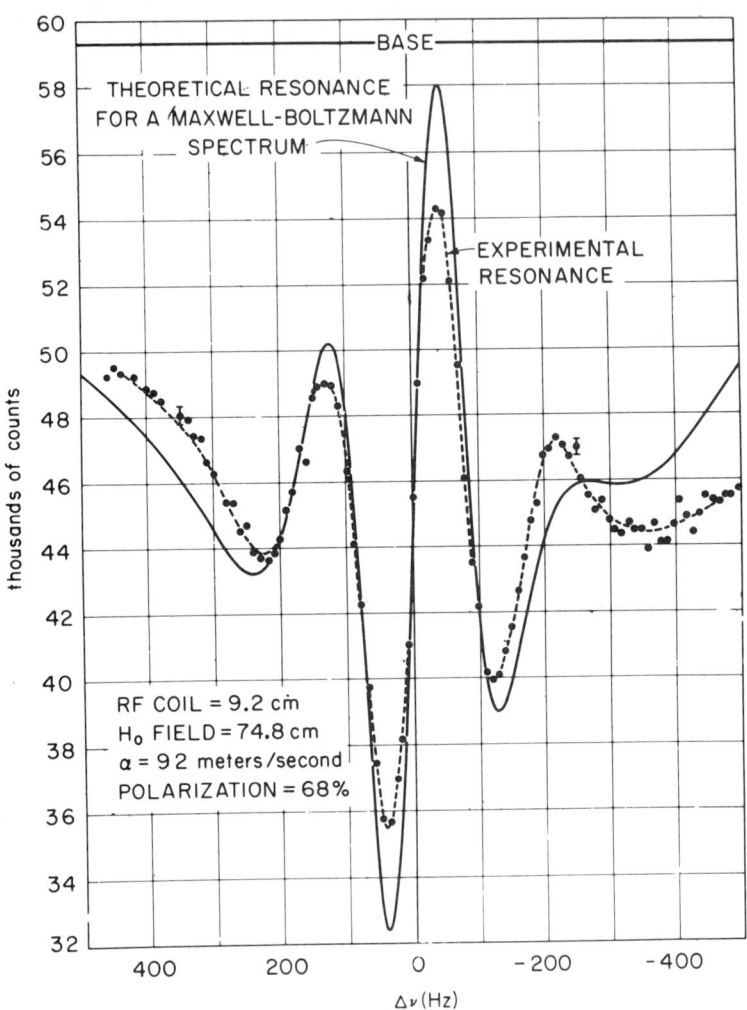

FIGURE 2. Typical magnetic resonance with a phase shift of $\pi/2$ between the two oscillatory fields. The calculated transition probability for a Maxwell-Boltzmann distribution characterized by a temperature of 1°K is shown in the solid curve. The departure of the experimental curve from the theoretical one when far from resonance is to be expected from the known departure of the beam velocity from a Maxwell-Boltzmann distribution.

The electric field is applied over a length of 196 cm and typically has a value of about 100 kV/cm. The static magnetic field was about 17 G and the neutron beam was 89% polarized.

Great care in the experiment must be taken to avoid spurious effects, which could either simulate a nonexistent electric dipole moment or mask an existing one. Fortunately, a number of things can be done to eliminate or minimize such spurious effects. The relative phase of the two oscillatory fields, can be shifted from +90 degrees to −90 degrees in which case the slope of the curve at the resonance position is reversed with a consequent reversal of the effect of the electric field on the detected neutron beam intensity. This reversal in the electric dipole moment effect eliminates many possible spurious effects. The phase was reversed once per second. Fortunately, in addition, many of the possible spurious effects cancel themselves because of the parity or time reversal symmetry of the effect. For example, there can be an effect of the electric field upon the frequency due to the force from the electric field pulling the magnets together and thereby changing the magnetic field. However, this effect and many others go as E^2 and consequently cancel on subtracting of results with reversed electric fields. A check on the existence of such E^2 effects can also be obtained from observations at zero electric field. Likewise to detect magnetic effects from the field-reversing mechanism, the leads to the source of potential are reversed at intervals. In addition, measurements are made when no potential is present but when the reversing switches are successively changed. The importance of such control measurements is illustrated by the fact that for some months there was a very small residual effect when the switches were reversed in the absence of any potential. This was eventually eliminated by moving the reversing switches still further from the apparatus and by increasing their magnetic shielding.

An important source of a spurious effect has been observed in recent runs of high sensitivity. Whenever there is a spark across the electric plates, the accompanying current produces a slight magnetic field, which in turn produces a very small residual change in the permanent magnetic field, due to the hysteresis of the iron. Even if the neutron counts during the period of the spark are excluded, the residual change in the permanent magnetic field can give a false result. This trouble, however, can be eliminated if the existence of sparks are recorded and if care is taken to assure that equal amounts of measurements with the fields in opposite directions are utilized in each interval between sparks.

One of the most bothersome spurious effects is that due to motion of the neutrons with a velocity **v** through the electric field **E** since such motion produces an effective magnetic field **E** × **v**/c. This effective magnetic field can then interact with the known neutron magnetic moment to produce an added precession frequency, which will look like that due to an electric dipole moment since it will reverse with the reversal of **E**. This effect is drastically reduced by making **E** parallel to **H**. If exact parallelism could be obtained the effect would be completely eliminated since this spurious magnetic field would be perpendicular to the initial magnetic field with the result that the effect would go as E^2 instead of **E**. However, because of the residual magnetism of ferromagnetic materials and magnetic shields, one can never be absolutely certain as to the direction of the magnetic field, with the result that **E** and **H** cannot be made exactly parallel, and the perpendicular component of E can produce an apparent electric dipole effect through the **E** × **v**/c effective magnetic field. The existence of such an effect, however, can be detected by changing the velocity of the neutrons since the spurious effects should be proportional to the neutron velocity. Consequently, all the data is analyzed in terms of an electric dipole moment and an apparent electric dipole proportional to the neutron velocity. The neutron velocity is altered in either of two ways. In some cases, the velocity is changed by changing the angle of neutron reflection from mirrors and in all cases the measurements are repeated many times with the direction of the neutrons through the apparatus reversed. For this reason, as can be seen in FIGURE 2, the basic neutron resonance apparatus is fastened to a turntable, which can be rotated to have the neutrons pass through the apparatus in opposite directions. The necessity for experiments at altered velocity greatly increases the running time of the experiment since the $E \times v/c$ effect must be measured with equal precision to that desired for the neutron electric dipole moment.

Results

The results of the present phase of measurements at the Institute Laue-Langevin are

$$\mu_E/e = (0.4 \pm 1.5) \times 10^{-24} \text{ cm}.$$

In other words, the neutron electric dipole moment, if it exists at all, is less than 3×10^{-24} cm. To emphasize the smallness of this result, I should emphasize to nuclear physicists that this is 10^{-24} cm not cm^2; it corresponds to 10^{-48} cm^2. If the neutron were expanded to the size of the earth this asymmetry would correspond to an incremental height of 0.01 cm in the northern hemisphere.

There have been numerous theoretical predictions as to the value of the neutron electric dipole moment. All theories that account for the CP-violating decay of the K_L^o meson[7] predict nonzero values for the neutron electric dipole moment. The predictions of these theories[18-38] are shown in FIGURE 3. Each lettered block

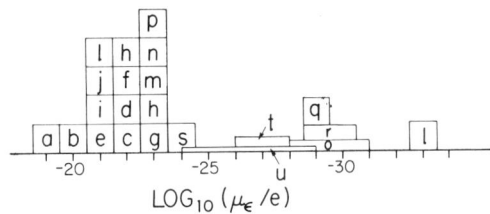

FIGURE 3. Theoretical predictions of the neutron electric dipole moment. Each lettered block corresponds to a different theory with the letter referring to a specific reference below. The basis of the different theories are indicated in square brackets below with EM indicating a theory that attributes the time reversal violation to the electromagnetic theory, W attributes it to the weak forces, MW to a milliweak force, SW to a new superweak force. Normally, the rectangle indicating each theory is a square spanning one decade; where the authors propose a wider spread, the rectangle is adjusted accordingly but with the same area as for other theories. Key: (a) Feinberg,[18] EM; (b) Salzman & Salzman,[19] EM; (c) Barton & White,[20] EM2; (d) Broadhurst,[21] EM2; (e) Babu & Suzuki,[22] MWΔS = 0; (f) Meister & Rhada,[23] MWΔS = 0; (g) Gourishankar,[24] MWΔS = 0; (h) McNamee & Pati,[25] MWΔS = 0; (i) Nishijima & Swank,[26] MWΔS = 0; (j) Nishijima,[27] MWΔS = 0; (k) Boulware,[28] MWΔS = 0; (l) Wolfenstein,[29] SWΔS = 2; (m) Pais & Primack,[30] MW; (n) Lee,[31] MW; (o) Okun,[32] SW; (p) Mohapatra,[33] MW; (q) Frenkel & Ebel,[34] MW; (r) Wolfenstein,[35] SW; (s) Weinberg,[36] MW; (t) Pakvasa & Tuan,[37] MW; (u) Mohapatra & Pati,[38] MW.

in the figure corresponds to the prediction of a different theory. One source of interest of the present experimental limit is that it provides extreme difficulties for many of the theoretical predictions and significant difficulty for most of the theories except those that attribute the time reversal asymmetric interaction to a new super weak force.

Neutron Electric Dipole Moment Experiments with Bottled Ultra-Cold Neutrons

The neutron electric dipole moment experiment so far described depends upon the fact that neutrons at a velocity of 80 m/sec will be totally reflected by many materials at glancing angles of approximately 5 degrees. As the velocity of the neutrons diminish, the glancing angle for total reflection increases until finally at a velocity of 6 m/sec total reflection can be obtained even at normal incidence on many surfaces. Under such circumstances, it is possible in principle to store neutrons in an enclosed bottle. For many years,[39-41] we have been anxious to do such experiments that would provide for the neutrons many of the advantages of the successive oscillatory field experiments with stored atoms,[39,40] including those with the hydrogen maser.[40] Up until recently, however, we have had no prospect of obtaining access to such ultracold neutrons.

Zeldovitch,[42] Vladimirski,[43,44] and the late Dr. F. L. Shapiro have discussed bottled ultracold neutrons, and Shapiro and his collaborators[45] have shown that ultracold neutrons can be stored in bottles for up to 20 seconds. Improvements in the techniques with ultracold neutrons and storage bottles have been made by Steyerl,[46] Ageron,[47] Lobashov,[48] Taran,[49,50] Pendlebury, Gollub and Smith,[51,52] and Miller, Dress, and Ramsey,[52,53] and others. Specific experiments have been discussed by Ramsey, Miller, and Dress,[39,41,52,53] by Taran,[49,50] and by Pendlebury, Smith, and Gollub.[51,52]

The experiment at ILL now being prepared to use ultracold bottled neutrons to set a limit to the neutron electric dipole moment is a collaboration between J. Byrne, R. Gollub, J. M. Pendlebury, K. F. Smith, N. F. Ramsey, W. B. Dress, P. D. Miller, A. Steyerl, P. G. H. Sandars, P. Ageron, and P. Perrin of the University of Sussex, Harvard University, Oak Ridge National Laboratory, Munich, Oxford University, ILL, and CENG. Ultracold neutrons at approximately 6 meters per second will be led by a neutron-conducting pipe into the apparatus shown in FIGURE 4. The neutrons will be stored in a cylinder approximately 15 cm in diameter and 10 cm high with the top plates being metallic—probably beryllium —and the sides of the cylinder being of beryllia insulator. The oscillatory field is applied to the admission and exit tubes so the resonance can be observed by the previously described successive oscillatory field technique.[39] The resonance will be observed in a similar fashion to our present neutron beam experiment and observation will be at the steepest point of the resonance curve. The change in beam intensity correlated with the application of an electric field will then be examined to set a limit to the neutron electric dipole moment.

The use of stored ultracold neutrons possesses two particularly important advantages. The resonance curve for the 18-second storage time of the neutron should be approximately 800 times narrower than in the present experiment with a corresponding increase in sensitivity. Furthermore, as mentioned earlier, a large fraction of running time in the present experiment must be devoted to eliminating the $\mathbf{E} \times \mathbf{v}/c$ effect. Since it is the average value of $\mathbf{v}$ that is important, this effect is drastically diminished when the neutrons enter and leave by the same exit hole with an 18-second storage time instead of passing through the apparatus at a velocity of 80 m/sec. As a result of the reduced effective magnetic field from $\mathbf{E} \times \mathbf{v}/c$, it should also be possible to use a much weaker static magnetic field with an accompanying reduction in the field stability problem.

Although the new experiment being planned will have the above-marked advantages, it must be recognized that it will still be an extremely difficult one. The limit has by now been pushed to such a low value that care must be taken to avoid

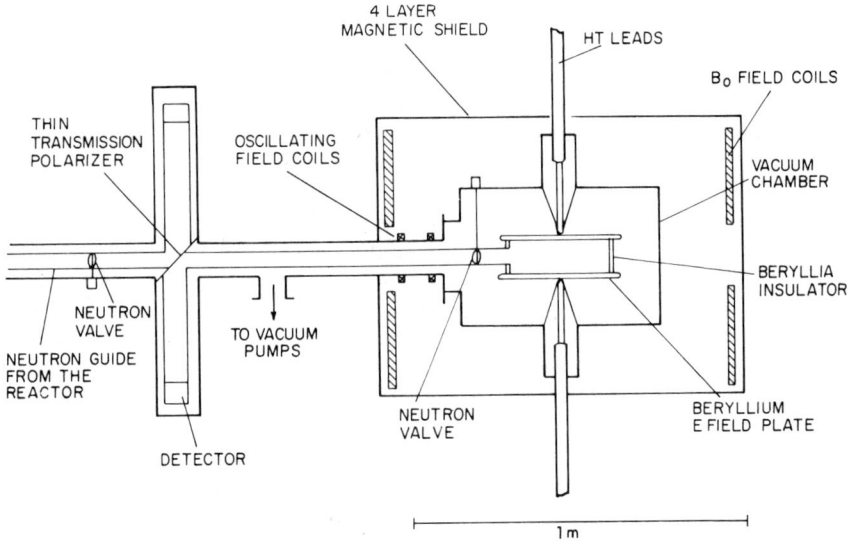

FIGURE 4. Schematic diagram of apparatus for conducting the neutron magnetic resonance experiment with bottled neutrons.

all possible systematic effects. Although some of these are intrinsically reduced in an experiment with bottled neutrons, other serious problems will remain. For example, problems due to stray magnetic fields (especially when associated with reversals of the electric field) and to magnetic field changes resulting from electrical sparks can be just as serious with bottled neutrons as with neutron beams. These problems have already caused much difficulty in the beam version of the experiment and should be even more formidable in the bottled neutron experiment, which seeks to lower the limit for the neutron electric dipole moment by a factor of 100 to 1000.

The apparatus is planned to be capable of being operated in either of two fashions. In one, pulsed ultracold neutrons will be admitted for a few seconds and then stored with the neutron valve closed for approximately 30 seconds before the valve is reopened so the neutrons can escape past the oscillatory field for a second time.

In the second mode of operation, the neutrons will continuously be introduced and permitted continuously to bounce out of the neutron bottle with a mean storage time of approximately 18 seconds. The precision of the two methods of observation are comparable and the two procedures should mutually compliment each other.

With an electric field of 30 kV/cm and a multilayer Mumetal or Moly-Permalloy magnetic shield, it should be possible to achieve a limit on the electric dipole moment of 10^{-26} cm. To go to a lower limit will probably require superconducting magnetic shields. These are currently contemplated but decisions on a subsequent phase of the experiment will not be taken until later. With superconducting shields and sufficiently long observation times, it should be possible to lower the limit to 10^{-27} cm. With a larger cell diameter and other improvements, sensitivity of the order of 10^{-28} cm might ultimately be reached.

Other Neutron Beam Magnetic Resonance Experiments

Since it will take more than a year before the ultracold neutron beam can be available at ILL and before the apparatus for the bottled neutron experiment can be ready, and since the new apparatus will be required to achieve a significant improvement in the present limit, the collaborators of the present experiment plan to modify the present apparatus so it can be used during the coming year markedly to improve the accuracy of the measurements of the neutron *magnetic* dipole moment. At present, the neutron magnetic dipole moment is the least accurately known of all the nucleon and lepton magnetic moments. The magnetic moments of the negative electron, the positive muon, and the proton are all known to a fractional error less than 3×10^{-8} whereas the fractional error in the neutron moment is 1000 times greater or 3×10^{-5}. Although the present apparatus was designed with the neutron electric dipole moment exclusively in mind, by coincidence it turns out to be an appropriate design for measuring the magnetic moment of the neutron. Although the magnetic field is low and probably cannot easily be raised much above 800 gauss, this disadvantage is more than offset by the large magnetic gap, which permits an accurate calibration of the magnetic field because of the smaller inhomogeneities that result from the increased gap. In the previous most accurate experiment[54] the precision of the result was primarily limited by this field inhomogeneity and the consequent difficulty in calibrating the magnetic field accurately.

The magnetic moment will be measured in an apparatus that is essentially the same as that now being used in the neutron electric dipole moment experiment. Permanent magnets, however, can be added to increase the magnetic field from 15 oerstead to 800 oerstead. The magnetic field can be calibrated in several alternative ways. One is by the use of a proton NMR magnetometer and another is by the use of a rubidium magnetometer. A still different alternate is to pump water at high speed through a high magnetic field storage region to polarize the protons and then to have the water pass through the neutron beam pipe at high velocity, with the resonance being observed by the separated oscillatory field method[17]; in this case the second oscillatory field region has many of the characteristics of a volume filled with molecules in "super-radiant" states. The flowing-water method has the advantage of a close similarity between the averagings done by the neutrons and by the protons, as each are confined to the neutron pipe. It is anticipated that all three methods will be used. The greatest possible care must be devoted to assuring that the magnetic field at the time of the proton calibration is the same as that during the measurements with the neutron.

With this technique, it appears that it should be relatively easy to improve the accuracy of the measurement of the neutron magnetic moment by at least a factor of 100 and hopefully by a somewhat larger factor.

There are two other interesting neutron beam experiments that we would very much like to do as soon as we can fit in the time without significant interference with the primary priority we attach to the neutron electric dipole moment experiments. From the experimental point of view, the two experiments are closely related, since both involve the measurement of a small parity-violating reorientation of the neutron spin when it passes through matter.

Kabir, Karl, and Obryk[55] have pointed out that when a neutron passes through an optically active medium, (one whose constituents are parity asymmetric, as solutions of levulose), the transverse component of the neutrons polarization should precess about the direction of propagation about 10^{-5} radians in traversing a centimeter of a representative optically active medium. Since an electric dipole moment of 10^{-24} cm provides a precession of only 0.3×10^{-5} radians, there should be sufficient sensitivity in an adaptation of the apparatus for that experiment to observe such a precession; among the apparatus changes required in such an adaptation would be the provision of a weak longitudinal magnetic field instead of a weak transverse field. If sensitivity alone were the only requirement, the measurement should be relatively easy. Unfortunately, the primary difficulty will arise from the existence of the neutron magnetic moment and the necessity of demonstrating that there is no small magnetic perturbation associated with insertion of the interacting material that produces a comparable or greater recession. This can be seen from the fact that a magnetic field of 3×10^{-8} gauss would produce 10^{-5} radians of precession while a 100 m/sec neutron traversed the 1.9 meter length of our present apparatus. Some benefit could be obtained by making the apparatus as short as possible, but there is still a severe requirement for eliminating any change in magnetic field associated with the change in the sample. A great improvement in this problem can be obtained by simultaneously running neutron beams through the apparatus in opposite directions. The parity-violating effect depends on the direction of the neutron velocity while the magnetic precession is independent of the neutron direction so by simultaneously measuring the precession for opposite directions of neutron motion the two effects can be distinguished.

A second experiment of even more fundamental interest is to look for a similar parity-violating precession of the neutron in passing through a medium that is not optically active. Michel[56] and Stodolsky[57] have pointed out that a precession of 1.4×10^{-6} radians should occur if cold neutrons passed through at about 1 meter of say bismuth. The source of the parity-violating rotation in this case would not be the optical activity of the medium but instead would be the parity-violating weak interaction of the neutron. It would be of great interest in this way to observe directly the weak interactions of neutrons with matter, and the parity-violating character of the weak interactions provides a unique signature through the spin precession to distinguish this interaction from the strong forces that usually dominate the interaction of neutrons with matter. The problem of magnetic effects in this case is of course even more severe than in the case of an optically active medium because the precession angles from the desired effects are even smaller. However, as in the previous experiment, great benefit could be obtained in distinguishing the precession due to the weak forces from those due to magnetic fields by simultaneously making observations on neutrons that pass through the apparatus in opposite directions.

Although the possibility of doing these experiments was first discussed as a neutron beam magnetic resonance experiment, they can also be done by Mezei's interesting neutron spin echo technique.[58] To avoid unnecessary duplication, we have discussed a combined effort with Mezei to observe these two interesting effects probably using a modification of his neutron spin echo technique. The principal problem is finding time to fit these experiments into our respective programs, particularly in view of the primary priority attached to further lowering the limit on the electric dipole moment of the neutron.

References

1. PURCELL, E. M. & N. F. RAMSEY. 1950. Phys. Rev. **78**: 807.
2. SMITH, J. H. 1950. Thesis. Harvard University. Unpublished.
3. SMITH, J. H., E. M. PURCELL & N. F. RAMSEY. 1957. Phys. Rev. **108**: 120.
4. LEE, T. D. & C. N. YANG. 1957. Phys. Rev. **105**: 1671.
5. WU, C. S., E. AMBLER, R. W. HAYWARD, D. D. HOPPES & R. P. HUDSON. 1957. Phys. Rev. **105**: 1413.
6. LANDAU, L. 1957. Zh. Eksperim. i Teov. Fiz. **32**: 405. Translation: 1971. Sov. Phys. JETP **5**: 336.
7. RAMSEY, N. F. 1958. Phys. Rev. **109**: 225.
8. CHRISTENSON, J. H., J. W. CRONIN, V. L. FITCH & R. TURLAY. 1964. Phys. Rev. Lett. **13**: 138.
9. MILLER, P. D., W. B. DRESS, J. K. BAIRD & N. F. RAMSEY. 1967. Phys. Rev. Lett. **19**: 381. References to many of the theoretical predictions for nuclear electric dipole moments are given in references 12-21 cited herein.
10. BAIRD, J. K., P. D. MILLER, W. B. DRESS & N. F. RAMSEY. 1969. Phys. Rev. **179**: 1285.
11. SHULL, C. & R. NATHAN. 1967. Phys. Rev. **19**: 384.
12. COHEN, V. W., E. LIPWORTH, R. NATHAN, N. F. RAMSEY & H. B. SILSBEE. 1969. Phys. Rev. **177**: 1942.
13. DRESS, W. B., P. D. MILLER & N. F. RAMSEY. 1973. Phys. Rev. **D7**: 3147.
14. DRESS, W. B., J. K. BAIRD, P. D. Miller & N. F. RAMSEY. 1968. Phys. Rev. **170**: 1200.
15. SMITH, K. & M. PENDLEBURY. 1968. Private communication.
16. APOSTOLESCU, IONESCU, IONESCU-BUJOR, MEITERTS & PETROSCU. 1970. Rev. Romaine Phys. **15**: 343.
17. RAMSEY, N. F. 1956. Molecular Beams. Oxford University Press. New York, N.Y.
18. FEINBERG, G. 1965. Phys. Rev. **B140**: 1402.
19. SALZMAN, F. & G. SALZMAN. 1965. Phys. Rev. Lett. **15**: 91.
20. BARTON, G. & E. D. WHITE. 1969. Phys. Rev. **184**: 1660.

21. BROADHURST, D. J. 1970. Nucl. Phys. **B20**: 603.
22. BABU & SUZUKI. 1967. Phys. Rev. **162**: 1359.
23. MEISTER, N. T. & T. K. RHADA. 1964. Phys. Rev. **B135**: 769.
24. GOURISHANKAR, G. R. 1968. Cana. J. Phys. **46**: 1843.
25. MCNAMEE, P. & J. C. PATI. 1969. Phys. Rev. **178**: 2273.
26. NISHIJIMA, K. & L. J. SWANK. 1967. Nuc. Phys. **B3**: 565.
27. NISHIJIMA, K. 1969. Prog. Theor. Phys. **41**: 739.
28. BOULWARE, D. G. 1969. Nuovo Cimento **40A**: 1041.
29. WOLFENSTEIN, L. 1964. Phys. Rev. Lett. **13**: 562.
30. PAIS, A. & J. PRIMACK. 1973. Phys. Rev. **D8**: 625, 3036.
31. LEE, T. D. 1973. Phys. Rev. **D8**: 1226. 1974. Phys. Reports **9**: 143.
32. OKUN, L. B. 1969. Comments on Nuclear and Particle Physics **3**: 135.
33. MOHAPATRA, R. N. 1972. Phys. Rev. **D6**: 2026.
34. FRENKEL, J. & M. E. EBEL. Complex Cabbibo Angle and CP Violation in a Class of Gauge Theories. Univ. of Wisconsin Preprint; and Reference 35 below.
35. WOLFENSTEIN, L. 1974. Nucl. Phys. **B77**: 375.
36. WEINBERG, S. Private Communication. The predicted value of this gauge theory which attributes CP nonconservation purely to the exchange of Higgs bosons is $\mu_E/e = 2.3 \times 10^{-24}$ cm.
37. PAKVASA, S. & S. F. TUAN. 1975. Phys. Rev. Lett. **34**: 553.
38. MOHAPATRA, R. N. & J. C. PATI. 1975. Phys. Rev. **D11**: 569.
39. RAMSEY, N. F. 1957. Rev. Sci. Inst. **28**: 57.
40. KLEPPNER, D., N. RAMSEY, P. FIJILSTADT & M. GOLDENBERG. 1960. Phys. Rev. Lett. **1**: 232. 1960. Phys. Rev. **8**: 361.
41. RAMSEY, N. F. 1969. Proposal to ILL for Neutron Electric Dipole Moment Experiment.
42. ZELDOVITCH, Y. B. 1959. Zh. Eksp. Zero. Fiz. **36**: 1952. 1959. Sov. Phys. [JETP **9**: 1389.]
43. VLADIMIRSKII, V. 1961. Sov. Phys. JETP **12**: 740.
44. DROSHKEVICH, A. 1963. Sov. Phys. JETP **16**: 56.
45. LUSCHIKOV, V. I., YU N. POKOTILIOSKY, A. V. STVELKOV & F. L. SHAPIRO. 1969. Zh. Eksp. Teov. Fiz., Pis mo v Redaktsiya **9**: 40. [1969. Sov. Phys. JETP Lett. **9**: 23.]
46. STEYERL, A. & W. B. TRUSTEDT. 1973. Private communication.
47. AGERON, P., J. M. ASTRUS & J. VERDIER. Preliminary Project of an Ultra-Cold Neutron Source at the High Flux Reactor, Institute Laue-Langevin, Grenoble. Unpublished.
48. LOBASTOV, V. M., G. D. PORSEV & A. B. SEREBROV. 1973. Preprint No. 37, Konstantinov Institute of Nuclear Physics of the Academy of Science of the USSR, Leningrad.
49. Elementarnaya Teoria Metoda Electricheskova Kipolnova Momenta Neitrona C. Pomoshyn UXN. I Protochini.
50. Variant. Yu V. Taran, Dubna, Preprint P3-7147, Elementarnaya Teoris Metoda Opredeleria Electricheskova Dipolnova Momenta Neitrona C. Pomoshyn UXN. II Nakopitelnu Variant, Yu V. Taran, Dubna Preprint P3-71479.
51. GOLLUB, R. & J. M. PENDLEBURY. 1972. Contemp. Phys. **13**: 519.
52. BYRNE, J., R. GOLLUB, J. M. PENDLEBURY & K. F. SMITH in collaboration with N. F. RAMSEY, W. B. DRESS, P. D. MILLER, A. STEYERL, P. G. H. SANDARS, P. AGERON & P. PERRIN. A Proposal to Search for the Electric Dipole Moment of the Neutron Using Bottled Neutrons, Institute Laue-Langevin (1974) and Rutherford Laboratory (1975).
53. MILLER, P. D. 1974. Measurements of the Electric Dipole Moment of the Neutron. Second International School on Neutron Physics, Alushta, Crimea, USSR (1974).
54. COHEN, V. W., N. R. CORNGOLD & N. F. RAMSEY. 1958. Phys. Rev. **104**: 283.
55. KABIR, P. K., KARL G. & E. OBRYK. 1974. Phys. Rev. **D10**: 1471.
56. MICHEL, F. C. 1964. Phys. Rev. **133**: B329.
57. STODOLSKY, L. 1974. Max Planck Institute. MPI-PAE/PTh9.
58. MEZEI, F. 1972. Z. Phys. **255**: 146.

SUPERFLUID MOTIONS

Mario Rasetti

*Istituto di Fisica del Politecnico
Torino, Italy*

Tullio Regge*

*Institute for Advanced Study
Princeton, New Jersey 08540*

Introduction

We would like to develop in this paper a number of intuitive concepts related to the theory of superfluid motions. Although the discussion may prove useful also for Fermi fluids, we shall restrict ourselves for simplicity to Bose Superfluids, therefore essentially to He II.

In the superfluid state He II is known to undergo motions with no apparent viscosity, and obeying the Euler-Bernoulli equations for potential flow

$$\frac{\partial \rho}{\partial t} + \mathrm{div}(\rho \mathbf{v}) = \quad m \frac{\partial \varphi}{\partial t} + \frac{d}{d\rho} e(\rho)\rho + \frac{mv^2}{2} = 0. \quad (1)$$

$$\mathbf{v} = \nabla \varphi$$

On the other hand, besides the macroscopic description of the motion given by Equation 1, we have a microscopic theory of the fluid through the N-body Schrödinger equation:

$$\left[-\frac{\hbar^2}{2m} \sum_{\alpha=1}^{N} \Delta_N + \sum_{\alpha>\beta} V(x_\alpha - x_\beta) \right] \Psi = i\hbar \frac{\partial \Psi}{\partial t}. \quad (2)$$

The theory of elementary excitations of He II, as originally developed by Landau,[1] rests on the assumption that the lowest lying states of He II can be adequately described by quantizing hydrodynamics in the form of Equation 1. The approach by Feynman and Cohen[2] instead is directly related to Equation 2 and can be regarded as a proper microscopic theory. Here however the intuitive picture of the motion of the fluid is less evident. Furthermore there seems to be no clear connection between the Feynman-Cohen theory and the beautiful concept of quantized vortices introduced by Feynman and Onsager. We intend to present here an heuristic derivation of Equation 1 from the wave equation. There is nothing in our methods that has not already appeared in the literature and we feel that each step is familiar to some of the experts in the field. However, we have never seen the whole sequence in its entirety; this fact and the occasional appearance of unclear statements on the subject have convinced us of the need for the present discussion. Our starting point is the Feynman wave function for a vortex.

*Sponsored in part by the National Science Foundation, Grant No. GP-40768X.

The most important feature of the Feynman Ansatz is that it exists, that is, that it should be possible at all to speak of the wave function of a fluid in motion. The existence of this wave function is an equivalent way of stating the "coherence" of the superfluid motion. The initial data is, of course, the ground state wave function for Equation 2:

$$\Phi = \Phi(x_1 \ldots x_N). \tag{3}$$

Intuitive reasons and numerical calculations show that Φ can be chosen to be everywhere non negative. Φ is supposed to vanish very rapidly when atoms come so close as to feel their repulsive core. However, if the atoms are evenly spaced, Φ is practically a constant. A superfluid is known to perform only irrotational motions, that is motions in which the velocity field is locally a gradient:

$$\mathbf{v}(x) = \nabla \varphi(x). \tag{4}$$

A vortex or a system of vortices is characterized by a velocity potential $\varphi(x)$ which is not single valued. $\varphi(x)$ therefore has a branching locus, the filament of the vortex, in which φ is undefined. In He II the thickness of the filament is very small, of the order of 1Å. In what follows we shall be discussing really what happens outside the filament. When one heats the sample lots of vortices are excited and they ultimately fill every available space so that φ becomes less and less useful and finally undefined at the λ point. At $T = 0$, however, it was proposed by Feynman that the wave function Ψ for the moving fluid should be given by

$$\Psi = \exp\left[i\frac{m}{\hbar}\sum_{\alpha=1}^{N}\varphi(x_\alpha)\right]\Phi. \tag{5}$$

Indeed if we apply the momentum operator $P_\alpha = (\hbar/i)\partial_\alpha$, we find

$$\mathbf{P}_\alpha \exp\left[i\frac{m}{\hbar}\varphi(x_a)\right] = m\mathbf{v}(x_\alpha)\exp\left[i\frac{m}{\hbar}\varphi(x_\alpha)\right],$$

where $m\mathbf{v}(x_\alpha)$ is the local momentum of the αth particle moving with velocity $\mathbf{v}$. It is clear that this analogy holds only if $v(x)$ varies very slowly within the fluid. For simplicity we suppose that there is only one vortex whose filament is the z axis. In this case if the fluid can be regarded as incompressible the velocity potential is given by

$$\varphi(x) = c\chi \quad x = \sqrt{x^2 + y^2}\cos\chi \quad y = \sqrt{x^2+y^2}\sin\chi \quad c = \text{const.} \tag{6}$$

This potential is not single-valued and increases by $2\pi c$ each turn around the z axis. However, the Ansatz must remain single-valued; therefore the change in φ must induce a trivial phase change in the wave function. It follows

$$2\pi c \cdot \frac{m}{\hbar} = 2n\pi \quad c = \frac{n\hbar}{m}, \quad n \text{ integer}$$

which is the Feynman-Onsager (FO) quantization of the vortex. This quantization is actually observed, and Equation 5 must be taken very seriously. Yet it is not free of objections. It was remarked already by Lin[5] that this wave function does not satisfy the Schrödinger equation in any obvious way. We should expect that the Ansatz is valid also for any potential motion, including sound waves in the fluid. Yet there seems no obvious way to link the Ansatz either with the Euler-Bernoulli equation or with the Feynman-Cohen theory. In what follows we plan to amend these shortcomings and to show that with minor changes Equation 5 becomes a very good wave function for the moving fluid as long as we limit ourselves to those motions that are slow with respect to the roton frequency and where φ does not vary appreciably over an interparticle distance. The amended Feynman wave function is really a "secular" Ansatz in which a sufficiently small volume of the fluid can be considered as a modulated local ground states corresponding to a slowly moving local density and phase. If we attempt to treat high frequency phonons with this method we find that it corresponds essentially to the Feynman-Cohen theory without the backflow term. The appearance of this term is then really a signal that the conventional hydrodynamical picture becomes inadequate.

Hydrodynamics

Let us first introduce a compact notation. We use a two-index label $a \equiv (\alpha, i)$ where $\alpha = 1, \ldots, N$ and $i = x, y, z$, for the generic degree of freedom of the N-particle system. Therefore we denote by x a $3N$ dimensional vector of components $x_a = (\mathbf{x}_\alpha)_i$, equal to the ith Cartesian coordinate of the αth particle. Also $dx \equiv dx_{11} dx_{12} \ldots dx_{N3}$ and $\Delta = \sum_{\alpha=1}^{N} \Delta_\alpha$, where Δ_α is the Laplace operator for the αth coordinate. The ground state wave function $\Phi(x)$ has a few qualitative features that are worth keeping in mind. The most important of these is that Φ is grossly insensitive to the actual configuration of the atoms as long as each atom is allotted the same specific volume on the average (for the ground state, 45.8 Å^3). This property implies that to some extent Φ is invariant under sufficiently smooth diffeomorphisms. Each differentiable map of R^3 into R^3 produces locally some stretching and compression of a small volume element in addition to a translation and a rotation, which are not interesting. If the map preserves the measure of the element our statement is then that Φ remains approximately invariant. It would be quite desirable to restate this property in a more rigorous mathematical language, but at the moment this seems to be still far away.[9] The net result of the invariance is that it costs remarkably little energy to deform Φ with a differentiable measure preserving map. This is the main reason why He II is a fluid at $T = 0$. The same deformation attempted on a solid would produce an enormous increase in energy. If the map causes a local scaling of the volume element, then Φ is transformed into the ground state with a different energy corresponding to the different density. It should be noted that all these properties are really only approximate and valid for very smooth maps and small deviations from the ground state density.

A diffeomorphism of this kind often arises by considering the flow defined by the differential equations

$$\frac{dx_a}{dt} = v_a(x, t) = \frac{\partial \Lambda}{\partial x_a}$$

$$x(\xi, t_0) = \xi \tag{7}$$

$$\Theta = \frac{m}{\hbar} \Lambda, \qquad (8)$$

where the set of coordinates ξ_a defines the initial conditions at $t = t_0$. In Equation 6 we shall suppose that the "total" velocity potential Λ is really the sum of all potentials of each particle. In this case the flow equations really decouple into N copies of the same equation. Without this assumption we would admit the presence of backflow terms and the set of Equation 6 would become exceedingly difficult to deal with.

From Equation 7 we can define the inverse functions by solving for the ξs,

$$\xi_a(x, t).$$

Our amended Feynman (F) wave function is resting on a Eulerian point of view of the hydrodynamics of the system. We assume in fact that the wave function depends on x through the ξ. With this assumption we propose the following wave function for the coherent motion under the flow F:

$$\Psi_F(x) = \exp\left[i \frac{m}{\hbar} \sum_{\alpha=1}^{N} \varphi(\mathbf{x}_\alpha)\right] \sqrt{J} \Phi(\xi) = \exp(i\Theta) \Phi_F. \qquad (9)$$

The Jacobian

$$J = \frac{\partial(\xi_{1x}, \xi_{1y}, \ldots, \xi_{Nz})}{\partial(x_{1x}, x_{1y}, \ldots, x_{Nz})} \qquad (10)$$

in Equation 9 naturally guarantees that the wave function is properly normalized for any diffeomorphism, i.e., for any macroscopic motion of the fluid. Differentiating Φ_F with respect to t, and noticing that $\Phi_F |_{t=t_0} = \Phi$, we have

$$\frac{\partial \Phi_F}{\partial t} = \left\{ \sqrt{J} \frac{\partial \Phi}{\partial t} + \Phi \frac{\partial \sqrt{J}}{\partial t} \right\}$$

$$= \left\{ \sqrt{J} \frac{\partial \Phi}{\partial \xi_a} \frac{\partial \xi_a}{\partial t} + \Phi \frac{\partial \sqrt{J}}{\partial t} \right\}. \qquad (11)$$

Now, for any η independent of t, by very definition Equation 7,

$$\xi_a[x(\eta, t), t] = \eta_a. \qquad (12)$$

Hence

$$\frac{\partial \xi_a}{\partial t} + \frac{\partial \xi_a}{\partial x_b} \frac{\partial x_b}{\partial t} = 0, \qquad (13)$$

and finally

$$\frac{\partial \xi_a}{\partial t} = -v_b \frac{\partial \xi_a}{\partial x_b}. \qquad (14)$$

Equation 11 can be written as

$$\frac{\partial \Phi_F}{\partial t} = -v_a \sqrt{J} \frac{\partial \Phi}{\partial x_a} + \Phi \frac{\partial \sqrt{J}}{\partial t}. \qquad (15)$$

On the other hand J can be factored out explicitly as

$$\sqrt{J} = \prod_{\alpha=1}^{N} \sqrt{J_\alpha}.$$

Therefore we need to calculate only the derivative of J_α for the αth particle. The derivative is taken by keeping x constant. It is, however, clear that we have the equality (we drop the index α temporarily)

$$J\rho(0) \equiv J(x,t)\rho[\xi(x,t),0] = \rho(x,t),$$

because the same number of particles initially in the volume element $d^3\xi$ and given by $\rho(0)d^3\xi$ is later found at d^3x and is given there by $d^3x\rho(x,t)$. The continuity equation yields

$$-\operatorname{div}(\rho v) = -\frac{\partial}{\partial x_i}[\rho(0)Jv_i] = \frac{\partial \rho}{\partial t}$$

$$= \frac{\partial J}{\partial t}\rho(0) + J\frac{\partial \rho(0)}{\partial \xi_i}\frac{\partial \xi_i}{\partial t} = \frac{\partial J}{\partial t}\rho(0) - J\frac{\partial \rho(0)}{\partial x_i}v_i, \quad (16)$$

because of Equation 14. Therefore for the αth atom

$$\frac{\partial J_\alpha}{\partial t} + \operatorname{div}(J_\alpha v_\alpha) = 0,$$

and in general

$$\frac{\partial \sqrt{J}}{\partial t} = -\frac{1}{2\sqrt{J}}\frac{\partial}{\partial x_a}(Jv_a). \quad (17)$$

It follows that the definition 9 is equivalent to the differential equation

$$\frac{\partial \Phi_F}{\partial t} = -\frac{1}{2}\sum_{\alpha=1}^{N}\left[v(x_\alpha)\frac{\partial}{\partial x_\alpha} + \frac{\partial}{\partial x_\alpha}v(x_\alpha)\right]\Phi_F. \quad (18)$$

Notice also here, although in the previous discussion the time $t = 0$ plays a special role, any other initial condition (in particular, $t_0 = -\infty$) could have been selected. The meaning of the entire procedure is then that the ground state drifts downstream along the velocity field, while acquiring the Feynman-Onsager phase Θ.

A Variational Principle

We are now able to set up a variational principle for Ψ_F. It is interesting to note how such a procedure copies the one followed by E. Schrödinger in his classical paper.[6] We write down an action integral A of the form

$$A\{\varphi\} = \int dt\, dx\, \Psi^*\left(i\hbar\frac{\partial}{\partial t} - H\right)\Psi, \quad (19)$$

where the argument within the curly brackets is intended as a reminder that A is indeed to be considered as a functional in the velocity potential φ. H is the Hamiltonian.

We intend to use Ψ_F as a trial function in the variational equation

$$\delta A \equiv 0. \tag{20}$$

Notice that in performing the variation, $\varphi(x)$ is to be considered as independent variable, while

$$\rho(\mathbf{x}) = \int \Phi_F^2(\mathbf{x}_1 \mathbf{x}_2, \ldots, \mathbf{x}_N) d^3\mathbf{x}_2 \ldots d^3\mathbf{x}_N$$

$$= \frac{\partial[\xi(x,t)]}{\partial[x]}\bigg|_{t=t_0} \rho(0) = J\rho(0) \tag{21}$$

is a dependent parameter connected to φ by the continuity equation 17.

Before performing the variation explicitly, we state an auxiliary identity, namely,

$$\left(i\hbar \frac{\partial}{\partial t} - H\right)\Psi_F = -\exp(i\Theta)\left\{\sum_{\alpha=1}^{N}\left[m\frac{\partial\varphi(\mathbf{x}_\alpha)}{\partial t} + \frac{1}{2}mv^2(\mathbf{x}_\alpha) + H\right]\right\}\Phi_F. \tag{22}$$

The proof of Equation 22 implies some nontrivial analysis, but can be easily sketched.

First, observe that

$$-\frac{\hbar^2}{2m}\Delta(e^{i\Theta}\Phi_F) = -\frac{\hbar^2}{2m}\Delta\Phi_F e^{i\Theta} - \frac{\hbar^2}{2m}\Phi_F \Delta e^{i\Theta} - \frac{\hbar^2}{m}\nabla\Phi_F \cdot \nabla e^{i\Theta}. \tag{23}$$

Now $a = (\alpha, i)$,

$$\nabla_a e^{i\Theta} = i\nabla_a \Theta e^{i\Theta} = i\frac{m}{\hbar}e^{i\Theta}\nabla_a\varphi(\mathbf{x}_\alpha) = i\frac{m}{\hbar}e^{i\Theta}v(\mathbf{x}_\alpha), \tag{24}$$

and

$$\Delta e^{i\Theta} = \text{div}(i\nabla\Theta e^{i\Theta}) = i\Delta\Theta e^{i\Theta} - (\nabla\Theta)^2 e^{i\Theta}$$

$$= \sum_{\alpha=1}^{N}\left[i\frac{m}{\hbar}\Delta\varphi(\mathbf{x}_\alpha) - \left(\frac{m}{\hbar}\right)^2 v^2(\mathbf{x}_\alpha)\right]e^{i\Theta}$$

$$= \frac{m}{\hbar^2}\sum_{\alpha=1}^{N}\left[i\hbar \,\text{div}\, v(\mathbf{x}_\alpha) - mv^2\mathbf{x}_\alpha\right]e^{i\Theta}. \tag{25}$$

On the other hand

$$i\hbar\frac{\partial}{\partial t}(e^{i\Theta}\Phi_F) = i\hbar\frac{\partial\Phi_F}{\partial t}e^{i\Theta} - m\Phi_F e^{i\Theta}\sum_{\alpha=1}^{N}\frac{\partial\varphi(\mathbf{x}_\alpha)}{\partial t}, \tag{26}$$

and writing H as

$$H = -\frac{\hbar^2}{2m}\Delta + V(x), \tag{27}$$

where V is simply a multiplicative operator, Equation 22 is finally easily obtained.

Besides consideration of the identity 22, another observation is in order before proceeding to the variation.

First notice that any differentiable map, i.e., any flow F, can be locally decomposed into a direct product of a rotation a translation and a volume preserving dilation accompanied by a local change of density. Φ_F then can be considered as a "modulated" ground state where the local density is a smooth function over distances of the order of the hard core diameter.

Since Φ has no nodes, for the above reasons Φ_F must have no zeroes and we assume for the modulated ground state—for obvious smoothness and continuity considerations—the following generalization

$$H \Phi_F = \sum_{\alpha=1}^{N} e[\rho(\mathbf{x}_\alpha, t)] \Phi_F \tag{28}$$

of the conventional equation for the ground state energy at constant density ρ

$$H \Phi = \sum_{\alpha=1}^{N} e(\rho) \Phi = N e(\rho) \Phi, \tag{29}$$

where $e(\rho)$ is the chemical potential of the ground state. Inserting Equation 22 and Equation 28 into Equation 19, in which Φ_F is the trial function, the action integral assumes the form

$$A\{\varphi\} = \int dt\, d^3\mathbf{x} \left\{ -m\rho(\mathbf{x}) \frac{\partial \varphi(\mathbf{x})}{\partial t} - \frac{1}{2} m\rho(\mathbf{x}) [\text{grad } \varphi(\mathbf{x})]^2 - \rho(\mathbf{x}) e[\rho(\mathbf{x})] \right\}, \tag{30}$$

where of course, Equations 17 and 21 imply

$$\frac{\partial \rho}{\partial t} = -\text{div}[\rho(\mathbf{x}) \text{ grad } \varphi(\mathbf{x})]. \tag{31}$$

Notice that the action integral 30 is identical with the classical Euler-Bernoulli action.[6]

Indeed, if we perform the variation $\varphi \to \varphi + \delta\varphi$, with the usual suitable boundary condition $\delta\varphi = 0$ on $\partial\Omega$,

$$\delta_\varphi A = \int dt\, d^3\mathbf{x}\, \delta\varphi \cdot \left\{ \frac{\partial \rho}{\partial \varphi} \left[-m \frac{\partial \varphi}{\partial t} - e(\rho) - \rho \frac{de}{d\rho} - \frac{1}{2} m v^2 \right] + \left[m \frac{\partial \rho}{\partial t} + m \text{ div}(\rho \text{ grad } \varphi(\mathbf{x})) \right] \right\}. \tag{32}$$

The last square brackets term on the right hand side of Equation 32 vanishes because of Equation 31 and inserting 32 into 20 one recovers the Bernoulli equation

$$m\dot{\varphi} + \frac{1}{2} m v^2 + \frac{d\rho e(\rho)}{d\rho} = 0, \tag{33}$$

which appears now as a variational approximation to the time-dependent wave equation.

Coherent States

The discussion in the previous chapters can be completed by analyzing velocity potentials of the form

$$\varphi(x,t) = \lambda a^+(x) e^{-i\omega t} + \lambda^* a^-(x) e^{i\omega t}, \tag{34}$$

where λ is a small parameter, $a^+ = (a^-)^*$. We assume that this potential corresponds to a transient wave with absence of vortices. In this case we may form the approximate wave function

$$\Psi(x,t) = \Phi[\xi(x,t)] \sqrt{J} \exp\left[i\frac{m}{\hbar} \sum_{\alpha=1}^{N} \varphi(x_\alpha)\right], \tag{35}$$

which should become a good solution of the wave equation,

$$\left(-\frac{\hbar^2}{2m}\Delta + V\right)\Psi = i\hbar\frac{\partial \Psi}{\partial t}, \tag{36}$$

when ω is small, that is, in the long wavelength limit.

The relation between this solution and the wavefunction of the phonon of frequency ω is quite similar to the one existing between coherent states of the electromagnetic field and the multiphoton states. The result is obtained by expanding Ψ in a series in powers of the parameter λ. We have not been able to carry out this expansion in general; the result up to second order is given by

$$\Psi = \Phi + \frac{i2m\lambda}{\hbar} a^+ e^{-i\omega t} \Phi - \frac{2m^2\lambda^2}{\hbar^2}(a^+)^2 e^{-2i\omega t} \Phi$$

$$- \frac{m^2|\lambda|^2}{\hbar\omega}(H - 2\hbar\omega) a^+ a^- \Phi + O(\lambda^3). \tag{37}$$

But

$$a^+ a^- \Phi = (a^+ a^- - \langle a^+ a^- \rangle)\Phi + \langle a^+ a^- \rangle \Phi,$$

and the first term in our approximation is very close to a 2-phonon state of energy $2\hbar\omega$. The second term yields

$$\Psi \sim \Phi + \frac{2im\lambda}{\hbar} a^+ e^{-i\omega t} \Phi - \frac{2m^2\lambda^2}{\hbar^2}(a^+)^2 e^{-2i\omega t} \Phi$$

$$- \frac{2m^2|\lambda|^2}{\hbar^2} \langle a^+ a^- \rangle \Phi + O(\lambda^3)$$

and our expansion bears a strong similarity to a coherent state. The quantity $|\lambda|^2$ plays here the role of the sound intensity. This intensity is of course related to the average number of phonons and not directly to the phonon occupation number. At higher frequencies the Ansatz becomes progressively inadequate and the anharmonicity of the medium introduces phonon–phonon interactions, which require a more sophisticated approach.

At this stage our approximate wave function provides a unified physical picture of the different phenomena like phonon propagation and vortices but provides no new information on any one of them.

We hope however that the amended Feynman Ansatz may provide some new insight into the behavior of He II and associated nonlinear phenomena, whose treatment through the conventional phonon–phonon interaction often lacks direct physical clarity. Also in Reference 3 we have derived an approximate formula for the Gruneisen constant, having some similarity with the conventional virial theorem.

References

1. LANDAU, L. D. 1941. J. Phys. (U.S.S.R.) **5:** 71.
2. FEYNMAN, R. P. 1954. Phys. Rev. **94:** 262. Progr. Low Temp. Phys. **1:** 17 (1955).
3. FEYNMAN, R. P. 1955. Progr. Low Temp. Phys. **1:** 17.
4. RASETTI, M. & T. REGGE. 1973. J. Low Temp. Phys. **13:** 249.
5. LIN, C. C. 1963. Proc. Intl. School of Physics E. Fermi. XXI Course. :93. Academic Press. New York, N.Y.
6. SCHRÖDINGER, E. 1926. Ann. der Phys. **79:** 361.
7. LAMB, M. 1932. Hydrodynamics. Dover Publications. New York, N.Y.
8. WILKS, J. 1967. The Properties of Liquid and Solid Helium. Clarendon Press. Oxford, England.
9. RASETTI, M. & T. REGGE. 1975. Physica **80A:** 217.

THE MAJORANA FORMULA

Julian Schwinger

*Department of Physics
University of California at Los Angeles
Los Angeles, California 90024*

This retrospective paper is submitted in witness of my appreciation and affection for I. I. Rabi.

A natural consequence of my early association with Rabi and his atomic beam school was a protracted fascination with atomic and nuclear moments and, more generally, the quantum theory of angular momentum. At the heart of the attitude that I adopted toward angular momentum theory was the celebrated theorem and formula of Majorana. These established, qualitatively and quantitatively, how the behavior of an arbitrary magnetic moment in a time varying magnetic field is related to that of a spin $\frac{1}{2}$ system. The original Majorana paper[1] was baffling ["... ogni stato sarà rappresentato da $2j$ punti sulla sfera unitaria...." Where did that come from?] and it was obligatory to find a quantum mechanical derivation of the formula, in particular. My answer was only hinted at in a 1937 paper[2] ["The evaluation of the matrix element (43) may be carried out, for an arbitrary j, by a method which will not be given here. The results are in complete agreement with those obtained from Majorana's general theorem" (by which I meant the formula).] That method was, of course, the explicit construction of an arbitrary angular momentum j as a superposition of $2j$ spin $\frac{1}{2}$ systems. What was thus left implicit (did I then think it of so little consequence?) was actually the most important result in this work, although the technique of transformation to rotating coordinates acquired a certain conceptual utility as later reviewed in a joint paper with Rabi and Ramsey.[3]

The subject remained quiescent until 1945 (the year for renewed attention to basic physics) when Bloch and Rabi[4] remarked on the derivation of Majorana's formula from the spin $\frac{1}{2}$ representation, and I began to write a paper supporting the thesis that the expression of symmetry concepts in quantum mechanics does not require the injection of group theory as an independent mathematical discipline. A major piece of evidence was to be the derivation of Majorana's theorem and formula, thereby finally making explicit the veiled reference of 1937. For some reason this paper was never completed; it is reproduced here in APPENDIX 1 in its final, unfinished form, including the absence of some equation numbers.

In 1951 I wrote a monograph on angular momentum (unpublished), in which the methods of second quantization were used to describe the superposition of spin $\frac{1}{2}$ systems. With this technique all of angular momentum theory, including the important theorems of Racah, were reconstructed and some new results obtained. The paper was later included in an angular momentum collection.[5] Still later, the publication[6] of yet another uncompleted manuscript (1955) disclosed that the operator construction used in this angular momentum representation appears naturally, at a more elementary level than the multiparticle viewpoint of second quantization.

A publication[7] in 1958 by A. Meckler made me aware, to my chagrin, that some aspects of the Majorana formula had not been brought out previously, although they were implicit in the 1951 monograph. I asserted this, along with some further consequences, in a paper that was submitted to *The Physical Review* in the fall of 1959. That paper was rejected by Editor S. Goudsmit for reasons that I then found so incomprehensible that I cannot now recall them. The paper appears here in APPENDIX 2 without alteration.

Finally, let me remark on the expectation value statements near the end of this paper, the first of which is $<J_z>_{m'} = m' \cos \theta$. What I want to emphasize is not their derivation from the quantum theory of angular momentum, but the converse —the intuitive simplicity of these statements, especially the first one, which permits an elementary induction of the general mathematical framework of quantum mechanics, in contrast with the deductive approach, which is so destructive of physical insight. For a short sketch of the argument, consider an atom about which it is known experimentally (Stern-Gerlach) that the magnetic moment, or associated angular momentum, takes only two values along any line; symbolize these by + and −. If atoms are initially selected in a + state, say, and a subsequent measurement is made along another line inclined at the relative angle Θ, the average value found will obviously be the geometrical projection, $+ \cos \Theta$, or in terms of probabilities (which are inescapable given the discreteness of such responses to measurement in the three-dimensional continuum):

$$\cos \Theta = p(+,+) - p(-,+), \quad 1 = p(+,+) + p(-,+),$$

and thus

$$p(+,+) = \cos^2 \tfrac{1}{2} \Theta, \quad p(-,+) = \sin^2 \tfrac{1}{2} \Theta.$$

But more than this appears if Θ is regarded as the angle between two independent directions that are characterized by the angles θ, ϕ and θ', ϕ',

$$\cos \Theta = \cos \theta \cos \theta' + \sin \theta \sin \theta' \cos(\phi - \phi'),$$

namely,

$$p(+,+) = \left| \cos \frac{\theta}{2} \cos \frac{\theta'}{2} + \sin \frac{\theta}{2} e^{-i\phi} \sin \frac{\theta'}{2} e^{i\phi'} \right|^2,$$

$$p(-,+) = \left| -\sin \frac{\theta}{2} e^{i\phi} \cos \frac{\theta'}{2} + \cos \frac{\theta}{2} \sin \frac{\theta'}{2} e^{i\phi'} \right|^2.$$

Here exhibited are the constructions of probabilities as absolute squares of complex number-valued probability amplitudes, and the manner of multiplicative composition of such amplitudes to produce other probability amplitudes. It is a modest beginning (for a complete development in somewhat this spirit see Reference 6) but one that goes directly to the heart of the quantum experience. I would hope that Rabi finds such an attitude more congenial than the conventional mathematical dogma, about which he evinced some skepticism (to which, at the time, I turned an uncomprehending ear—but perhaps it had its effect after all?).

References

1. MAJORANA, E. 1932. Atomi orientati in campo magnetico variabile. Nuovo Cimento **9**: 43–50.
2. SCHWINGER, J. 1937. Phys. Rev. **51**: 648.
3. RABI, I. I., N. F. RAMSEY & J. SCHWINGER. 1954. Rev. Mod. Phys. **26**: 167.
4. BLOCH, F. & I. I. RABI. 1945. Rev. Mod. Phys. **17**: 237.
5. BIEDENHORN, L. C. & H. VAN DAM, eds. 1965. Quantum Theory of Angular Momentum. Academic Press. New York, N.Y.
6. Quantum Kinematics and Dynamics. 1970. :242–245. W. A. Benjamin. New York, N.Y.
7. MECKLER, A. 1958. Phys. Rev. **111**: 1447.

APPENDIX 1

A NOTE ON GROUP THEORY AND QUANTUM MECHANICS*

Introduction

Symmetry concepts are of great importance in quantum mechanics. For this reason, it is often claimed that the mathematical discipline concerned with the analysis of symmetry—group theory—must be employed in order to deal satisfactorily with the symmetry problems of quantum mechanics. Opposed to this sentiment, which is advanced by most of those possessed with a knowledge of the subject, is the feeling, shared by the vast majority of physicists, that group theory is an unnecessary addition to the already heavy mathematical burden of the working theoretical physicist. In support of the latter group, it may be argued that the mathematical methods of group theory are too general for the purposes of quantum mechanics, that quantum mechanics is adequate to describe, within its own framework, those symmetry operations that arise from physical considerations.

It is the purpose of this note to bolster the last contention by deriving, with the aid of elementary quantal operator methods, a number of results that have thus far been considered striking examples of the power of group theoretic methods applied to quantum mechanics. These are: Majorana's theorem; the matrices representing the finite rotation operators (in group theoretic language, the irreducible representations of the rotation group); the unitary transformation describing the composition of angular momenta; and the selection rules and magnetic quantum number dependence of tensor operator matrices.

All of these questions relate to the properties of angular momenta, and our methods will require little more than the elementary consequences of the communtation relation

$$\mathbf{J} \times \mathbf{J} = i\mathbf{J} \quad (1)$$

satisfied by the angular momentum vector $\mathbf{J}$ (in units of $\hbar$). In particular we shall

---

*This material was originally started as a paper in 1945, but was never completed. It is presented here in its unfinished form.

constantly employ the operators

$$\mathbf{J}_+ = \mathbf{J}_x + i\mathbf{J}_y, \mathbf{J}_- = \mathbf{J}_x - i\mathbf{J}_y \tag{2}$$

applied to the eigenvectors $\Psi(j,m)$, which are defined by

$$\begin{aligned}J_z\Psi(j,m) &= m\Psi(j,m) \qquad m = j, j-1, \ldots, -j \\ J^2\Psi(j,m) &= j(j+1)\Psi(j,m) \quad j = 0, \tfrac{1}{2}, 1, \ldots.\end{aligned} \tag{3}$$

It is well known that

$$\begin{aligned}\mathbf{J}_+\Psi(j,m) &= \sqrt{(j-m)(j+m+1)}\,\Psi(j,m+1), \\ \mathbf{J}_-\Psi(j,m) &= \sqrt{(j+m)(j-m+1)}\,\Psi(j,m-1).\end{aligned} \tag{4}$$

By repeated application of these operators, one obtains

$$\begin{aligned}\mathbf{J}_+^{m-m'}\sqrt{\frac{(j+m')!}{(j-m')!}}\Psi(j,m') &= \sqrt{\frac{(j+m)!}{(j-m)!}}\Psi(j,m), \\ \mathbf{J}_-^{m'-m}\sqrt{\frac{(j-m')!}{(j+m')!}}\Psi(j,m') &= \sqrt{\frac{(j-m)!}{(j+m)!}}\Psi(j,m),\end{aligned} \tag{5}$$

of which an important specialization is

$$\begin{aligned}\Psi(j,m) &= \sqrt{\frac{1}{(2j)!}\frac{(j-m)!}{(j+m)!}}\,\mathbf{J}_+^{j+m}\Psi(j,-j), \\ \Psi(j,m) &= \sqrt{\frac{1}{(2j)!}\frac{(j+m)!}{(j-m)!}}\,\mathbf{J}_-^{j-m}\Psi(j,j).\end{aligned} \tag{6}$$

Majorana's Theorem

The physical problem to which this theorem applies is that of a quantum mechanical system, possessing a magnetic moment, in a time varying magnetic field. It was shown by Majorana, and independently by Güttinger, that the change in state of a system with total angular momentum j could be obtained from that of the simpler system $j = \tfrac{1}{2}$.

To construct a quantum mechanical derivation of this result, we first note that an angular momentum vector $\mathbf{J}$, associated with the angular momentum quantum number, or spin, j, can be regarded as the superposition of $2j$ individual angular momenta with spin $\tfrac{1}{2}$, i.e.,

$$\mathbf{J} = \sum_{k=1}^{2j} \mathbf{J}_k = \frac{1}{2}\sum_{k=1}^{2j} \sigma_k, \tag{7}$$

on employing the Pauli spin vector σ_k to represent the component angular momenta of spin $\tfrac{1}{2}$. To verify this statement, we observe that the commutation relation for **J** is fulfilled in consequence of that for $\mathbf{J}_k$,

$$\mathbf{J} \times \mathbf{J} = \sum_{k,l} \tfrac{1}{2}(\mathbf{J}_k \times \mathbf{J}_l + \mathbf{J}_l \times \mathbf{J}_k) = \sum_k \mathbf{J}_k \times \mathbf{J}_k = i\sum_k \mathbf{J}_k = i\mathbf{J}$$

while the magnetic quantum number m, the eigenvalue of $\mathbf{J}_z$,

$$m = \sum_{k=1}^{2j} m_k$$

has as its maximum value, j, since the possible values of m_k are $\pm \tfrac{1}{2}$. The eigenvector $\Psi(j,j)$, associated with $m = j$, is easily constructed in terms of the eigenvectors of the component spins, for $m = j$ can be achieved only with $m_1 = m_2 = \ldots m_{2j} = \tfrac{1}{2}$. Hence,

$$\Psi(j,j) = \Phi(\tfrac{1}{2}, \ldots, \tfrac{1}{2}), \tag{8}$$

where, generally, $\Phi(m_1, \ldots, m_{2j})$ denotes the simultaneous eigenvector of the operators J_{kz}, associated with the indicated eigenvalues. In a similar way,

$$\Psi(j,-j) = \Phi(-\tfrac{1}{2}, \ldots, -\tfrac{1}{2}). \tag{9}$$

The entire set of $2j + 1$ eigenvectors $\Psi(j,m)$ can then be obtained from either of the operator Equations 6. Of course, the system of $2j$ independent angular momenta will have states with spin values that are smaller than j, but all such states are excluded by our method of construction, since the state $m = j$ can not be realized with a spin less than j.

The Schrödinger equation for a system with a magnetic moment $\mu = g\mu_0 \mathbf{J}$, in a magnetic field $\mathbf{H}(t)$, is

$$i\hbar \frac{\partial}{\partial t} \Psi(t) = -g\mu_0 \mathbf{J} \cdot \mathbf{H}(t)\Psi(t), \tag{10}$$

where μ_0 is the magneton appropriate to the system, and g is the gyromagnetic ratio. The eigenvector representing the state of the system at time t can be regarded as the result of a unitary operator applied to the eigenvector describing the initial state of the system:

$$\Psi(t) = U(t)\Psi(0). \tag{11}$$

The Schrödinger equation may then be replaced by an equation of motion for the operator $U(t)$, namely,

$$i\hbar \frac{\partial}{\partial t} U(t) = -g\mu_0 \mathbf{J} \cdot \mathbf{H}(t)U(t) = -\sum_{k=1}^{2j} g\mu_0 \mathbf{J}_k \cdot \mathbf{H}(t)U(t). \tag{12}$$

In the latter form, it is evident that the component spins are dynamically independent, which is the essential basis of Majorana's theorem. To exhibit the dynamical independence of the component angular momenta explicitly, we write

$$U(t) = u_1(t)u_2(t) \ldots u_{2j}(t), \tag{13}$$

where the unitary operator $u_k(t)$ obeys the equation of motion

$$i\hbar \frac{\partial}{\partial t} u_k(t) = -g\mu_0 \cdot \mathbf{J}_k \cdot \mathbf{H}(t) u_k(t) \quad k = 1, 2, \ldots, 2j. \tag{14}$$

and commutes with the angular momentum operators of all other component spins. The original problem has thus been reduced to that of a system with spin $\frac{1}{2}$ and gyromagnetic ratio g, in the magnetic field $\mathbf{H}(t)$.

The quantity of physical interest is the probability that the system shall be found with magnetic quantum number m at the time t, if initially it was prepared in the state with quantum number m'. The desired probability, $W(m, m'; t)$, is obtainable from the matrix elements of the operator $U(t)$, for, if $\Psi(0) = \Psi(j, m')$, it follows from Equation 11 that

$$\Psi(t) = U(t) \Psi(j, m') = \sum_{m=-j}^{j} \Psi(j, m) U_{m,m'}(t), \tag{15}$$

whence

$$W(m, m'; t) = |U_{m,m'}(t)|^2, \tag{16}$$

according to the probability interpretation of quantum mechanics. Our problem is the explicit calculation of the reorientation probability $W(m, m'; t)$, in terms of the corresponding probability $w(m, m'; t)$ for the simple system with spin $\frac{1}{2}$.

To evaluate the matrix element $U_{m,m'}(t)$, we employ the scalar product representation

$$U_{m,m'}(t) = (\Psi(j, m), U(t) \Psi(j, m')). \tag{17}$$

The eigenvectors $\Psi(j, m)$ and $\Psi(j, m')$, according to Equations 6 and 8, can be conveniently written in the form

$$\Psi(j, m) = \sqrt{\frac{1}{(2j)!} \frac{(j+m)!}{(j-m)!}} \left(\frac{d}{d\lambda}\right)^{j-m} e^{\lambda \mathbf{J}_-} \Phi(\tfrac{1}{2}, \ldots, \tfrac{1}{2}),$$

$$\Psi(j, m') = \sqrt{\frac{1}{(2j)!} \frac{(j+m')!}{(j-m')!}} \left(\frac{d}{d\mu}\right)^{j-m'} e^{\mu \mathbf{J}_-} \Phi(\tfrac{1}{2}, \ldots, \tfrac{1}{2}), \tag{18}$$

in which it is understood that λ and μ are to be placed equal to zero after performing the indicated differentiations. The utility of this representation follows from the elementary property

$$e^{\lambda \mathbf{J}_-} = e^{\lambda \mathbf{J}_{1-}} e^{\lambda \mathbf{J}_{2-}} \cdots e^{\lambda \mathbf{J}_{2j-}} = \prod_{k=1}^{2j} e^{\lambda \sigma_{k-}/2}$$

which, combined with Equation 13, the analogous factorization of $U(t)$, enables us to write

$$U_{m,m'}(t) = \frac{1}{(2j)!} \sqrt{\frac{(j+m)!}{(j-m)!} \frac{(j+m')!}{(j-m')!}} \left(\frac{d}{d\lambda}\right)^{j-m} \left(\frac{d}{d\mu}\right)^{j-m'}$$

$$\left(\Phi(\tfrac{1}{2}, \ldots, \tfrac{1}{2}), \prod_{k=1}^{2j} e^{1/2\lambda \sigma_{k+}} u_k(t) e^{\mu \sigma_{k-}/2} \Phi(\tfrac{1}{2}, \ldots, \tfrac{1}{2})\right) \tag{19}$$

In deriving this result, the Hermitian property of the scalar product has been used, as expressed by

$$(A\Psi, \Psi) = (\Psi, A\dagger\Psi), \qquad (20)$$

where $A\dagger$ is the operator adjoint to A. In particular

$$\mathbf{J}_-\dagger = \mathbf{J}_+, \mathbf{J}_+\dagger = \mathbf{J}_-. \qquad (21)$$

Since the operator contained in Equation 19 is a product of operators referring to independent systems, the matrix element is a corresponding product of matrix elements for the individual systems. Further, all of the latter quantities are equal, for the $2j$ spin systems are dynamically equivalent. Thus

$$\left(\Phi(\tfrac{1}{2},\ldots,\tfrac{1}{2}), \prod_{k=1}^{2j} e^{\lambda\sigma_k+/2} u_k(t) e^{\mu\sigma_k-/2} \Phi(\tfrac{1}{2},\ldots,\tfrac{1}{2})\right)$$

$$= [(\Phi(\tfrac{1}{2}), e^{\lambda\sigma_+/2} u(t) e^{\mu\sigma_-/2} \Phi(\tfrac{1}{2}))]^{2j}$$

$$= [(e^{\lambda\sigma_-/2}\Phi(\tfrac{1}{2}), u(t) e^{\mu\sigma_-/2}\Phi(\tfrac{1}{2}))]^{2j}. \qquad (22)$$

Now, the Pauli spin operator $\tfrac{1}{2}\sigma_-$ has the properties

$$\tfrac{1}{2}\sigma_-\Phi(\tfrac{1}{2}) = \Phi(-\tfrac{1}{2}), \quad \tfrac{1}{2}\sigma_-\Phi(-\tfrac{1}{2}) = 0,$$

according to Equation 4. Therefore

$$e^{\lambda\sigma_-/2}\Phi(\tfrac{1}{2}) = \left[1 + \lambda(\tfrac{1}{2}\sigma_-) + \frac{\lambda^2}{2!}(\tfrac{1}{2}\sigma_-)^2 + \cdots\right]\Phi(\tfrac{1}{2})$$

$$= \Phi(\tfrac{1}{2}) + \lambda\Phi(-\tfrac{1}{2}),$$

and

$$(e^{\lambda\sigma_-/2}\Phi(\tfrac{1}{2}), u(t) e^{\mu\sigma_-/2}\Phi(\tfrac{1}{2})) = ([\Phi(\tfrac{1}{2}) + \lambda\Phi(-\tfrac{1}{2})], u(t)[\Phi(\tfrac{1}{2}) + \mu\Phi(-\tfrac{1}{2})])$$

$$= u_{1/2,1/2}(t) + \lambda u_{-1/2,1/2}(t) + \mu u_{1/2,-1/2}(t) + \lambda\mu u_{-1/2,-1/2}(t).$$

Hence

$$U_{m,m'}(t) = \frac{1}{(2j)!}\sqrt{\frac{(j+m)!\,(j+m')!}{(j-m)!\,(j-m')!}}\left(\frac{d}{d\lambda}\right)^{j-m}\left(\frac{d}{d\mu}\right)^{j-m'}$$

$$\times [u_{1/2,1/2}(t) + \lambda u_{-1/2,1/2}(t) + \mu u_{1/2,-1/2}(t) + \lambda\mu u_{-1/2,-1/2}(t)]^{2j}$$

which, in symbolic form, is the solution to our problem, for the probability amplitudes $U_{m,m'}(t)$ of the system with arbitrary spin j are hereby constructed from those of the system with spin $\tfrac{1}{2}$.

The further simplification of this result is facilitated by the use of a more specific form for the matrix elements of the operator u. We shall first demonstrate that the determinant of the unitary matrix $U(t)$ is unity,

$$\det U(t) = 1. \qquad (23)$$

This statement is true for any j and therefore applies to the operator $u(t)$, describing a system of spin $\frac{1}{2}$. The theorem will be proved under the following assumptions, that $U(t)$ obeys an equation of motion of the form

$$\frac{\partial}{\partial t} U(t) = K(t) U(t),$$

that the diagonal sum (spur) of the operator K is zero,

$$\text{sp } K = 0,$$

and that, at $t = 0$, $U(t)$, reduces to the unit operator

$$U(0) = 1.$$

In the problem under consideration, the operator K is

$$K(t) = \frac{i}{\hbar} g\mu_0 \mathbf{J} \cdot \mathbf{H}(t),$$

which indeed has a vanishing spur, since the diagonal sum of any angular momentum matrix necessarily vanishes, in consequence of the commutation relations 1. That $U(0)$ is the unit operator is an immediate result of Equation 11.

The proof of the theorem involves the following lemma, which is established in Appendix 1,†

$$\frac{\partial}{\partial t} \det U = (\det U)\left(\text{sp } \frac{\partial}{\partial t} U \cdot U^{-1}\right).$$

Now

$$\text{sp } \frac{\partial}{\partial t} U \cdot U^{-1} = \text{sp } K = 0,$$

whence

$$\frac{\partial}{\partial t} \det U = 0.$$

Therefore $\det U$ is independent of t and is equal to its value at $t = 0$, which, by Equation 11, is unity, thus completing the proof of Equation 23.

It is easily shown that the restrictions

$$u\dagger u = 1, \quad \det u = 1,$$

imposed on the elements of the matrix u, lead to the following relations:

$$u_{-1/2, 1/2} = -u^*_{1/2, -1/2}, u_{-1/2, -1/2} = u^*_{1/2, 1/2}, \quad |u_{1/2, 1/2}|^2 + |u_{1/2, -1/2}|^2 = 1.$$

In accordance with the last restriction, we may write, in all generality,

$$u_{1/2, 1/2} = \cos \tfrac{1}{2} \alpha \, e^{i(\beta + \gamma)/2}, u_{1/2, -1/2} = i \sin \tfrac{1}{2} \alpha \, e^{i(\beta - \gamma)/2}$$

with α, β, and γ designating arbitrary angles. Thus the general form of the matrix

†The Appendix referred to here was not completed.

u is

$$u = \begin{pmatrix} \cos \tfrac{1}{2}\alpha \, e^{i(\beta+\gamma)/2}, & i \sin \tfrac{1}{2}\alpha \, e^{i(\beta-\gamma)/2} \\ i \sin \tfrac{1}{2}\alpha \, e^{-i(\beta-\gamma)/2}, & \cos \tfrac{1}{2}\alpha \, e^{-i(\beta+\gamma)/2} \end{pmatrix}.$$

It may be noted that the reorientation probabilities $w(m, m'; t)$ depend only on the angle α

$$w(\tfrac{1}{2},\tfrac{1}{2};t) = w(-\tfrac{1}{2},-\tfrac{1}{2};t) = \cos^2 \tfrac{1}{2}\alpha,$$
$$w(-\tfrac{1}{2},\tfrac{1}{2};t) = w(\tfrac{1}{2},-\tfrac{1}{2};t) = \sin^2 \tfrac{1}{2}\alpha.$$

The result obtained by substituting the matrix elements of u into Equation 19 may be simplified by changing the variables λ and μ in accordance with

$$\lambda \to -i \cot \tfrac{\alpha}{2} e^{i\beta} \lambda, \quad \mu \to -i \cot \tfrac{\alpha}{2} e^{i\gamma} \mu,$$

whence

$$U_{m,m'} = \frac{1}{(2j)!} \sqrt{\frac{(j+m)!\,(j+m')!}{(j-m)!\,(j-m')!}} \, i^{2j-m-m'} \sin^{2j}\tfrac{\alpha}{2} \cot^{m+m'}\tfrac{\alpha}{2} \, e^{i(m\beta+m'\gamma)}$$
$$\times \left(\frac{d}{d\lambda}\right)^{j-m} \left(\frac{d}{d\mu}\right)^{j-m'} \left[1 + \lambda + \mu - \lambda\mu \cot^2\tfrac{\alpha}{2}\right]^{2j}$$

The differentiation with respect to either λ or μ is now easily performed, giving the two alternative forms:

$$U_{m,m'} = \sqrt{\frac{(j+m)!}{(j-m)!} \frac{1}{(j+m')!(j-m')!}} \, i^{2j-m-m'} \sin^{2j}\tfrac{\alpha}{2} \cot^{m+m'}\tfrac{\alpha}{2} \, e^{i(m\beta+m'\gamma)}$$
$$\times \left(\frac{d}{d\lambda}\right)^{j-m} (1+\lambda)^{j+m'}\left(1 - \lambda \cot^2\tfrac{\alpha}{2}\right)^{j-m'},$$

$$U_{m,m'} = \sqrt{\frac{(j+m')!}{(j-m')!} \frac{1}{(j+m)!(j-m)!}} \, i^{2j-m-m'} \sin^{2j}\tfrac{\alpha}{2} \cot^{m+m'}\tfrac{\alpha}{2} \, e^{i(m\beta+m'\gamma)}$$
$$\times \left(\frac{d}{d\mu}\right)^{j-m'} (1+\mu)^{j+m}\left(1 - \mu \cot^2\tfrac{\alpha}{2}\right)^{j-m}$$

The final differentiations are equivalent to the problem of picking out the coefficient of t^c in $(1+t)^a (1-xt)^b$, namely,

$$\sum_r (-1)^r \frac{a!\,b!}{(b-r)!(c-r)!(a-c+r)!} \frac{x^r}{r!},$$

where $x = \cot^2(\alpha/2)$, and either $a = j+m$, $b = j-m$, $c = j-m'$, or $a = j+m'$, $b = j-m'$, $c = j-m$. The end result, obtained from either form, is

$$U_{m,m'} = \sqrt{(j+m)!(j-m)!(j+m')!(j-m')!} \; i^{2j-m\times m'} \sin^{2j}\frac{\alpha}{2} \cot^{m+m'}\frac{\alpha}{2}$$

$$\times \; e^{i(m\beta+m'\gamma)} \sum_r (-1)^r$$

$$\times \; \frac{1}{(j-m-r)!(j-m'-r)!(m+m'+r)!r!} \cot^{2r}\frac{\alpha}{2} \quad (25)$$

in which the summation is to be extended over all non-negative integers that lead to meaningful factorials. The transition probability, $W(m,m';t)$, is then

$$W(m,m';t) = (j+m)!(j-m)!(j+m')!(j-m')! \; \sin^{4j}\frac{\alpha}{2} \cot^{2(m+m')}\frac{\alpha}{2}$$

$$\times \left[\sum_r (-1)^r \frac{1}{(j-m-r)!(j-m'-r)!(m+m'+r)!r!} \cot^{2r}\frac{\alpha}{2} \right]^2 \quad (26)$$

where the angle α is to be related to the solution of the problem with spin $\frac{1}{2}$ by means of Equation 16. This is the formula obtained by Majorana.

A few simple properties of the operator U may be noted here. If $U_{m,m'}$ is regarded as a function of the three angles α, β, and γ, it is seen from Equation 25 that

$$U_{m,m'}(\alpha,\beta,\gamma) = U_{m',m}(\alpha,\gamma,\beta) \quad (25')$$

whence

$$W(m,m';t) = W(m',m;t)$$

since the transition probability is independent of β and γ. Further, on placing $m + m' + r = s$ in Equation 25, one finds

$$U_{m,m'} = \sqrt{(j+m)!(j-m)!(j+m')!(j-m')!}$$

$$\times i^{2j+m+m'} \sin^{2j}\frac{\alpha}{2} \cot^{-(m+m')}\frac{\alpha}{2} \; e^{i(m\beta+m'\gamma)} \sum_s (-1)^s$$

$$\times \frac{1}{(j+m-s)!(j+m'-s)!(-m-m'+s)!s!} \cot^{2s}\frac{\alpha}{2},$$

which shows that

$$U_{m,m'}(\alpha,\beta,\gamma) = U_{-m,-m'}(\alpha,-\beta,-\gamma),$$

and

$$W(m,m';t) = W(-m,-m';t).$$

As a last property, we assert that

$$U_{m,m'}(\alpha,\beta,\gamma) = (-1)^j U_{m,-m'}(\pi-\alpha, \pi+\beta, \pi+\gamma), \quad (27)$$

which can be proved by placing $r = j - m - s$ in Equation 25. This gives

$$U_{m,m'} = \sqrt{(j + m)!(j - m)!(j - m')!(j + m')!}$$

$$\times i^{2j-m+m'} \cos^{2j}\frac{\alpha}{2} \tan^{m-m'}\frac{\alpha}{2} e^{i(m\beta + m'\gamma)}(-1)^{j-m+m'} \sum_s (-1)^s$$

$$\times \frac{1}{(j - m - s)!(j + m' - s)!(m - m' + s)!s!} \tan^{2s}\frac{\alpha}{2},$$

which is equivalent to Equation 27. The result states that $W(m, -m'; t)$ is the same function of $w(\frac{1}{2}, -\frac{1}{2}; t)$ as $W(m, m'; t)$ is of $w(\frac{1}{2}, \frac{1}{2}; t)$. In view of these relations, it is sufficient to know $U_{m,m'}$ for nonnegative values of m and m'.

The series that occurs in Equation 25 may be recognized as that of the hypergeometric function. If a and b are nonnegative intergers, while c is a positive integer,

$$F(-a, -b; c; -x) = a!b!(c - 1)! \sum_r (-1)^r \frac{1}{(a - r)!(b - r)!(c + r - 1)!} \frac{x^r}{r!},$$

whence

$$U_{m,m'} = \frac{1}{(m + m)!} \sqrt{\frac{(j + m)!\,(j + m')!}{(j - m)!\,(j - m')!}} i^{2j-m-m'} \sin^{2j}\frac{\alpha}{2} \cot^{m+m'}\frac{\alpha}{2} e^{i(m\beta + m'\gamma)}$$

$$\times F\left(m - j, m' - j; m + m' + 1; -\cot^2\frac{\alpha}{2}\right),$$

provided $m + m' \geq 0$. Having thus established contact with the well-developed theory of the hypergeometric function, we may avail ourselves of the various known transformations to cast Equation 25 into different forms. For example, the transformations

$$F(-a, -b; c; -x) = (1 + x)^a F\left(-a, b + c; c; \frac{x}{1 + x}\right)$$

$$= (1 + x)^b F\left(-b, a + c; c; \frac{x}{1 + x}\right),$$

enable us to rewrite Equation 25 in the two equivalent forms:

$$U_{m,m'} = \frac{1}{(m + m')!} \sqrt{\frac{(j + m)!\,(j + m')!}{(j - m)!\,(j - m')!}}$$

$$\cdot i^{2j-m-m'} \sin^{m-m'}\frac{\alpha}{2} \cos^{m+m'}\frac{\alpha}{2} e^{i(m\beta + m'\gamma)}$$

$$\times F\left(m - j, j + m + 1; m + m' + 1; \cos^2\frac{\alpha}{2}\right),$$

$$U_{m,m'} = \frac{1}{(m+m')!} \sqrt{\frac{(j+m)!\,(j+m')!}{(j-m)!\,(j-m')!}}$$

$$\cdot i^{2j-m-m'} \sin^{m'-m}\frac{\alpha}{2} \cos^{m+m'}\frac{\alpha}{2} e^{i(m\beta+m'\gamma)}$$

$$\times F\left(m'-j, j+m'+1; m+m'+1; \cos^2\frac{\alpha}{2}\right), \quad (28)$$

which are related to each other by Equation 25'.

The hypergeometric function that occurs in Equation 28 is a polynomial in $\cos^2\alpha/2$ of degree $j - m$, and may be recognized as the Jacobi polynomial,

$$J_n(a,b;x) = F(-n, a+n; b; x)$$

$$= \frac{(b-1)!}{(b+n-1)!} x^{1-b}(1-x)^{b-a}\left(\frac{d}{dx}\right)^n x^{b-1+n}(1-x)^{a-b+n},$$

with

$$n = j - m, a = 2m + 1, b = m + m' + 1, \text{ and } x = \cos^2\frac{\alpha}{2}.$$

Hence,

$$U_{m,m'} = \sqrt{\frac{(j+m)!}{(j-m)!}} \frac{1}{(j+m')!(j-m')!} i^{2j-m-m'} \sin^{m'-m}\frac{\alpha}{2} \cos^{-(m+m')}\frac{\alpha}{2}$$

$$\times e^{i(m\beta+m'\gamma)} \left[\left(\frac{d}{dx}\right)^{j-m} x^{j+m'}(1-x)^{j-m'}\right]_{x=\cos^2(\alpha/2)}.$$

It is interesting that this form of the result is obtained immediately from Equation 24 on placing

$$\lambda = -1 + \frac{x}{\cos^2(\alpha/2)}.$$

APPENDIX 2

The Majorana Formula‡

The relation of a recent version of Majorana's formula to known angular momentum theorems is pointed out.

The celebrated formula of Majorana characterizes the dynamical properties of the magnetic moment associated with an angular momentum vector, in a time-dependent magnetic field. For an arbitrary time variation the net effect of the interaction is to produce a rotation of the angular momentum vector through a

‡This material was submitted as a paper to *The Physical Review* in 1959 but was not published.

definite angle, and the dynamical problem is fully equivalent to the kinematical one of relating angular momentum states associated with different directions in space. The latter problem has naturally received much attention and a variety of equivalent forms are known for the matrix elements of the unitary rotation operator,

$$\langle jm | U(\omega) | jm' \rangle = U_{mm'}^{(j)}(\varphi \vartheta \psi),$$
$$= e^{-im\varphi} U_{mm'}^{(j)}(\vartheta) e^{-im'\psi},$$

from which the required probabilities are obtained:

$$p(m, m'; \vartheta) = [U_{mm'}^{(j)}(\vartheta)]^2.$$

In a recent note§ an interesting formula is stated for this probability as a Legendre polynomial expansion. We wish to point out that this and other results in that note are immediate consequences of known theorems in the theory of angular momentum. Let us begin with the general multiplication property of the rotation matrices, expressive of the composition of angular momenta,

$$U_{m_1 m_1'}^{(j_1)}(\omega) U_{m_2 m_2'}^{(j_2)}(\omega)$$
$$= \sum_{jmm'} (2j + 1) X(j_1 j_2 j, m_1 m_2 m) U_{mm'}^{(j)}(\omega)^* X(j_1 j_2 j, m_1' m_2' m'),$$

and, after equating j_1 and j_2, pick out the products that are independent of the Eulerian angles φ and ψ. With the aid of the relations

$$U_{-m-m'}^{(j)}(\vartheta) = (-1)^{m-m'} U_{mm'}^{(j)}(\vartheta), \ U_{00}^{(l)}(\vartheta) = P_l(\cos \vartheta),$$

we obtain immediately the desired form,

$$[U_{mm'}^{(j)}(\vartheta)]^2$$
$$= \sum_{l=0}^{2j} (2l + 1)(-1)^{j-m} X(jjl, m-m0) P_l(\cos \vartheta)(-1)^{j-m'} X(jjl, m' - m'0).$$

Some symmetry properties implied by those of the X coefficients are

$$p(m, m'; \vartheta) = p(m', m; \vartheta) = p(-m, -m'; \vartheta) = p(-m, m'; \pi - \vartheta).$$

Another aspect of this formula emerges in terms of the solid harmonic functions of the angular momentum vector **J** that are defined by the generating function

---

§Not unnaturally, the present author is particularly aware of his own contribution to the subject, prepared in late 1951 as a report to Nuclear Development Associates, White Plains, N. Y. All results cited appear in this report. The relation of the notation to that of other writers in the field is described by A. R. Edmonds (1957. Angular Momentum in Quantum Mechanics. Princeton, N.J.). It should also be mentioned that an identity among Racah coefficients, which is usually ascribed to Biedenharn (1953. J. Math. Phys. **31**: 287.), is contained quite explicitly in this report as Equation 5.36.

$$\frac{1}{2^l l!} \left(\frac{2l+1}{4\pi} \right)^{1/2} [-z_+^2 (J_x + iJ_y) + z_-^2 (J_x - iJ_y) + 2z_+ z_- J_z]^l$$

$$= \sum_{m=-l}^{l} \frac{z_+^{l+m} z_-^{l-m}}{[(l+m)!(l-m)!]^{1/2}} Y_{lm}(\mathbf{J}).$$

It has been shown that

$$< jm \mid Y_{l0}(\mathbf{J}) \mid jm >$$

$$= \left[\frac{(2l+1)(2j+1)}{4\pi} \{j(j+1)\}^l \right]^{1/2} (-1)^{j-m} X(jjl, m-m0),$$

where

$$\{\mathbf{J}^2\}^l = \prod_{n=0}^{l-1} \left[\mathbf{J}^2 - \frac{n}{2}\left(\frac{n}{2} + 1\right) \right].$$

Accordingly, if we introduce the Legendre polynomial operator $P_l(\mathbf{J})$ as the spherical harmonic function

$$P_l(\mathbf{J}) = \left[\frac{2l+1}{4\pi} \{\mathbf{J}^2\}^l \right]^{-1/2} Y_{l0}(\mathbf{J}),$$

the probability formula reads

$$p(m, m'; \vartheta) = \frac{1}{2j+1} \sum_{l=0}^{2j} (2l+1) P_l(j, m) P_l(\cos \vartheta) P_l(j, m'),$$

in which

$$P_l(j, m) = < jm \mid P_l(\mathbf{J}) \mid jm >.$$

The first few Legendre operators are

$$P_0(\mathbf{J}) = 1$$

$$P_1(\mathbf{J}) = [\mathbf{J}^2]^{-1/2} J_z$$

$$P_2(\mathbf{J}) = [\mathbf{J}^2(\mathbf{J}^2 - \tfrac{3}{4})]^{-1/2} \tfrac{1}{2}(3J_z^2 - \mathbf{J}^2).$$

The observation¶ that the matrix elements $P_l(j, m)$ are related to the Tchebichef polynomials of a discrete variable,**

$$2^l [\{j(j+1)\}^l]^{1/2} P_l(j, m) = t_l(j+m),$$

is in accordance with the asymptotic connection of these polynomials with Legendre functions. The detailed equivalence with the Tchebichef polynomials is

---

¶MECKLER, A. 1958. Phys. Rev. **111**: 1447.
**We use the notation of Higher Transcendental Functions (1953. Vol. II. Bateman Manuscript Project. McGraw-Hill. New York, N.Y.).

required by the orthogonality properties of the Legendre operators, as contained in

$$\delta_{ll'} = \frac{2l+1}{2j+1} \text{tr}^{(j)} P_l(\mathbf{J}) P_{l'}(\mathbf{J})$$

$$= \frac{2l+1}{2j+1} \sum_{m=-j}^{j} P_l(j,m) P_{l'}(j,m),$$

and

$$\sum_{l=0}^{2j} \frac{2l+1}{2j+1} P_l(j,m) P_l(j,m') = \delta_{mm'}.$$

As an application of the first orthogonality property, note that the expectation value of the function of m given by $P_l(j,m)$ is

$$<P_l(\mathbf{J})>_{m'} = \sum_{m=-j}^{j} P_l(j,m) p(m,m';\vartheta)$$

$$= P_l(\cos \vartheta) P_l(j,m').$$

The first two examples of this relation are

$$<J_z>_{m'} = m' \cos \vartheta$$
$$<3J_z^2 - j(j+1)>_{m'} = (3m'^2 - j(j+1))\tfrac{1}{2}(3\cos^2\vartheta - 1).$$

Since the second orthogonality property enables us to recover the probability formula, the complete set of $2j$ expectation values is fully equivalent to the knowledge of the individual probabilities. A direct derivation of the expectation value form is obtained from the rotational transformation properties of the operator harmonic functions,

$$U^{-1} Y_{lm}(\mathbf{J}) U = \sum_{m'} U_{mm'}^{(l)}(\omega)^* Y_{lm'}(\mathbf{J}),$$

for

$$<P_l(\mathbf{J})>_{m'} = <jm'|U^{-1}P_l(\mathbf{J})U|jm'>$$
$$= P_l(\cos\vartheta) P_l(j,m').$$

THE PROBLEM OF MASS

Steven Weinberg*

Lyman Laboratory of Physics
Harvard University
Cambridge, Massachusetts 02138

I recently heard Rabi describe the early calculations of atomic spectra, which were carried out before the development of a systematic quantum mechanics in 1925-6, as being a blend of "artistry and effrontery." Since those days, we have of course learned a great deal, and many things that were previously obscure now seem perfectly simple, while new obscurities have descended upon us. Over the last decade, we have seen a satisfying synthesis of the theories of weak, electromagnetic, and strong interactions, which has provided explanations for many of the ad hoc hypotheses that had been previously introduced into particle physics on chiefly empirical grounds. However, one essential element of this systematic theory has remained obscure: we must take the masses of the leptons and quarks as input parameters, without any real idea of where they came from.

In this article I would like to offer some observations on the determination of the quark masses, and the problem of their origin. The discussion of the determination of quark masses in the first three sections goes over familiar ground, though I believe that some of this material is new. As to the discussion in the last section on the origin of the masses, I can only hope in the spirit of Rabi's remark, that whatever these speculations lack in artistry, they will make up in effrontery.

What Are the Quark Mass Ratios?

Very likely, no one will ever see a free quark. It is therefore necessary to take some care in saying precisely what we mean by a quark mass. For the purposes of this article, the quark masses are to be understood as the mass parameters appearing in the effective quantum field theory of strong interactions, known as quantum chromodynamics.[1-3] The relation between such masses and the quark masses used in phenomenological "atomic" models of physical hadrons is somewhat indirect, and will be taken up briefly in the next section.

Quantum chromodynamics is governed by the two principles of renormalizability and "color" guage invariance. The color guage group is taken as SU(3): quarks of each "flavor" (u, d, s, c, etc.), come in three colors, say R, W, B. It is also assumed that the elementary hadronic fields consist solely of the quark fields plus the eight "gluon" gauge fields of color SU(3), but no strongly interacting scalars. Under these assumptions, the Lagrangian of the strong interactions

---

*On leave in 1976-77 at the Physics Department, Stanford University, Stanford, Calif. 94305.

necessarily takes the extremely simple form

$$\mathcal{L} = \overline{q}_{ck} Z_{kl} \gamma_\mu D_{cd}^\mu q_{dl} - \overline{q}_{ck} m_{kl} q_{cl} - \tfrac{1}{4} F_c^{\mu\nu} F_{c\mu\nu}. \qquad (1)$$

Here q_{ck} is the field of a quark, with color index c running over the three values R, W, B, and flavor index k running over an unknown number of values u, d, s, c, and so forth; D^μ is the covariant derivative for the color gauge group SU(3); $F_{c\mu\nu}$ is the color-gauge-covariant curl of the gluon field; and Z and m are unknown color-invariant matrices.

Apart from color gauge invariance itself, the global symmetries of such a theory are just the symmetries of the matrices Z and m. We can always redefine the quark fields, by subjecting their left and right-handed parts to independent matrix transformations,† in such a way as to bring Z and m to the form

$$Z = 1 \qquad m = \begin{bmatrix} m_u & 0 & 0 & \cdots \\ & m_d & & \\ 0 & & m_s & \\ 0 & & & \cdots \\ \vdots & & & \end{bmatrix}$$

with the diagonal elements $m_u, m_d, m_s, \ldots$ all real. The Lagrangian therefore necessarily conserves the numbers of quarks of each type in this "mass basis," and also respects parity, charge conjugation, and time-reversal invariance. In addition, if n of the eigenvalues of m happen to be small, the Lagrangian will have an approximate chiral SU(n) ⊗ SU(n) symmetry,‡ consisting of all color-conserving unitary unimodular position-independent transformations on the left-handed and (independently) the right-handed parts of the n quark fields of low mass.

Now, the success of the soft-pion theorems[6-8] of the mid-1960s indicates that the strong interactions do in fact respect a nearly exact SU(2) ⊗ SU(2) symmetry. In the framework of quantum chromodynamics, there must then be two quarks u, d, with the charge of the u one unit greater than that of the d, and with masses m_u, m_d that are very small. This symmetry is spontaneously broken, leaving over a subgroup, the isospin group SU(2), which is not broken spontaneously, but only violated by electromagnetism and by the u–d quark mass difference.

The strong interactions are also known to respect a less exact symmetry, the "eightfold way"[9] of flavor SU(3). To the extent that the strong-interaction Lagrangian respects both SU(2) ⊗ SU(2) and SU(3) invariance, it must also obey an approximate SU(3) ⊗ SU(3) invariance. It follows that there must be a third quark s, of the same charge as d, and of low mass, though not so low as u and d.

---

†The argument here is identical to that used long ago in a different context by G. Feinberg, et al.[4] The nonperturbative effects associated with "instantons" require special treatment here, but the conclusions are not changed.

‡The Lagrangian will also have a chiral U(1) symmetry, which leads to unacceptable results, such as the existence of a pseudoscalar isoscalar meson with mass less than $\sqrt{3} m_\pi$. This problem seems to be eliminated through the discovery of nonperturbative effects associated with the Adler-Bell-Jackiw anomaly in gauge theories. See G. t'Hooft.[5]

The SU(3) ⊗ SU(3) symmetry is spontaneously broken down to SU(3), with SU(3) violated chiefly by the s–u and s–d quark mass difference.

We are thus led to consider a theory of the strong interactions that is globally SU(3) ⊗ SU(3) invariant, except for a "small" perturbation

$$\mathcal{H}_m = m_u \bar{u}u + m_d \bar{d}d + m_s \bar{s}s. \tag{3}$$

Here u, d, and s stand for the Dirac fields of the quarks of definite mass, with color indices suppressed. Our first problem is to determine the ratios of the masses in Equation 3.

This problem can be attacked by the standard methods of current algebra.§ If we were to ignore effects of virtual photons, the mass of any pseudoscalar Goldstone boson Π_a would be given to first order in $\mathcal{H}_m$ by

$$\mu_m^2(\Pi_a) = -F_\pi^{-2} \int d^4x\, d^4y \langle [A_a^0(x), [A_a^0(y), \mathcal{H}_m(0)]] \rangle_0, \tag{4}$$

where $F_\pi \cong 190$ MeV is the fundamental constant of current algebra, and $A_a^\mu(x)$ is the partially-conserved axial-vector current $-i\bar{\psi}\gamma^\nu\gamma_5\lambda_a\psi$ associated with Π_a. Using Equation 3 in Equation 4 gives the masses

$$\mu_m^2(\pi^+) = \mu_m^2(\pi^0) = 4F_\pi^{-2}[\langle \bar{u}u \rangle_0 m_u + \langle \bar{d}d \rangle_0 m_d], \tag{5}$$

$$\mu_m^2(K^+) = 4F_\pi^{-2}[\langle \bar{u}u \rangle_0 m_u + \langle \bar{s}s \rangle_0 m_s], \tag{6}$$

$$\mu_m^2(K^0) = 4F_\pi^{-2}[\langle \bar{d}d \rangle_0 m_d + \langle \bar{s}s \rangle_0 m_s], \tag{7}$$

Since SU(3) is not spontaneously broken, to the accuracy of this result we can take the vacuum expectation values equal

$$\langle \bar{u}u \rangle_0 = \langle \bar{d}d \rangle_0 = \langle \bar{s}s \rangle_0 \equiv \langle \bar{q}q \rangle_0. \tag{8}$$

To these values of $\mu_m^2(\Pi_a)$ we must add the contribution $\mu_\gamma^2(\Pi_a)$ of virtual photons, to obtain a total mass¶

$$\mu^2(\Pi_a) = \mu_m^2(\Pi_a) + \mu_\gamma^2(\Pi_a). \tag{9}$$

The photon terms cannot be expressed as simply as in Equations 5–7, but they are subject to the relations[13]

$$\mu_\gamma^2(\pi^0) = \mu_\gamma^2(K^0) = 0, \tag{10}$$

$$\mu_\gamma^2(K^+) = \mu_\gamma^2(\pi^+). \tag{11}$$

From Equations 5–11 we now find

$$\mu^2(\pi^0) = 4F_\pi^{-2} \langle \bar{q}q \rangle_0 (m_u + m_d) \tag{12}$$

---

§The derivation of Equations 5–7 is essentially that given by Gell-Mann et al.[10] Equivalent results were derived by Glashow and Weinberg,[11] but were not related to quark mass ratios. In contrast to these authors, we make no attempt here to take the effects of SU(3) breaking into account, beyond terms of first order in m_s.

¶For pions, Equation 5 shows that the mass difference is purely electromagnetic. It is very satisfying that this is the one mass splitting that has been successfully calculated using photon exchange alone. See Das et al.[12]

$$\mu^2(K^0) = 4F_\pi^{-2} <\bar{q}q>_0 (m_s + m_d) \qquad (13)$$

$$\mu^2(K^+) - \mu^2(K^0) - \mu^2(\pi^+) + \mu^2(\pi^0)$$
$$= 4F_\pi^{-2} <\bar{q}q>_0 (m_u - m_d) \qquad (14)$$

It follows that the quark masses stand in the ratios**

$$m_d/m_u = \frac{\mu^2(K^0) - \mu^2(K^+) + \mu^2(\pi^+)}{2\mu^2(\pi^0) + \mu^2(K^+) - \mu^2(K^0) - \mu^2(\pi^+)} = 1.80, \qquad (15)$$

$$m_s/m_d = \frac{\mu^2(K^0) + \mu^2(K^+) - \mu^2(\pi^+)}{\mu^2(K^0) - \mu^2(K^+) + \mu^2(\pi^+)} = 20.1. \qquad (16)$$

It is striking that the quark masses are not at all degenerate: Equation 15 shows that d is almost twice as heavy as u, while Equation 16 shows that s is very much heavier than either u or d. Why then is isotopic spin a very good symmetry of nuclear physics, and the eightfold way a fairly good symmetry? The answer has been anticipated in our previous discussion: it must be that, although the quark masses are nowhere near equal, they are all in some sense small.†† If the u and d quark masses were precisely zero, the strong interactions would have an exact SU(2) ⊗ SU(2) symmetry, which would presumably be spontaneously broken, leaving isospin as an unbroken subgroup. With d and u very light, the real world is close to this ideal situation, and isospin is very well conserved, whatever the ratio of the d and u quark masses may be.‡‡ Similarly, if s were also massless, there would be an exact SU(3) ⊗ SU(3) symmetry, which must be spontaneously broken down to SU(3); with m_s sufficiently small, SU(3) remains as an approximate symmetry, whatever the values of the s, d, and u quark mass ratios may be. But on the level of the Lagrangian of the strong interactions, isospin is no better a symmetry than chiral SU(2) ⊗ SU(2), and SU(3) is no better than chiral SU(3) ⊗ SU(3). We return to the implications of this remark in the third section of this paper.

What Are the Quark Masses?

We have only had to deal so far with ratios of quark masses. Indeed, as we have defined them the quark masses m_u, m_d, m_s contain a common infinite numeri-

---

**Similar results have been obtained before; see, e.g., footnote 8 of Halprin et al.[25] The one feature of Equations 15 and 16 that may be new is that effects of photon exchange have been extracted here from the observed pseudoscalar meson masses; otherwise the d-u mass ratio would be smaller, about 1.5 instead of 1.8. An independent calculation of this ratio based on the neutron-proton mass difference gives a value of about 1.5. See Gasser and H. Leutwyler.[39]

††This point of view may be widely held. I first heard it advocated in a conversation with H. Fritzsch.

‡‡An interesting illustration is provided by the two-dimensional Schwinger model: it has been shown by Coleman[14] that for arbitrary ratios of bare quark masses, the ratio of physical fermion masses approaches unity as the bare quark masses approach zero.

cal factor,§§ which is needed to cancel a corresponding factor in matrix elements of the quark bilinears $\bar{u}u$, $\bar{d}d$, and $\bar{s}s$. In order to give meaning to the absolute values of the quark masses, it is necessary to renormalize the quark operator products, so that they have some definite matrix elements in some definite hadron states.

This is to some extent a matter of arbitrary definition, but we can formulate a reasonable approximation scheme which is useful in its own right, and leads to a natural normalization of the quark masses. Suppose that in the limit of zero quark mass, the expectation value of the operator $\bar{q}_k q_k$ for quarks of flavor k in any hadron state h is simply proportional to the number N_{hk} of these quarks that are supposed to be in the hadron h:

$$<\bar{q}_k q_k>_h \simeq Z_m N_{hk} \tag{17}$$

with Z_m a universal renormalization factor. (Equation 17 is consistent with the SU(3) symmetry of physical states in the limit of zero quark mass, but it is of course a considerably stronger assumption.) The "renormalized" quark masses are then naturally defined by

$$m_k^* \equiv Z_m m_k \tag{18}$$

so that the expectation value of $m_h \bar{q}_k q_k$ in a hadron state h is simply m_k^* times the number N_{hk} of quarks of flavor k in that state.

In order to check assumption 17 and to determine the m_k^*, let us consider the consequences of this assumption for the mass splittings within unitary multiplets. These splittings are almost entirely due to the term $m_s \bar{s} s$ in Equation 3, so by use of first-order perturbation theory, the mass of any hadron h should be given by

$$m_h \simeq m_0 + m_s <\bar{s}s>_h \tag{19}$$

where m_0 is a common mass for all members of a given unitary multiplet, but may vary from multiplet to multiplet. Together with Equations 17 and 18, this implies that

$$m_h \simeq m_0 + m_s^* N_{hs}, \tag{20}$$

where N_{hs} is the number of strange quarks in the hadron h. Equation 20 incorporates the familiar Gell-Mann-Okubo mass formula,[9] but it has the further consequences, that there is no splitting between isosinglets and isotriplets in unitary octets, and that the splittings between members of unitary multiplets which differ in strangeness by one unit should all be the same.

These rules are known to be fairly well satisfied. The average splittings in the 1^- meson octet, the 2^+ meson octet, the $\frac{1}{2}^+$ N octet, and the $\frac{3}{2}^+$ Δ decimet are, respectively, 120, 110, 190, and 150 MeV, while the $\rho - \omega$ and $\Sigma - \Lambda$ mass differences are 10 and 75 MeV. As a rough compromise, we can take the coefficient m_s^*

§§The fact that ratios of bare masses are finite is shown by the renormalization procedures of G. 't Hooft[15] and Weinberg.[16]

in Equation 20 as ¶¶

$$m_s^* \simeq 150 \text{ MeV}. \quad (21)$$

Using the mass ratios derived in the preceding section, we have then

$$m_d^* \simeq 7.5 \text{ MeV}, \quad (22)$$

$$m_u^* \simeq 4.2 \text{ MeV}. \quad (23)$$

It must be emphasized that these are not the quark masses to be used in "naive" quark models of physical hadrons. Rather, it seems likely that the so-called constituent quarks appearing in the naive quark models can be treated as ordinary hadrons, which happen to be trapped, but which receive most of their mass from the spontaneous breakdown of $SU(3) \otimes SU(3)$ rather than from the mass terms in the Lagrangian. This suggests that the masses of the constituent quarks should satisfy Equation 20, with m_0 of the order of $m_N/3$ or $m_\rho/2$; hence the constituent quark masses might be about 350 MeV for u and d, and 500 MeV for s. (However, the mass difference of the u and d constituent quarks is *not* just equal to $m_d^* - m_u^*$, because electromagnetic effects contribute terms to the constituent quark masses of order αm_0.[17]) In any case, these familiar mass values, which are commonly used in quark model calculations, are totally different from the masses in the strong interaction Lagrangian.

It is even more speculative to extend these considerations to quarks heavier than the s quark; these are almost certainly too heavy to allow the use of current algebra to calculate quark mass ratios. If we were to extend Equation 20 to hadrons that are supposed to contain charmed quarks, we would estimate the renormalized charmed quark mass as

$$m_c^* \simeq 1500 \text{ MeV} - 350 \text{ MeV} = 1150 \text{ MeV} \quad (24)$$

so that m_c/m_s would be around 7.7. However there are two independent arguments, based, respectively, on spectral function sum rules[32] and on asymptotic freedom and observed masses,[33] which indicate that m_c/m_s is in the range of 3.5 to 4. The question of the mass of the c quark appears at this time to be surrounded with confusion.

The results of this section have potentially important applications to the phenomenology of the Higgs bosons. In the simplest unified gauge theory of the weak and electromagnetic interactions, there is one elementary weakly coupled neutral scalar boson H^0, with couplings

$$2^{1/4} G_F^{1/2} H^0 \bar{q} m q \quad (25)$$

where $G_F = 1.02 \times 10^{-5} m_p^{-2}$ is the Fermi coupling constant. Hence the coupling of the Higgs boson to a quark of flavor k is

$$g_{Hk} = 2^{1/4} G_F^{1/2} m_k^* = 4 \times 10^{-6} m_k^* \text{ (MeV)}. \quad (26)$$

Thus, in calculating effects of Higgs boson exchange in ordinary or muonic

¶¶ For other attempts to set a scale for these quark masses, see Leutwyler,[34,35] Testa,[36] Gava *et al.*,[37] and Furlan *et al.*[38]

atoms,*** the coupling of the H^0 to a proton or neutron can be estimated as

$$g_{Hp} \simeq 2g_{Hu} + g_{Hd} \simeq 6.4 \times 10^{-5},$$

$$g_{Hn} \simeq 2g_{Hd} + g_{Hu} \simeq 7.7 \times 10^{-5}.$$

Do Strong Interactions Always Conserve Isospin?

We have concluded in the first section that isospin is observed to be a good symmetry of strong interactions only because most measured quantities are quite insensitive to the very small d and u quark masses. This is presumably the case for quantities like the nucleon mass, the ρ mass and width, the πN cross sections at 1 GeV, and so on. However, there are a few measurable quantities which vanish in the limit of vanishing d and u quark masses. Since d and u are so far from being degenerate, we expect that quantities of this sort, which *are* sensitive to m_d and m_u, should show a gross violation of isospin conservation.

Can we test this remarkable conclusion? One thinks first of the pion-pion scattering lengths, which are well known to depend sensitively on the chiral transformation properties of the symmetry breaking terms in the Lagrangian.††† Unfortunately, there is a theorem, which states that the d-u quark mass difference will produce no isospin violation in any purely pionic low-energy matrix element. Any such matrix element can be calculated using an effective Lagrangian[22] constructed from a pion field which transforms according to a 3-dimensional non-linear realization of chiral SU(2) ⊗ SU(2). For one convenient such realization, the chiral-invariant "kinematic" Lagrangian is

$$\mathcal{L}_{KIN} = -\tfrac{1}{2} \partial_\mu \pi \partial^\mu \pi (1 + \pi^2/F_\pi^2)^{-2}. \tag{27}$$

To this, we must add terms with the same chiral transformation properties as the symmetry breaking term (3). This term consists of an SU(2) ⊗ SU(2) invariant $m_s \bar{s}s$, which does not concern us here; plus an isoscalar $\tfrac{1}{2}(m_u + m_d)(\bar{u}u + \bar{d}d)$, which transforms like the fourth component of a chiral four-vector; plus an isovector term $\tfrac{1}{2}(m_u - m_d)(\bar{u}u - \bar{d}d)$ which transforms like the third component of a chiral four-vector. Now, we can construct a term in the effective Lagrangian with the same transformation properties as the isoscalar $\bar{u}u + \bar{d}d$:

$$\mathcal{L}_4 = -\tfrac{1}{2} m_\pi^2 \pi^2 (1 + \pi^2/F_\pi^2)^{-1}. \tag{28}$$

Because $\mathcal{L}_4$ involves no derivatives, $\mathcal{L}_4$ and $\mathcal{L}_{KIN}$ make contributions of the same order of magnitude to processes with pion four-momenta of order m_π, such as pion-pion scattering at threshold. However, it is not possible to construct a term in the effective Lagrangian with the same transformation properties as the iso-

---

***The shift of energy levels by Higgs exchange in muonic atoms was considered by Jackiw and Weinberg.[18] The conversion of muons into electrons in atomic orbits by Higgs exchange with the nucleus was recently considered by Bjorken and Weinberg.[19]

†††See Reference 20 and Reference 21, Section IV. Note that the assumption that SU(2) ⊗ SU(2) is violated in the Lagrangian by a chiral four-vector term (which was used to derive the ratio of the $\pi\pi$ scattering lengths) is now seen as a direct consequence of quantum chromodynamics.

vector perturbation $\bar{u}u - \bar{d}d$, using only the pion field without derivatives. (In order to have the same G-parity as $\bar{u}u - \bar{d}d$, such a term would have to be odd in the pion field, but it would then be a pseudoscalar, and could not appear in the Lagrangian.) Thus, aside from true electromagnetic effects, all purely pionic low energy processes automatically conserve isospin. One example of this theorem is provided by the pion masses themselves: Equation 5 shows that the u and d quark masses contribute equal values to $\mu^2(\pi^+)$ and $\mu^2(\pi^0)$, proportional to $m_u + m_d$, so that the pion mass difference is a purely electromagnetic effect.

Since purely pionic processes do not allow observation of gross isospin violations, let us turn to pion-nucleon scattering. In general, the scattering length for pion-nucleon scattering at threshold is[23]

$$a(\pi_a N \to \pi_b N') = \frac{2i\pi^2 m_N}{(m_N + m_\pi)F_\pi^2} \int d^4z \, \exp(-im_\pi z^0)$$

$$< N' | \{m_\pi^2 T[A_b^0(0), A_a^0(z)]_n + \delta(z^0)[A_b^0(0), \partial_\mu A_a^\mu(z)]$$

$$- im_\pi \delta(z^0)[A_b^0(0), A_a^0(z)]\} | N > . \quad (30)$$

Here a and b are the initial and final pion isovector indices; $|N'>$ and $|N>$ are the states of the final and initial nucleon at rest; $A_a^\mu(x)$ is the axial-vector current; and the subscript n indicates that the time-ordered product excludes terms in which either axial-vector current is dominated by a single pion pole. In the gauge theory of strong interactions, the axial current is

$$A_a^\mu(z) = -i \begin{pmatrix} \bar{u} \\ \bar{d} \end{pmatrix} \gamma_5 \gamma^\mu \tau_a \begin{pmatrix} u \\ d \end{pmatrix} \quad (31)$$

with τ_a the Pauli isospin vector.

For reasons that will be made clear below, let us now restrict our attention to the case of π^0-nucleon scattering; for which $a = b = 3$. In this case, the commutator in the last term in Equation 30 vanishes. The first term is dominated by a one-nucleon pole, and yields a contribution to the scattering length that is just what would be found in the Born approximation with pseudovector coupling

$$a^{\text{BORN}}(\pi^0 n \to \pi^0 N) = -\frac{g_A^2 m_\pi^2}{4\pi(m_N + m_\pi)F_\pi^2} = -1.05 \times 10^{-15} \, \text{cm} \quad (32)$$

where $g_A \simeq 1.2$ is the familiar axial-vector coupling constant. We note that this term is the same for the $\pi^0 n$ and $\pi^0 p$ scattering lengths, as would be expected from isospin conservation.

This leaves the middle term in Equation 30, known as the "σ-term." The divergence of the axial-vector current here is

$$\partial_\mu A_3^\mu = 2i[m_u \bar{u}\gamma_5 u - m_d \bar{d}\gamma_5 d], \quad (33)$$

and its commutator with the axial-vector current is

$$\delta(z^0)[A_3^0(0), \partial_\mu A_3^\mu(z)] = -4i\delta^4(z)\{m_u \bar{u}(0)u(0) + m_d \bar{d}(0)d(0)\} \quad (34)$$

Hence this contribution to the scattering length takes the form

$$a^\sigma(\pi^0 N \to \pi^0 N) = \frac{m_N}{\pi(m_N + m_\pi)F_\pi^2} \langle m_u \bar{u}u + m_d \bar{d}d \rangle_N, \quad (35)$$

where $\langle \cdots \rangle_N$ denotes the expectation value

$$\langle \cdots \rangle_N \equiv \langle N | \cdots | N \rangle (2\pi)^3. \quad (36)$$

The expectation values appearing in Equation 35 can be estimated by using the approximations introduced in the preceding section. Those give

$$\langle \bar{u}u \rangle_n = \langle \bar{d}d \rangle_p = Z_m,$$

$$\langle \bar{u}u \rangle_p = \langle \bar{d}d \rangle_n = 2Z_m,$$

$$m_d Z_m \equiv m_d^*, \quad m_u Z_m \equiv m_u^*,$$

so Equation 3 becomes

$$a^\sigma(\pi^0 n \to \pi^0 n) = \frac{m_N}{\pi(m_N + m_\pi)F_\pi^2} (2m_d^* + m_u^*), \quad (37)$$

$$a^\sigma(\pi^0 p \to \pi^0 p) = \frac{m_N}{\pi(m_N + m_\pi)F_\pi^2} (2m_u^* + m_d^*). \quad (38)$$

Using the numerical values in Equations 22, 23, and including the Born term in 30, we now have the total scattering lengths

$$a(\pi^0 n \to \pi^0 n) = 1.9 \times 10^{-15} \text{ cm}, \quad (39)$$

$$a(\pi^0 p \to \pi^0 p) = 1.4 \times 10^{-15} \text{ cm}. \quad (40)$$

Isospin conservation would say that these should be equal—instead, we see that they differ by over 30%.

These results serve to illustrate the point of this section, that the u-d quark mass difference can produce gross violations of isospin conservation. However, in view of the crude approximations used to calculate the expectation values in Equation 35, the precise values in 39 and 40 should not be given too much weight.

As an alternative to the above calculation, we might try to give up the approximations of the previous section and use instead only SU(3) symmetry, under the assumption that SU(3) is exactly conserved except for the strictly first-order effects of quark mass differences. We would then have the SU(3) relations

$$\langle \bar{u}u \rangle_p = \langle \bar{d}d \rangle_n = \langle \bar{s}s \rangle_\Xi, \quad (41)$$

$$\langle \bar{u}u \rangle_n = \langle \bar{d}d \rangle_p = \langle \bar{s}s \rangle_\Sigma. \quad (42)$$

Also, Equation 19 would still be valid, and would yield the $\bar{s}s$ expectation values

$$m_s \langle \bar{s}s \rangle_\Xi = m_\Xi - m_N(1 - \xi), \quad (43)$$

$$m_s \langle \bar{s}s \rangle_\Sigma = m_\Sigma - m_N(1 - \xi), \quad (44)$$

where ξ is the fraction of the nucleon mass that is due to the s-quark mass

$$\xi \equiv m_s \langle \bar{s}s \rangle_N / m_N. \quad (45)$$

Inserting Equations 41–44 in 35 gives the σ-term contribution to the scattering lengths as

$$a^\sigma(\pi^0 n \to \pi^0 n) = \frac{m_N}{\pi(m_N + m_\pi)F_\pi^2}$$

$$\times \left[\frac{m_u}{m_s}[m_\Sigma - m_N(1 - \xi)] + \frac{m_d}{m_s}[m_\Xi - m_N(1 - \xi)] \right], \quad (46)$$

$$a^\sigma(\pi^0 p \to \pi^0 p) = \frac{m_N}{\pi(m_N + m_\pi)F_\pi^2}$$

$$\times \left[\frac{m_u}{m_s}[m_\Xi - m_N(1 - \xi)] + \frac{m_d}{m_s}[m_\Sigma - m_N(1 - \xi)] \right]. \quad (47)$$

We can now obtain numerical results for the total scattering lengths by using the quark mass ratios of the first section in Equations 46 and 47 and adding the result to the Born term 32. In this way, we find that

$$a(\pi^0 n \to \pi^0 n) = (2.9 + 6.1\xi) \times 10^{-15} \text{ cm}, \quad (48)$$

$$a(\pi^0 p \to \pi^0 p) = (2.5 + 6.1\xi) \times 10^{-15} \text{ cm}. \quad (49)$$

The unknown parameter ξ is expected to be fairly small, but it could easily be as large as $\frac{1}{6}$, and we do not even know its sign. It appears that only the *difference* of the $\pi^0 n$ and $\pi^0 p$ scattering lengths can be predicted with any confidence, as about 0.4 to 0.5 $\times 10^{-15}$ cm.

Similar considerations apply to the elastic scattering of charged pions, and to the charge-exchange processes $\pi^- p \to \pi^0 n$ and $\pi^+ n \to \pi^0 p$. However, the analysis here is very much more complicated. For one thing, the commutator in the last term of Equation 30 is now no longer zero and makes an isospin-conserving contribution that is larger than the isospin-violating σ terms by a factor of order m_N/m_π. Even worse, the exchange of virtual photons violates the soft-pion theorems for charged pions, and therefore makes an isospin-violating contribution to the πN scattering lengths of the order of α/m_N, not much smaller than the σ term in Equation 30. In contrast, photon exchange respects the soft-pion theorems for neutral pions (as shown for example by Equation 10) so its effect on the π^0 scattering lengths is only of order α times the scattering lengths themselves. I will not pursue this matter here, but it should be kept in mind that measurements of the σ terms in $\pi^\pm N$ scattering may be seriously affected by isospin violations due to photon exchange, as well as to the quark mass difference.

Unfortunately, it is not so easy to measure the $\pi^0 N$ scattering lengths. Perhaps the only hope would be to measure the relative phases of various partial wave amplitudes for the photoproduction processes $\gamma N \to \pi^0 N$, near threshold, and use Watson's theorem to relate these phases to the relative phases of the $\pi^0 N$ scattering amplitudes.[24]

One can think of other places where a gross violation of isospin conservation might show up:‡‡‡ in πK scattering lengths, in the differences between pion-hadron coupling coupling constants and their values as given by the Goldberger-Treiman formula; and at Adler zeros in decay amplitudes. All these possibilities deserve further theoretical study, but none will be easy to deal with experimentally. It is frustrating that nature should have done such a good job in hiding the isospin violation in the quark masses from our view.

Where Do the Small Masses Come From?

In renormalizable gauge theories, the quark and lepton masses arise in lowest order from the vacuum expectation values of scalar fields, and in special cases also from a true bare mass term in the Lagrangian. However, the masses of the electron and the u and d quarks are so small, that one is led to suspect that they are naturally zero in lowest order; they would then arise from radiative corrections of some sort, rather than from bare masses or vacuum expectation values of scalar fields. (The adverb "naturally" here has a technical meaning[26]: the e, u, and d masses are supposed to vanish to zeroth order in gauge couplings as a result of the symmetry and representation-content of the theory, for all values of the parameters of the theory in at least some finite range.) This section will explore the origin of the u, d, and e masses, and will consider a possible modification in this point of view for the d quark.

For the sake of definiteness, we shall concern ourselves here only with a "minimal" model: four quark fields and four lepton fields, whose left-handed parts transform as doublets $(u_o, d_o)_L$, $(c_o, s_o)_L$, $(\nu_e, e)_L$, $(\nu_\mu, \mu)_L$ under the weak and electromagnetic gauge group $SU(2) \otimes U(1)$, and with all right-handed quark and lepton fields singlets. (The label o here means that these quark fields are defined by their weak interaction properties, not by the diagonalization of the quark mass matrix. The subscripts L and R indicate the parts of the fermion fields proportional to $(1 + \gamma_5)/2$ or $(1 - \gamma_5)/2$, respectively.) Let us tentatively suppose that the vacuum expectation values of the scalar fields give the s_o, c_o, and μ_o nonzero masses, while leaving the u_o, d_o, and e_o and neutrino massless. We will further suppose that this is what is called a "type 1" zeroth-order mass relation,§§§ which follows from the representation content of the fermion and scalar fields, irrespective of the values of the scalar field vacuum expectation values, or scalar-fermion coupling constants. The point of this assumption is that it guarantees that the masses of the u, d, and e arise only from emission and reabsorption of gauge vector bosons by the fermions, and not from graphs involving virtual scalar bosons, whose properties are almost entirely unknown. That part of the change in the fermion mass matrix which contains corrections to "type 1" zeroth-order mass

---

‡‡‡The possibility that quark mass differences produce second class currents in beta decay has been considered by Halprin et al.[25] They find that because the currents are not sensitive to the small quark masses, the resulting second-class terms are very small.

§§§This class of zeroth order mass relation was described in Reference 27. Also see Reference 26.

relations is in general given by¶¶¶

$$\delta m = - \frac{1}{16\pi^2} \sum_{N \neq \gamma} [-m_o t_N^2 + 4\gamma_4 t_N \gamma_4 m_o t_N] \ln \mu_N^2$$
$$- \frac{3}{16\pi^2} e^2 m_o \left[-\frac{1}{2} + \ln m_o^2 \right]. \tag{50}$$

Here m_o is the zeroth-order mass matrix; N labels the gauge vector bosons of definite mass μ_N; and t_N is the Hermitian matrix representation of the associated gauge group generator, including a gauge coupling constant factor. This expression does not depend on the unit of mass used to calculate the logarithms, because the assumption of a "natural" zeroth-order mass relation (of *any* type) implies in general that****

$$\sum_{N \neq \gamma} m_o t_N^2 + m_o e^2 = 0, \tag{51}$$

$$\sum_{N \neq \gamma} \gamma_4 t_N \gamma_4 m_o t_N + m_o e^2 = 0. \tag{52}$$

In the case of interest here, where the zeroth order mass relation takes the form of a vanishing of some fermion masses, Equation 50 may be simplified to

$$\delta m = - \frac{1}{4\pi^2} \sum_{N \neq \gamma} \gamma_4 t_N \gamma_4 m_o t_N \ln \mu_N^2 \tag{53}$$

except between eigenstates of m_o with nonzero mass.

Now, it is quite clear that the observed weak and electromagnetic interactions cannot in this picture give masses to the u, d, or e, because the $W^\pm$, Z^0, and γ only connect the left-handed fields of massless fermions to each other. We are therefore compelled here to suppose that the weak and electromagnetic gauge group $SU(2) \otimes U(1)$ is part of a larger gauge group Γ, which includes transformations that take heavy and light fermions into each other. We do not see effects of the gauge bosons associated with such transformations, so we must suppose that they are very heavy,†††† and therefore that Γ must be much more strongly broken than $SU(2) \otimes U(1)$. Fortunately these vector bosons can be almost arbitrarily heavy, and still produce the necessary mass shifts, because it is only logarithms of vector boson mass ratios that enter in Equation 53.

Since $SU(2) \otimes U(1)$ is less strongly broken than Γ, we can use it to sort out the

---

¶¶¶See Reference 26, Equation 4.33. We assume here that the quark masses are all much less than the intermediate vector boson masses.

****See the discussion after Equation 4.22 in Reference 26.

††††For instance, a vector boson of mass μ_N which couples with full strength to both $\bar{\mu} e$ and $\bar{s} d$ would produce the decay $K_L^0 \rightarrow \mu^\pm e^\mp$, with a rate less than that of $K^+ \rightarrow \mu^+ \nu$ by a factor of order m_w^4 / μ_N^4. In order to keep the rate within known bounds, we would need $\mu_N > 2 \times 10^2 m_w$. Such enormous masses can be avoided only if we assume that the heavy vector boson masses and couplings conserve strangeness, charm, and muon number.

properties of the very heavy vector bosons which produce the mass shifts. Such vector bosons must be coupled to both the right- and left-handed parts of the fermion fields, but since all right-handed fermion fields are assumed here to be SU(2) singlets, the relevant heavy vector bosons must be SU(2) singlets. Thus, the masses of the u, d, and e can only arise from terms in Equation 53 corresponding to the transitions

$$u_L \to c \to u_R \quad d_L \to s \to d_R \quad e_L \to \mu \to e_R. \tag{54}$$

We therefore expect that in order of magnitude

$$m_u/m_c \approx m_d/m_s \approx m_e/m_\mu \approx \frac{g^2}{4\pi^2} \times \text{logarithms}, \tag{55}$$

where g is a typical coupling constant of the gauge group Γ. In the special case where Γ is the direct product of SU(2) ⊗ U(1) with some other group Γ_1, and the doublets (u_o, c_o), (d_o, s_o), and (e_o, μ_o) form (for each helicity) identical representations of Γ_1, we even expect that

$$m_u/m_c = m_d/m_s = m_e/m_\mu \tag{56}$$

to first order in g^2.

The order-of-magnitude estimates in Equation 55 seem to work out very well for the u-c and e-μ mass ratios. Indeed, using the mass values derived in the second section, we find that $m_u/m_c = 0.0037$, not very different from the ratio $m_e/m_\mu = 0.0048$, and also not very different from the value $e^2/4\pi = 0.0023$ expected if $g \approx e$. Given the uncertainty in an estimate of m_c, it is not even ruled out that m_u/m_c is precisely equal to m_e/m_μ, as would be required by Equation 56.

However, the ratio m_d/m_s is very different, more than an order of magnitude greater than m_u/m_c or m_e/m_μ or $e^2/4\pi^2$. It does not appear likely that the d quark could get its mass from transitions to any new sort of heavy quark, because there apparently is no quark of charge $-\frac{1}{3}$ anywhere near the needed mass $(m_\mu/m_e)m_d$. Some other mechanism is needed to produce the d quark mass.

Another problem with the general picture described so far is that it does not allow for an appreciable Cabibbo angle. It is true that the heavy vector bosons of Γ might produce off-diagonal terms in the quark mass matrix through transitions like $u_{L,R} \to c \to c_{R,L}$ and $d_{L,R} \to s \to s_{R,L}$. However, these off-diagonal masses would be of order $g^2 m_c/4\pi^2$ or $g^2 m_s/4\pi^2$, so that the Cabibbo angle would be of order $g^2/4\pi^2$, which as we have seen is of order 2 to 4 × 10^{-3}, instead of the observed value[28-30] $\theta_c = 0.229 \pm 0.003$.

One way out of both these difficulties is to suppose that the Cabibbo angle and the d mass are not effects of radiative corrections, but arise already in the lowest order of perturbation theory.‡‡‡‡ For an arbitrary charge $-\frac{1}{3}$ quark mass matrix, we can always perform a unitary transformation on the right-handed quark fields

‡‡‡‡ The idea that the d quark mass arises from the Cabibbo mixing with the s quark, and is therefore of order $\theta_c^2 m_s$, was suggested to be by J. D. Bjorken. A number of authors have attempted to derive the relations $\theta_c^2 \approx m_d/m_s$ or $\theta_c^2 \approx m_\pi^2/2m_K^2$ from considerations of finiteness for weak radiative corrections. See Gatto et al.,[40] Cabibbo and Maiani,[41] and Jackiw and Schnitzer.[42]

s_{oR} and d_{oR}, so that the mass matrix connects d_{oL} only with s_{oR}, not d_{oR}. The lowest order mass term in the effective Lagrangian is then

$$\mathcal{H}_{om} = -a\bar{d}_{oL}s_{oR} - b\bar{s}_{oL}d_{oR} - c\bar{s}_{oL}s_{oR} + \text{H.C.} \tag{57}$$

The left-handed parts of the quark fields s, d of definite mass may be expressed in terms of s_{oL} and d_{oL} by the unitary transformation:

$$d_L = e^{i\alpha} \cos\theta_c d_{oL} + e^{i\beta} \sin\theta_c s_{oL},$$

$$s_L = -e^{i\gamma} \cos\theta_c d_{oL} + e^{i\delta} \sin\theta_c s_{oL},$$

where θ_c is the Cabibbo angle, and α, β, γ, and δ are real phases, with $\alpha - \gamma = \beta - \delta$. In general, θ_c and m_d/m_s are functions of the two ratios $|a/c|$ and $|b/c|$, so there is no necessary relation between them. In order to see what sort of relation might hold, let us suppose that parity is conserved *in this basis,* so that

$$|a| = |b| \tag{58}$$

It is straightforward then to eliminate the unknown ratio $|a/c|$, and show that

$$\frac{1}{2}\tan 2\theta_c = \frac{\sqrt{R}(1 + R)}{(1 - R^2)} \tag{59}$$

where $R \equiv m_d/m_s$. Using the result $R = 0.0498$ from Equations 21 and 22, we find $\theta_c = 0.219$ radians. (For such small values of R and θ_c, Equation 59 can be adequately approximated by $\theta_c = \sqrt{R}$.) If we first subtracted from m_d an amount $(m_e/m_\mu)m_s$, which might be due to "radiative" corrections, we would have $R = 0.0448$, and Equation 59 would give $\theta_c = 0.222$ radians. Either result is in quite remarkable agreement with the experimental value[28-30] $\theta_c = 0.229 \pm 0.003$ radians, certainly within the uncertainty in our use of current algebra to calculate R.

It should be emphasized that the rather small value of the d quark mass appears from this point of view as an accident. The only small number that must be introduced here in an ad hoc way is the ratio $|a/c| \simeq \theta_c = 0.23$. More generally, if instead of Equation 58 we only assumed that $|a|$ and $|b|$ were of the same order of magnitude and somewhat smaller than $|c|$, then we would have θ_c of order $|a/c|$ or $|b/c|$, and m_d/m_s would be of order θ_c^2.

The important question here is one of naturalness: are there theories in which Equation 58 follows directly from the gauge group and the representation content of the fields? If we ignore all other problems, the answer is yes. For instance, we might suppose that the gauge group of the weak and electromagnetic interactions is $SU(2) \otimes SU(2) \otimes U(1)$, with parity conserved, and with quark fields forming the multiplets§§§§

$$(\tfrac{1}{2}, 0) : (u_o, d_o)_L (c_o, s_o)_L,$$

$$(0, \tfrac{1}{2}) : (u_o, d_o)_R (c_o, s_o)_R.$$

---

§§§§ Models based on an $SU(2) \otimes SU(2) \otimes U(1)$ gauge group have also been previously considered by Weinberg,[27] Pati and Salam,[31] Mohapatra and Pati,[43] Senjanovic and Mohapatra,[44] Fritzsch and Minkowski,[45] Mohapatra and Sidhu,[46] Georgi,[47] and de Rújula *et al.*[48]

The scalar fields that can enter into renormalizable interactions with such quarks are necessarily of the $(\frac{1}{2},\frac{1}{2})$ type. Suppose these form two non-Hermitian 2×2 matrices Φ and Ω, and that there is a discrete symmetry under which the fields are multiplied with the factors

$$(u_o, d_o)_L : +i \quad (u_o, d_o)_R : -i$$
$$(c_o, s_o)_L : +1 \quad (c_o, s_o)_R : +1$$
$$\Phi : +1 \quad\quad \Omega : +i.$$

The most general scalar-quark interaction is then of the form

$$f(\overline{c,s})_L \Phi \begin{pmatrix} c \\ s \end{pmatrix}_R + f'(\overline{c,s})_L \Omega \begin{pmatrix} u \\ d \end{pmatrix}_R + f''(\overline{u,d})_L \Omega \begin{pmatrix} c \\ s \end{pmatrix}_R$$

$$+ f'''(\overline{c,s})_L \tau_2 \Phi^* \tau_2 \begin{pmatrix} c \\ s \end{pmatrix}_R + \text{H.C.}$$

This is invariant under a space inversion transformation

$$q_L \leftrightarrow q_R \quad \Phi \to \Phi\dagger \quad \Omega \to \Omega\dagger$$

provided we take

$$f = f^* \quad f'' = f'^*$$

The vacuum expectation values of the neutral scalar fields then automatically yield effective mass terms of the form 57, with $|a| = |b| = |f' < \Omega_{22}^0 >_o|$, so that Equation 59 is a "natural" zeroth-order relation.

The trouble with this particular example is that it offers no explanation of the vanishing of the lowest-order uu, uc, and cu mass terms, and it does not appear to have room for an enlarged gauge group Γ with vector bosons which could produce the small u quark mass. The search for a truly natural theory of the quark masses must continue. However, it may be a useful guide in this search to try to construct theories which preserve the successful relation $\theta_c = \sqrt{m_d/m_s}$.

I am grateful for frequent valuable conversations with colleagues at Stanford and SLAC, especially James Bjorken, Helen Quinn, and Marvin Weinstein. I have also benefited much from conversations with Sidney Coleman, Harald Fritzsch, and Benn Lee.

[NOTE ADDED IN PROOF: This paper was originally submitted for publication in the Rabi Festschrift in March 1977. Since then, two other papers have independently arrived at similar conclusions concerning the relation between the Cabibbo angle and the quark mass ratio in SU(2) ⊗ SU(2) ⊗ U(1) models; see A. De Rújula, H. Georgi, and S. Glashow, Harvard Preprint HUTP-77/A028, to be published, and F. Wilczek and A. Zee, Princeton preprint, to be published. For a generalization, see H. Fritzsch, to be published. For other recent work on SU(2) ⊗ SU(2) ⊗ U(1) models, see E. Ma, Oregon preprint OITS-68, to be published; J. Pati, S. Rajpoot, and A. Salam, Imperial College preprint ICTP/76/11, to be published; and H. Georgi and S. Weinberg, Harvard Preprint HUTP-77/A052, to be published. I wish to thank F. Wilczek for pointing out some algebraic errors in the original version of this paper, and to C. L. Ong and H. Leutwyler for informative communications.

References

1. FRITZSCH, H., M. GELL-MANN & H. LEUTWYLER. 1973. Phys. Lett. **B47**: 365.
2. GROSS, D. J. & F. WILCZEK. 1973. Phys. Rev. **D8**: 3497.
3. WEINBERG, S. 1973. Phys. Rev. Lett. **31**: 494.
4. FEINBERG, G., P. K. KABIR & S. WEINBERG. 1959. Phys. Rev. Lett. **3**: 527.
5. 'T HOOFT, G. 1976. Phys. Rev. Lett. **37**: 8.
6. WEINBERG, S. 1968. Proc. 14th Internat. Conf. High Energy Physics. :253. CERN. Geneva, Switzerland.
7. WEINBERG, S. 1970. *In* Lectures on Elementary Particles and Quantum Field Theory—1970 Brandeis Summer Institute in Theoretical Physics. S. Deser, M. Grisaru & H. Pendleton, Eds. MIT Press, Cambridge, Mass.
8. LEE, B. W. 1972. Chiral Dynamics. Gordon and Breach. New York, N.Y.
9. GELL-MANN, M. & Y. NE'EMAN. 1972. The Eightfold Way. W. A. Benjamin, Inc. New York, N.Y. A review with reprints of original articles.
10. GELL-MANN, M., R. J. OAKES & B. RENNER. 1968. Phys. Rev. **175**: 2195.
11. GLASHOW, S. L. & S. WEINBERG. 1968. Phys. Rev. Lett. **20**: 224.
12. DAS, T., G. S. GURALNIK, V. S. MATHER, F. E. LOW & J. E. YOUNG. 1967. Phys. Rev. Lett. **18**: 759.
13. DASHEN, R. 1969. Phys. Rev. **183**: 1245. Possible corrections have been discussed by LANGACKER, P. & H. PAGELS. 1973. Phys. Rev. **D8**: 4620.
14. COLEMAN, S. 1976. Ann. Phys. (N.Y.) **101**: 239.
15. 'T HOOFT, G. 1973. Nucl. Phys. **B61**: 455.
16. WEINBERG, S. 1973. Phys. Rev. **D8**: 3497.
17. LANE, K. & S. WEINBERG. 1976. Phys. Rev. Lett. **37**: 717.
18. JACKIW, R. & S. WEINBERG. 1972. Phys. Rev. **D5**: 2396.
19. BJORKEN, J. D. & S. WEINBERG. In Press. Phys. Rev. Lett.
20. WEINBERG, S. 1966. Phys. Rev. Lett. **17**: 616.
21. WEINBERG, S. 1968. Phys. Rev. **166**: 1568.
22. WEINBERG, S. 1967. Phys. Rev. Lett. **18**: 188. Also see References 6–8.
23. Reference 7: Equation 4.D.11.
24. BJORKEN, J. D. Personal communication.
25. HALPRIN, A., B. W. LEE & P. SORBA. 1976. Phys. Rev. **D14**: 2343.
26. WEINBERG, S. 1973. Phys. Rev. **D7**: 2337.
27. WEINBERG, S. 1972. Phys. Rev. Lett. **29**: 388.
28. WILKINSON, D. H. 1975. Nature **257**: 189.
29. WILKINSON, D. H. & D. E. ALBURGER. 1976. Phys. Rev. **D13**: 2517.
30. WILKINSON, D. H. Preprint. University of Sussex.
31. PATI, J. C. & A. SALAM. 1974. Phys. Rev. **D10**: 275.
32. ONG, C. L. 1977. Phys. Rev. To be published.
33. GEORGI, H. & H. D. POLITZER. 1976. Phys. Rev. **D14**: 1829.
34. LEUTWYLER, H. 1973. Phys. Lett. **48B**: 431.
35. LEUTWYLER, H. 1974. Nucl. Phys. **76B**: 413.
36. TESTA, M. 1975. Phys. Lett. **56B**: 53.
37. GAVA, E., F. LEGOVINI & N. PAVER. 1975. Nuovo Cimento Lett. **14**: 41.
38. FURLAN, G., N. PAVER & C. VERZEGNASSI. 1976. Nuovo Cimento **32A**: 75.
39. GASSER, J. & H. Leutwyler. 1975. Nucl. Phys. **B94**: 269.
40. GATTO, R., G. SARTORI & M. TONIN. 1968. Phys. Lett. **28B**: 128.
41. CABIBBO, N. & L. MAIANI. 1968. **28B**: 131.
42. JACKIW, R. & H. J. SCHNITZER. 1972. Phys. Rev. **d5**: 2008.

43. MOHAPATRA, R. N. & J. C. PATI. 1975. Phys. Rev. **D11**: 566, 2558.
44. SENJANOVIC, G. & R. N. MOHAPATRA. 1975. Phys. Rev. **D12**: 1502.
45. FRITZSCH, H. & P. MINKOWSKI. 1976. Nucl. Phys. **B103**: 61.
46. MOHAPATRA, R. N. & D. P. SIDHU. 1977. Phys. Rev. Lett. **38**: 667.
47. GEORGI, H. 1977. Harvard preprint HUTP-77/A009, to be published in the Proceedings of the 1977 Coral Gables Symposium.
48. DE RÚJULA, A., H. GEORGI & S. GLASHOW. Harvard preprint HUTP-77/A002, to be published.

ABOUT LIQUIDS

Victor F. Weisskopf

Department of Physics
Massachusetts Institute of Technology
Cambridge, Massachusetts 02139

Centre Européen pour la Récherche Nucleaire
Geneva
Switzerland

Introduction

Under ordinary terrestrial conditions matter appears in three states of aggregation: solids, liquids, and gases. The existence and the general properties of solids and gases are relatively easy to understand once it is realized that atoms or molecules have certain typical properties and interactions that follow from quantum mechanics. Liquids are harder to understand. Assume that a group of intelligent theoretical physicists had lived in closed buildings from birth such that they never had occasion to see any natural structures. Let us forget that it may be impossible to prevent them to see their own bodies and their inputs and outputs. What would they be able to predict from a fundamental knowledge of quantum mechanics? They probably would predict the existence of atoms, of molecules, of solid crystals, both metals and insulators, of gases, but most likely not the existence of liquids.

The essay does not show how the existence of liquids necessarily follows from quantum mechanics. Its aim is much less ambitious. It tries to present oversimplified models of the three states of aggregation that may help to get a better intuitive understanding of simple liquids, of the processes of melting and evaporation, their temperatures and energies, and, in particular, of the viscosity and of the self-diffusion coefficients of liquids.

Nothing in this essay is original, a lot is vastly oversimplified, but the author was happy when he could use these models in order to clarify a few points, which for him, before, were in the well-known gray area of "I should understand this better but I don't."

The Questions

In what follows we will consider simple substances only, that is, substances consisting of atoms or molecules that are exactly spherical and whose internal degrees of freedom do not participate in the heat exchange. We will refer to the constituents as "atoms," even when they may be molecules. The examples on which we will test our ideas are argon, oxygen, sodium, copper, mercury, methane, and water. The molecules of the last substance cannot be considered spherical nor can their internal degrees of freedom be neglected. Water, however, is such an important liquid that we nevertheless will try to use our simplified methods, which do indeed yield the right order of magnitude of the different properties.

We distinguish an energy ϵ_s which is the binding energy per atom of the solid at zero-temperature. This energy assumes very different values for different substances, depending on the type of binding between the "atoms." For example, for noble gases, $\epsilon_s \sim 0.1$ eV whereas for some metals it is of the order of several eV. We distinguish ϵ_B which is the boiling heat per atom at atmospheric pressure, and ϵ_M, the melting heat per atom. The following relation,

$$\epsilon_s \cong \epsilon_B + \epsilon_M + \tfrac{1}{2}kT_B, \qquad (2.1)$$

is a reasonable approximation under the assumption that the heat content of a solid and a liquid is $3kT$ per atom (Dulong-Petit law), whereas it is $(5/2)kT$ in the gas under constant pressure. We will never take into account any quantum effects in this paper. Usually kT_B and ϵ_M is much smaller than ϵ_s so that we may put roughly $\epsilon_s \sim \epsilon_B$. Indeed when we put

$$\epsilon_M = \frac{\epsilon_s}{a} \approx \frac{\epsilon_B}{a}, \qquad (2.2)$$

we find that, roughly speaking, a is of the order 10 to 30 for simple substances as seen in TABLE 1. We will give reasons for this relation. The melting temperature T_M depends somewhat but not very much on the pressure. We observe the following relation

$$kT_M \sim \epsilon_M, \qquad (2.3)$$

where k is Boltzmann's constant. Indeed the ratio ϵ_M/kT_M is always of the order unity as seen in TABLE 1. We will give reasons for this relation too.

The boiling temperature depends strongly on the pressure; in the vacuum it is zero; every substance evaporates in empty space. We define T_B as the boiling temperature at atmospheric pressure. There exists a relation referred to as Troutons rule:

$$kT_B = \frac{1}{b}\epsilon_B, \qquad (2.4)$$

where b is very near to 11 as seen in TABLE 1. We will explain this relation and indicate how the value of b can be calculated.

We also will apply our considerations to an estimate of the self-diffusion coefficient D of a liquid and its viscosity η. The coefficient D is defined as follows: consider an atom at $t = 0$ at $r = 0$; after a time t the average distance $\bar{r}$ of the atom from the point $r = 0$ is:

$$\bar{r} = \sqrt{Dt}. \qquad (2.5)$$

From purely dimensional reasons it follows that we can write

$$D = \frac{w_{\text{th}} d}{3} \cdot \zeta, \quad \zeta \sim \frac{1}{15} \text{ at } T \sim T_M \qquad (2.6)$$

where w_{th} is the thermal velocity of an atom, d is the average distance between atoms (there are d^{-3} atoms per unit volume). For simple liquids near the melting

TABLE 1
OBSERVED AND CALCULATED VALUES FOR VARIOUS SIMPLE LIQUIDS*

	Magnitude	Description, Equation Referred to	Ar	O_2	Na	Cu	Hg	CH_4	H_2O
1	ϵ_B	in eV	0.068	0.071	1.02	3.11	0.61	0.085	0.42
2	ϵ_M	in eV	0.012	0.0046	0.027	0.136	0.024	0.010	0.062
3	kT_B	in eV $\times 10^2$	0.75	0.78	9.9	24.5	5.43	0.96	3.22
4	kT_M	in eV $\times 10^2$	0.73	0.47	3.2	11.7	2.02	0.78	2.35
5	d	in 10^{-8} cm	3.4	3.3	3.4	2.3	2.9	4.0	3.1
6	$\log_{10} p^G$	obs†	2.35	−0.15	−7.1	−3.1	−5.7	1.96	+0.66
7	$\log_{10} p^G$	calc†, 3.10	2.50	+0.12	−6.7	−3.4	−6.0	1.83	[−0.31]
8	$x = \epsilon_M/kT_M$	‡	1.64	0.98	0.84	1.16	1.19	1.26	2.64
9	$b = \epsilon_B/kT_B$		9.1	9.1	10.3	12.7	11.2	8.85	13.0
10	b (theor.)	4.18	8.8	9.1	10.4	13.1	10.9	8.8	12.0
11	$\omega_E \times 10^{-12}$	3.2	6.7	6.9	12	26	5.5	9.6	14.6
12	$\omega_E \times 10^{-12}$	3.4	3.7	4.2	17.8	27.9	5.5	5.1	14.9
13	$\bar{\omega}_E \times 10^{-12}$	mean of the two above	5.2	5.6	14.9	27.0	5.5	7.4	14.6[30]§
14	$\delta_S(T=T_M)$	in 10^{-8} cm, 3.6	0.63	0.53	0.61	0.39	0.45	0.72	[0.30]
15	$\delta_L(T=T_M)$	in 10^{-8} cm, 4.3	0.92	0.69	0.76	0.52	0.60	0.98	[0.50]
16	$\delta_L^*(T=T_B)$	in 10^{-8} cm, 4.14	0.93	0.81	1.17	0.66	0.83	1.04	[0.54]
17	$d/\delta_L = \alpha$	4.1	3.7	4.5	4.5	4.4	4.8	4.1	[6.2]
18	d^3/v_L	obs, 3.5, 4.8	28	88	74	63	84	49	80
19	d^3/v_L	calc, 4.5, 4.6	35	91	70	68	95	56	[226]
20	d^3/v_L^*	obs, 4.19	49	49	13.2	18.4	47	53	585
21	d^3/v_L^*	calc, 4.15	34	50	13.4	28	28	41	[185]
22	ξ	obs, 2.6¶	1/15		1/18	1/12	1/11	1/22	1/65
23	κ	obs, 2.8**	6.4	26	11.1	7.3	9.4	10.6	28

*The observed values are taken from Landolt-Boernstein Numerical Data, 1965. Vol. II, 5, Springer Verlag, Berlin, and the A.I.P. Handbook, 3rd edit. McGraw Hill, New York, 1972.
†$p^{(G)}$ is the vapor pressure over the solid near melting temperature in millimeter mercury.
‡The calculated value is unity according to Equation 4.10.
§We introduce a fictitious value for water, $\omega_E = 30$, which gives better results. All values calculated for water with this value are put in brackets.
¶The theoretical value is 1/12 according to Equation 5.3.
**The theoretical value is 12 according to Equation 5.4.

point, we find that the pure number ζ is of the order of 1/15, as seen in TABLE 1. It will be understandable on the basis of our model.

The viscosity coefficient η of a liquid substance is defined as follows: assume that the substance moves with a velocity u in the x direction, but the magnitude of u changes in the z direction, so that there is a gradient (du/dz) of the overall velocity u. Then a certain amount P of momentum is transferred per unit time through a cm^2 of a plane perpendicular to z. P is proportional to du/dz:

$$P = \eta \frac{du}{dz} \qquad (2.7)$$

and the coefficient η is the viscosity. Again, for purely dimensional reasons we write

$$\eta = \frac{mw_{th}}{3d^2}\kappa, \quad \kappa \sim 12 \text{ at } T \sim T_M, \qquad (2.8)$$

where m is the mass of the atom and κ is a dimensionless number which, at or near melting temperature, is of the order 10 for simple liquids as indicated in TABLE 1. Also this relation will be made plausible by our model.

The size of the constant in Equation 7 is surprising when one compares it with the well-known formula for the viscosity of an ideal gas

$$\eta_{Gas} = \frac{nmw_{th}l}{3}, \quad l = \frac{1}{\sqrt{2}n\pi(2r)^2}. \qquad (2.9)$$

Here n is the number of atoms per unit volume, r is the radius of the atom, and l is the mean-free path. It is well known that this expression does not depend on the density. Surely, expression 8 is only valid for dilute gases but, since η_{Gas} is density independent, one would have thought that it should at least give the right order of magnitude also for a liquid. However, when we put $d \approx 2r$ for the liquid in which the atoms touch each other, we get formula 7 but with $\kappa = (\sqrt{2}\pi)^{-1} = 0.23$; that is about 50 times too small. We will be able to explain this discrepancy.

Gases and Solids

We begin with a discussion of solids and gases and of the equilibrium between these two states of aggregation. This will serve to fix our models and to describe the simplified methods that we will use in order to treat the equilibrium between two phases. We consider the gaseous state as a dilute ideal gas of N spherical atoms without any internal degrees of freedom, enclosed in a volume V. We introduce the volume per atom $v_G = V/N$ which fulfills the ideal gas equation:

$$v_G = \frac{kT}{p}, \qquad (3.1)$$

where p is the pressure.

We now describe our model of a solid (the Einstein model): Here we assume that each atom or molecule is a mass-point (mass m) tied elastically to its rest-position; the latter being fixed in place at cubic lattice points, with d as the nearest distance. The elastic bond has a frequency ω_E which we will determine as follows. The frequency ω_E is connected with the Debye temperature θ:

$$\omega_E = c\frac{k\theta}{h} = c\omega_D \quad c = \frac{1}{\sqrt{3}} \tag{3.2}$$

where c is a numerical factor. The choice of $c = 1/\sqrt{3}$ is based upon the following reasoning: ω_D is the highest frequency of the lattice in the Debye model, which assumes a frequency distribution proportional to $\omega^2 \, d\omega$. The value of c depends upon what average one wants to choose. We are interested mainly in the determination of the square of the average amplitude δ of the oscillators at a given thermal energy. δ^2 is proportional to ω^{-2} at a fixed energy and the average of ω^{-2} in the Debye model is $3\omega_D^{-2}$. Hence we get $\omega_E^2 = 3\omega_D^2$ and $c = 1/\sqrt{3}$.

The Debye frequencies are determined experimentally from the behavior of the specific heat at low temperatures. They are characteristic of the bond strength at small vibrational amplitudes. We get an idea of the behavior at large amplitudes by interpreting ω_E in the following way: In the Einstein model of a solid we would imagine that the energy ϵ_s necessary to lift the atom from its rest position within the solid to a rest position in empty space (we neglect quantum effects such as zero-point energies) must be given approximately by

$$\epsilon_s \cong \tfrac{1}{2} m\omega_E^2 (d/2)^2, \tag{3.3}$$

where d is the distance between the atoms in the solid. It is the potential energy when the atom is displaced by $d/2$. The binding energy ϵ_S is determined from the melting heat ϵ_M and the boiling heat ϵ_B according to relation **2.1**. Thus we get another way to determine ω_E:

$$\omega_E = [8(\epsilon_B + \epsilon_M + \tfrac{1}{2} kT_B)/md^2]^{1/2}. \tag{3.4}$$

We will use the arithmetic mean of the values calculated by **3.2** and **3.4**. The resulting values are found in TABLE 1. In the case of water it turned out that the agreement with facts is much better if a fictitious Einstein frequency is used which is two times higher than the mean of **3.2** and **3.4**.

In all our examples $h\omega_E$ turns out to be reasonably small compared to the melting temperatures. We therefore are allowed to neglect quantum effects when we are dealing with temperatures near melting or higher as we will do throughout this paper.

We now introduce the concept of "available volume" of an atom at a given temperature T. It is easy to visualize that concept in the Einstein-model of a solid. The available volume v_s is a measure of the volume in which the motion of the atom takes place; it will be of the order of the cube of the amplitude of vibration, that is proportional to $[kT/(m\omega_E^2)]^{3/2}$. A more exact definition of this volume is

$$v_S = \int dx^3 \exp\left[-\pi(x)/kT\right] = \left(\frac{2\pi kT}{m\omega_E^2}\right)^{3/2}, \tag{3.5}$$

where $\pi(x) = \tfrac{1}{2} m\omega_E^2 x^2$ is potential energy of the oscillator. This expression represents an integral over the volume weighted by the probability of binding the atom there.

It is useful to introduce a length δ_s indicating the linear dimension of the available volume:

$$\delta_s = v_s^{1/3} = \left(\frac{2\pi kT}{m\omega_E^2}\right)^{1/2}. \tag{3.6}$$

The length δ_s is of the order of 0.1 of the lattice distance d below the melting point. The volume v_s is of the order 10^{-3} of the cell volume d^3 and reaches about $1/250$ of d^3 at the melting point. Notice that δ_s is the cube root of the available volume and not the average amplitude of the oscillator vibrations. The latter amplitude R_s would be approximately equal to the radius of a sphere of that volume:

$$R_s \approx \left(\frac{3}{4\pi}\right)^{1/3} \delta_s. \tag{3.6a}$$

It probably is intuitively plausible that the corresponding available volume v_G for a gas is the volume per atom, namely, V/N as given by **3.1**:

$$v_G = V/N. \tag{3.7}$$

At atmospheric pressure this volume is about 1000 times larger than the cell volume d^3 of a solid and therefore several 10^5 times larger than the available volume in a solid.

We will use the available-volume concept in order to express the condition of equilibrium between two phases I and II. Such equilibrium exists when

$$\frac{v_\mathrm{I}}{v_\mathrm{II}} = e^{\Delta\epsilon/kT} \quad \Delta\epsilon = \epsilon_\mathrm{I} - \epsilon_\mathrm{II}. \tag{3.8}$$

Here v_i are the available volumes in the two phases and ϵ_i are the energies per atom of the two phases, but without counting the thermal energies, that is, when the atoms are at rest. So ϵ_i is zero in the gas phase and $\epsilon_i = -\epsilon_s$ in the solid phase. The relation **3.8** is intuitively plausible. Two phases can coexist if the difference between the available volumes is compensated by the Boltzmann factor corresponding to the difference in energy. The relation **3.8** is derived from the general laws of statistical mechanics in the Appendix.

Applying the relation **3.8** to the gas-solid equilibrium we get

$$\frac{v_G}{v_S} = e^{\epsilon_S/kT}. \tag{3.9}$$

In the gas, the atom has a much larger available volume to its disposition than in the solid, but the energy in the solid is lower by ϵ_S. We can transform **3.9** into an expression of the vapor pressure $p^{(G)}$ above a solid by means of **3.1** and get

$$p_G = \frac{kT}{v_s} e^{-\epsilon_S/kT}. \tag{3.10}$$

This equation says that, in our model, the vapor pressure is that of an ideal gas compressed so that the volume per atom is v_s, but then reduced by the Boltzmann factor exp $(-\epsilon_s/kT)$.

We will show a few examples of how our model works. Of course, our simplified model can only give very approximative results. As mentioned before, all quantum effects including zero-point energies are neglected.

All data, the ones that enter into our models, the results and the actual values, are assembled in TABLE 1. We see by comparing lines 6 and 7 that the values for the vapor pressures near the melting point are not too badly reproduced. Although the pressures differ by nine orders of magnitude, our results are good within a factor 2.5, except for water where we are off by a factor of 8.

Liquids

In an idealized solid the "location" x_i of each atom is fixed at the lattice points. We understand by "location" not its exact position but the position of the center of that volume $\delta_s^3 = v_s$ within which it performs harmonic vibrations (in the Einstein picture) with an amplitude of the other δ_s. This is long-range order. We now make the following assumption. When δ_S reaches a certain critical value δ_L the location x_i of the atoms do no longer remain fixed; the structure then is loosened such that the atoms do no longer oscillate around a fixed position; the locations change due to combinations of vibrations of neighboring atoms: the long-range order disappears. Then the individual atom rather performs a random motion with steps of the order δ_L instead of an exact oscillation around a fixed location. This will be our description of the liquid state. We picture it by imagining spherical atoms of size $\sim d$, almost closely packed but with spaces inbetween, allowing relative motions of neighbors against each other, of order δ_L. They move oscillator like for distances δ_L but do not necessarily return to the same point when the restoring force pushes them back. They perform a "hindered" random walk. The term "hindered" expresses the fact that the small ratio δ_L/d prevents them from moving freely in all directions.

It is very difficult to determine from first principles the critical distance δ_L at which the long-range order disappears. We will get from an analysis of the conditions that the transition from long-range order to the liquid state occurs when

$$\delta_L \geq \frac{d}{\alpha}, \quad \alpha \sim 4.5, \tag{4.1}$$

that is, the random displacements are of the order of one quarter to a fifth of the lattice distance d or larger.*

This loosening up of the solid state long-range order needs energy. It is the energy necessary to increase the δ_S given by 3.6 to δ_L as defined above. The melting heat ϵ_M is just that energy necessary to increase δ_S to δ_L at T_M. Assuming that the binding forces roughly have the character of a restoring force in an oscilla-

---

*Here and in the rest that follows, we neglect the difference in density of liquids and solids. We always define d as $d^3 = V/N$, where V is the volume of N atoms in the solid or liquid.

tor, the energy ϵ_M necessary to extend the displacements from δ_S to δ_L in the solid should be nearly equal to the difference in potential energy of an oscillator with those displacements. The potential energy is proportional to the square of the displacement. At the temperature T_M, the average potential energy is $(3/2)kT_M$. Thus we get

$$\frac{(3/2)kT_M + \epsilon_M}{(3/2)kT_M} = \frac{\delta_L^2}{\delta_S^2}. \tag{4.2}$$

So ϵ_M is the energy to stretch δ_S to δ_L at T_M. This equation allows us to calculate δ_L at the melting point from δ_S, which is given by 3.6:

$$\delta_L = \left(1 + \frac{2\epsilon_M}{3kT_M}\right)^{1/2} \left(\frac{2\pi kT}{m\omega_E^2}\right)^{1/2} \tag{4.3}$$

TABLE 1 shows the values δ_L according to 4.3 for a few simple liquids. Now we can determine empirically $\alpha = d/\delta_L$ at the melting point; the values are listed in the TABLE 1, line 17. They are the basis for our claim that α lies between 4 and 5.

The motion of the atom in the liquid is no longer exactly that of an oscillator since it does not return to the same point. But its motion still is "back and forth" with an amplitude corresponding to δ_L. If we want to approximate the force that drives it back by an oscillator-type restoring force, that oscillator would have a different (smaller) frequency ω_E'. It is easy to determine ω_E', since that new oscillator has an amplitude corresponding to δ_L at T_M, whereas the oscillator describing the solid state had an amplitude corresponding to δ_S. Since the amplitudes at fixed T are proportional to ω^{-1}, we get

$$\omega_E' = (\delta_S/\delta_L)\omega_E, \tag{4.4}$$

where δ_S/δ_L is given by 4.2. We emphasize again that the motion of the atom in our model of the liquid state is an oscillator motion only in a very approximative sense because it does not return to the same point of origin. Our model of the solid, however, deals with exact oscillatory motions of the atom.

The energy of the atom at rest in that oscillator which approximately describes the motion in the liquid, is not $-\epsilon_S$ as it is in the oscillator of the Einstein model of the solid, but $-\epsilon_L = -\epsilon_S + \epsilon_M$; it is higher by the melting heat. We therefore get

$$\epsilon_S - \epsilon_L = \epsilon_M. \tag{4.4a}$$

We now determine the increase of the available space v_S when a solid becomes liquid. In order to do this we must try to describe the liquid state a little more accurately. The liquid consists of atoms moving through distances δ_L before turning into another direction because of the presence of neighboring atoms. Since δ_L is rather much smaller than the average distance d between the atoms, the displacements of neighboring atoms cannot be completely independent of each other, they must be coordinated; one must make room for the other. Actually there will be a continuous transition from nearby atoms whose locations are tightly correlated and others further away that are less tightly correlated. We try to simplify the situation by introducing the concept of a "clump" of f neighboring atoms whose relative positions are as tightly correlated as in a solid (they form a lattice), whereas the positions of the atoms outside the clump are not at all correlated to those inside the clump. We do not assign definite atoms to definite clumps; whenever we look at the neighborhood of an atom we assume that f surrounding atoms form a clump.

The size of a clump is connected with the ratio $\alpha = d/\delta_L$ for the following reason: The position of an atom, α lattice distances away, should be independent of the position of a reference atom. Since our clump describes the continuously diminishing dependence by a sharp break from complete lattice to complete independence, we believe that the linear dimensions of a clump should be $\xi \alpha d$ where ξ is less than unity, say about 0.6. This choice will give reasonable results. We then have

$$f = \left(\xi \frac{d}{\delta_L}\right)^3 = (\xi \alpha)^3, \quad \xi \sim 0.6 \tag{4.5}$$

which, with the choice of $\alpha \approx 4.5$ would give $f \sim 20$. It should be noted that the values of α and f are connected; in principle, we have introduced only one arbitrary constant α to describe the transition from the solid to the liquid state.

This rough picture allows us to calculate the available space volume v_L in a liquid. We determine it on the basis of our clump model: if the position of all the atoms were completely free like in a gas, we would get $v_L \sim V/N = d^3$. Actually among the f atoms of a clump only one—we call it the reference atom—is completely free, whereas the other $(f - 1)$ atoms have only a volume δ_L^3 at their disposition since their locations are coordinated to that of the reference atom. Therefore the position phase space for the f atoms of the clump is

$$(v_L)^f = d^3 \delta_L^{3(f-1)},$$

and the available volume of a single atom in the liquid becomes

$$v_L = \delta_L^3 \left(\frac{d}{\delta_L}\right)^{3/f} = \delta_L^3 \alpha^{3/f}. \tag{4.6}$$

Note that the available volume is larger than the δ_L^3 by the factor $(d/\delta_L)^{3/f}$. This is because the motion of the atom is not really oscillatory in the liquid. The above factor expresses the increase in available volume due to the fact that the atom does not return to the same place in its oscillations within a volume δ_L^3.

The expression **4.6** allows us to determine directly the ratio v_L/d^3, from the ratio $\alpha = d/\delta_L$ at the melting point. We get

$$\frac{v_L}{d^3} = \alpha^{(-3+3/f)}. \tag{4.7}$$

As a first orientation, we may put $\alpha = 4.5$ and $f = 20$, and we get $v_L \approx d^3/73$.

We now proceed to calculate the equilibrium between the liquid and solid phase and use again expression **3.8**, which tells us that the ratio of available volumes must equal the Boltzmann-factor corresponding to the energy difference $\Delta \epsilon$ between the two phases. Of course, this difference is equal to the melting heat:

$\Delta\epsilon = \epsilon_M$.† We get

$$\frac{v_L}{v_S} = e^{\epsilon_M/kT_M}. \qquad (4.8)$$

We now can check our result **4.7** for v_L in terms of d^3. We determine v_L empirically from **4.8** by using **3.5** for v_S and the empirical values for d^3, ϵ_M, and T_M. The result is found in TABLE 1 line 18. We see that the values are not too far from $1/73$, which we would get by using **4.7** with $\alpha = 4.5$ and $f = 20$. If we use the actual values of α as listed in the table, the agreement with the empirical v_L is even better, (see line 19 of the table), except in the case of water.

The relation **4.8** allows us to get an estimate of the ratio between the melting heats ϵ_M and the melting temperature T_M. A very rough result is obtained by noting from TABLE 1 that $v_S \sim d^3/250$ whereas $v_L \sim (1/73)d^3$ according to **4.7** with $\alpha = 4.5$ and $f = 20$. Thus the ratio $v_L/v_S \sim 3$. Then it follows immediately from **4.8** that $\exp(\epsilon_M/kT_M) = 3$ or $\epsilon_M \approx kT_M$.

We can come to the same conclusion without making use of the empirical relation $v_S \sim d^3/250$. We use **3.5** and **4.6** in order to express the volumes v_S and v_L in terms of the lengths δ_S and δ_L, and we express the ratio of these two lengths in terms of ϵ_M/kT_M by means of **4.2**. We then find that **4.8** can be written in the form

$$\left(1 + \frac{2\epsilon_M}{3kT}\right)^{3/2} \alpha^{3/f} = \exp(\epsilon_M/kT_M). \qquad (4.9)$$

This relation determines ϵ_M/kT_M; it is equivalent to the following equation for $x \equiv \epsilon_M/kT_M$:

$$\alpha^{3/f}(1 + \tfrac{2}{3}x)^{3/2} = e^x. \qquad (4.10)$$

With the previously determined values $\alpha \sim 4.5$, $f \sim 20$ valid at the melting point, we find that the solution of **4.10** is $x \sim 1$.

Our model therefore predicts that $\epsilon_M \sim kT_M$, a relation that indeed is approximately fulfilled as seen in TABLE 1, line 8.

It is interesting to point out that our relations also contain the so-called Lindemann Melting-Point Formula. It connects the Debye temperature θ with the melting temperature T_M and usually is written in the following form:

$$\theta = D\left(\frac{T_M}{\mu V^{2/3}}\right)^{1/2}, \quad D = 120 \text{ cm g}^{1/2} \text{ deg}^{1/2}. \qquad (4.11)$$

Here μ is the molar weight in grams and V is the molar volume. In our notation $V = Ad^3$, where A is the Avogadro number and $\mu = Am$. Since $k\theta/h = \omega_D$ we may write **4.11** in the form

$$\omega_D = D'\left(\frac{kT_M}{md^2}\right)^{1/2}, \quad D' = 20.4. \qquad (4.12)$$

---

†Exactly speaking, $\Delta\epsilon$ should be the energy difference without counting the heat content. Since we describe both phases approximately by oscillators, the heat content is the same. The relation $\Delta\epsilon = \epsilon_M$ was shown to be correct in Equation 4.4a.

Now D' is a dimensionless constant. We can derive a similar relation from Equation 4.3 by putting $e_M \sim kT_M$. Then 4.3 becomes

$$\delta_L^2 = \frac{5}{3} \frac{2\pi k T_M}{m\omega_E^2}.$$

This can be written in the form

$$\omega_E^2 = \frac{5}{3} 2\pi \frac{kT_M}{md^2} \frac{d^2}{\delta_L^2}.$$

We use 3.2 to connect ω_E with ω_D and 4.1 to determine (d/δ_L). Then we obtain indeed Lindemann's formula 4.12 with $D' = 25.2$, which is near enough to Lindemann's value, considering the approximations made.

The same equations also permit an estimate of the melting heat in terms of the binding energy of the solid. The melting heat ϵ_M increases the average oscillator amplitude from the one corresponding to δ_S to the one corresponding to δ_L. We call these amplitudes R_L and R_S and they are related to δ_S and δ_L as indicated in 3.6a. The energy needed for this is

$$\epsilon_M = \frac{1}{2} m\omega_E^2 (R_L^2 - R_S^2) = \frac{1}{2} m\omega_E^2 \left(\frac{3}{4\pi}\right)^{2/3} \delta_L^2 \left(1 - \frac{\delta_S^2}{\delta_L^2}\right)$$

Equation 4.2 tells us that the ratio $(\delta_S/\delta_L)^2$ is $\frac{3}{5}$ with $x = 1$. Thus we obtain

$$\epsilon_M = \frac{1}{5} \left(\frac{3}{4\pi}\right)^{2/3} m\omega_E^2 \delta_L^2 = \frac{8}{5} \left(\frac{3}{4\pi}\right)^{2/3} \left(\frac{\delta_L}{d}\right)^2 \cdot \frac{1}{8} m\omega_E d^2.$$

in the previous section we already have interpreted $\frac{1}{8} m\omega_E^2 d^2$ as approximately equal to ϵ_S. We therefore get with $d = 4.5 \delta_L$:

$$\epsilon_M \approx \frac{1}{33} \cdot \frac{1}{8} m\omega_E d^2 \sim \frac{1}{33} \epsilon_S. \qquad (4.13)$$

We find that the melting heat is a small fraction of the binding energy or of the boiling heat according to 2.1. Actually the values of ϵ_M/ϵ_B lie between 1/7 and 1/40 as seen by comparing lines 1 and 2 of TABLE 1. Our considerations give the right order of magnitude. As expected, the first equality in 4.13 is somewhat better fulfilled; the values of $\epsilon_M/(1/8 m\omega_E^2 d^2)$ lie between 1/14 and 1/34. The deviations from the relation 4.13 come from the fact that the binding potential is not exactly an oscillator potential. It rises less steeply than with increasing distance. Equation 4.13 therefore probably is an underestimate of the ratio ϵ_M/ϵ_S.

We want to determine the available volume of a liquid not only at melting temperature but also at boiling temperature T_B. Let us distinguish the magnitudes at boiling by a star, such as v_L^* and δ_L^*. We make the simplest possible assumption, namely, that δ_L^* is the result of an extension of an oscillators amplitude at the temperature T_B when the melting heat ϵ_M is added, just as δ_L was the result of a similar extension of an oscillator at the temperature T_M. Then we get in analogy to **4.3**:

$$\delta_L^* = \delta_S^* \left(1 + \frac{2}{3} \frac{\epsilon_M}{kT_B}\right)^{1/2}, \delta_S^* = \left(\frac{2\pi kT_B}{m\omega_E^2}\right)^{1/2} \quad (4.14)$$

Together with δ_L the value of f will also change with temperature. Expression **4.5** tells us that f is proportional to δ_L^{-3}, so that we find

$$f^* = f\left(\frac{\delta_L}{\delta_L^*}\right)^3, v_L^* = (\delta_L^*)^3 \left(\frac{d}{\delta_L^*}\right)^{3/f^*} \quad (4.15)$$

As a first orientation we may put very approximately $T_B \sim 2T_M$ and $\epsilon_M \sim kT_M$. Then $\alpha^* = d/\delta_L^* \sim 3.6$ and $f^* \sim 10$. The available volume amounts to

$$v_L^* \sim (\alpha^*)^{-3+3/f^*} d^3 \sim \tfrac{1}{32} d^3. \quad (4.16)$$

We may compare this with the equally approximative value of v_L at the melting point: $v_L \sim (1/73) d^3$.

Now we construct the equation regulating the boiling process, in analogy to **3.9** and **4.8**:

$$\frac{v_G}{v_L^*} = e^{(\epsilon_S - \epsilon_M)/kT_B}, \epsilon_S - \epsilon_M = \epsilon_B + \frac{1}{2} kT_B \quad (4.17)$$

Here $\epsilon_S - \epsilon_M$ is the difference in energy between liquid and gas; it is equal to $\epsilon_B + \tfrac{1}{2}kT_B$ according to **2.1**. This relation allows us to get an idea of the ratio ϵ_B/kT_B. We again start with a rough estimate. Remember that gases at atmospheric pressure have a density of about 10^{-3} of solids; hence $v_G/d^3 \sim 10^3$. Furthermore **4.16** tells us that $v_G \sim d^3/32$; thus $v_G/v_L^* \sim 32000$ and $\epsilon_B/kT = ln\ 32000 + \tfrac{1}{2}$, which is just about 11. We have derived Trouton's rule!

A somewhat more detailed determination of $b = \epsilon_B/kT_B$ uses the ideal gas equation, $v_G = kT_B/p_0$, where p_0 is the atmospheric pressure, and **4.14** and **4.15** for the calculation of v_L^*. One then gets from **4.17**

$$b + \frac{1}{2} = \ln\left[\frac{kT_B}{p_0}/v_L^*\right]. \quad (4.18)$$

TABLE 1 shows that the values calculated with this expression agree quite well with the observed ones.

Trouton's rule can be used to get an upper limit on the melting heat ϵ_M. We previously found that $\epsilon_M \sim kT_M$. Obviously the boiling temperature must be higher than the melting temperature. Thus we get $\epsilon_M < kT_B$ and from Trouton's rule

$$\epsilon_M < \epsilon_B/11.$$

Since our previously derived expression was an underestimate, we conclude that ϵ_M should be somewhere between 1/11 and 1/33 of the boiling heat, which indeed is borne out by TABLE 1.

A more sensitive way to test **4.17** than the calculation of the Trouton coefficient b would be a comparison of v_L^* obtained "theoretically" from **4.14** and

4.15, with the "empirical value":

$$v_L^* = \frac{kT_B}{p_0} e^{-\epsilon'_B/kT} \text{ (empirical)}, \qquad (4.19)$$

where $\epsilon'_B = \epsilon_B + \tfrac{1}{2}kT_B$.

TABLE 1 gives the theoretical and the empirical values in terms of d^3 (lines 20 and 21). Here, as well as in the calculation of v_L at the melting point, the semiquantitative agreement is gratifying, particularly in view of the fact that the vL values depend on the heats of transformation and on the corresponding temperatures through the Boltzmann exponential, which, in particular in the case of the boiling process, is a strongly varying function. Note that v_L^* for sodium is abnormally small, a fact that is reproduced by the theory and is caused by the very high ratio between boiling and melting temperature. Our assumptions do not really apply to the case of water where we find factors of three between observed and calculated values.

Diffusion and Viscosity

Our model allows us to figure out the approximative values of the self-diffusion constant D and the viscosity η of a liquid. D is defined as follows: Consider an atom at $t = 0$ at $r = 0$; after a time t its average distance from the center is $\bar{r} = \sqrt{Dt}$. In our model of the liquid the atom performs a "hindered" random walk with δ_L as the length of the step. It is "hindered" because of the fact that δ_L is small compared to the distance d between the atoms. The direction of the next step is not completely independent of the previous one; it is more probable that it is in or near the opposite direction. This would decrease the diffusion coefficient compared to a true random motion. Were it a true random motion, we would get $D_r = \delta_L w_{th}/3$ where w_{th} is the average (thermal) velocity of the atom. Actually we write

$$D = F(f) \frac{\delta_L w_{th}}{3}, \qquad (5.1)$$

where the function F expresses the deviation from random. We expect F to be a decreasing function of the clump size f and, naturally $F = 1$ for $f = 1$. We may imagine that $F(f)$ has something to do with the ratio of the number of atoms at the clump surface to the total number in the clump, since the atoms at the surface have more freedom of moving. We therefore expect an approximate proportionality to $f^{-1/3}$ and we will tentatively put

$$F(f) \sim f^{-1/3}. \qquad (5.2)$$

We then get

$$D \approx \frac{\delta_L w_{th}}{3f^{1/3}} = \frac{dw_{th}}{3}\xi, \quad \xi = \frac{\delta_L}{df^{1/3}}. \qquad (5.3)$$

With $d/\delta_L \sim 4.5$ and $f \sim 20$ near the melting point we find $\xi \sim 1/12$ which is the right order of magnitude according to TABLE 1, line 21. Our formula gives a temperature dependence proportional to T^n with n lying between 1 and 3/2, since w_{th} is proportional to $T^{1/2}$ and δ_L and $f^{-1/3}$ are both proportional to a power slightly less than $\frac{1}{2}$ (see **4.14** and **4.15**). Actually, however, D rises at least as fast as T^2. The stronger temperature dependence can probably be explained by the anharmonicity of the atomic vibrations which makes the amplitude δ_L rise faster than **4.14**. In water the increase is even stronger and is caused by the breakage of hydrogen bonds with rising temperature.

We now try to determine the viscosity of a liquid. We have defined the viscosity coefficient η in Equation **2.7** as the factor between the momentum transfer P in the z direction per cm^2 and second when there is a gradient du/dz of the overall velocity u of the liquid in the x direction. We make use of the concept of "clump" and assume that all atoms in a clump have the same average velocity u. Let us look at an area in the x-y plane. There are d^{-2} atoms per cm^2 in or near (within d) that area, each of them performing vibrations with a frequency ω'_E which we have determined in **4.4**. One third of them vibrate perpendicular to the plane. Each of them moves (ω'_E/π) times per second up or down. Assume that the area separates one clump above from one below. Then each time an atom vibrates downwards it transfers the average momentum that it has in the upper clump to the lower clump and vice versa. The average momentum of each atom in the upper clump is larger by $M(du/dz)\Delta$ than the one in the lower clump, where $\Delta \sim f^{1/3}d$ is the linear dimension of a clump. We then get

$$P = \frac{1}{3} \frac{M}{d^2} \frac{\omega'_E}{\pi} f^{1/3} d \frac{du}{dz}$$

or

$$\eta = \frac{M\omega'_E}{3\pi d} f^{1/3}.$$

The atoms move back and forth over distances δ_L with an average velocity w_{th}. Hence $\omega'_E/\pi = w_{th}/\delta_L$. So we get:

$$\eta = \frac{Mw_{th}}{3d^2} k, \quad \kappa = f^{1/3} \frac{d}{\delta_L}. \tag{5.4}$$

We put again $d/\delta_L = \alpha \sim 4.5$ and $f = 20$ near the melting point; this gives us a value of $\kappa \approx 12$, in reasonably good agreement with the observed values found in TABLE 1, line 22. Again the temperature dependence of our expression is too weak. The coefficient κ goes as $f^{1/3}/\delta_L$ which would be somewhat less than T^{-1}, w_{th} goes with $T^{1/2}$. Altogether we would get T^m with m between 0 and $-\frac{1}{2}$ whereas, in fact, the viscosity diminishes strongly with increasing temperature with powers somewhat larger than unity. Simple liquids show a decrease of viscosity of about a factor two to four of the viscosity between melting and boiling. This effect can perhaps be understood qualitatively by the following corrections to our picture: the momentum delivered when "entering" in the adjacent clump might be larger than $M(du/dz)\Delta$, since the atom is to some extent "rigidly" tied to the clump.

This can be expressed by replacing M in **5.4** by $M' > M$. The ratio M'/M certainly decreases with temperature since the clumps become "looser" with rising T.

Water, as usual, is a special case again since the viscosity changes by a factor more than six between melting and boiling. Here the rising of temperature breaks more hydrogen bonds between the molecules; thus the momentum transfer is reduced considerably.

Complicated liquids, especially those with long chain molecules have a much larger viscosity, in particular near the melting point. Here the chains are mutually entangled; one chain winds itself around another. This is a phenomenon called "reptation" by L. DeGennes and causes a direct momentum transport over distances much longer than the chain length. The corresponding increase of viscosity may reach factors of a million at lower temperatures. Our simple expression represents only a lower bound for the viscosity in the case of complicated liquids.

Let us compare the expression **5.4** with the well-known formula of the viscosity of an ideal gas as given by **2.9**. This expression does not depend on the density, but it would be wrong to consider a liquid a highly compressed gas and use **2.9** for the viscosity. Indeed, the ratio of **5.4** to **2.9** is

$$\frac{\eta}{\eta_{gas}} = \sqrt{2}\,\frac{(2r)^2}{d\delta_L}\,f^{1/3}.$$

In the liquid we may put approximately $2r \approx d$, and we get with $\alpha = d/\delta_L \sim 4.5$ and $f = 20$, near the melting point:

$$\frac{\eta}{\eta_{gas}} = \sqrt{2}\pi\alpha f^{1/3} \sim 50.$$

The viscosity of the liquid is roughly two orders of magnitude higher than the one of a gas at the same temperature. The physical reasons for this factor are these: In the gas, the distance of momentum transfer is of the order of the mean free path l and the number of atoms crossing a unit area per second is $\nu_G \sim \frac{1}{3}\,w_{th}/d^3$. In the liquid the distance of momentum transfer is $\Delta \sim f^{1/3}d$, and the number of atoms crossing a unit area per second (by vibration) is $\nu_L \sim \frac{1}{3}\,w_{th}/(\delta_L d^2)$. At densities for which the atoms touch each other, $2r \sim d$, and the gas-kinetic expression $l = d^3/[\sqrt{2}\,\pi(2r)^2]$ becomes $l \sim d/\sqrt{2}\,\pi$ which, very roughly, is equivalent to $l \sim \delta_L$. Thus the viscosity of a liquid is larger because the distance of momentum transfer is $f^{1/3}d$ instead of $l \sim \delta_L$ (a factor about 12), and because there are ν_L crossings instead of ν_G (another factor about 4.5).

Thus we have shown that our oversimplified model of a liquid indeed gives rise to a viscosity and diffusion coefficient of the right order of magnitude. It is true, however, that the temperature dependences of these magnitudes are only rather qualitatively interpreted and probably insufficiently explained. Our model emphasizes the atomic rearrangements rather than the penetration of potential barriers in contrast to most of the models used in the current literature.‡

---

‡An excellent survey about the present state of our knowledge of liquids by John A. Barker and Doug Henderson has appeared recently in the *Review of Modern Physics*, 1976. **48**:587.

Acknowledgments

The author is grateful for most valuable advice from Richard Feynman and Kerson Huang and many others.

Appendix

In an equilibrium between two phases I and II, the Gibbs function G must not change when an atom is transferred from one phase to another and temperature, pressure and total number N of atoms is kept constant; $N = N_\mathrm{I} + N_\mathrm{II}$, where N_I and N_II are the number of atoms in the two phases:

$$\left(\frac{\partial G}{\partial N_\mathrm{I}}\right)_{T,p,N} = 0; \tag{A.1}$$

the Gibbs function of the two phase system can be expressed by the partition functions in the following way:

$$G = -kT \log (Q^{(\mathrm{I})} \cdot Q^{(\mathrm{II})}).$$

Here $Q^{(i)}$ is the partition function of the phase i containing N_i particles. In our simple models the partition function $Q^{(i)}$ can be written as the N_ith power of a magnitude $Z^{(i)}$:

$$Q^{(i)} = (Z^{(i)})^{N_i}, \tag{A.2}$$

where $Z^{(i)}$ is independent of N_i. We may call $Z^{(i)}$ the one-particle partition function. We then get

$$G = -kT(N_\mathrm{I} \log Z^{(\mathrm{I})} + (N - N_\mathrm{I}) \log Z^{(\mathrm{II})}).$$

Equation A.1 then becomes for $T \neq 0$,

$$\log \frac{Z^{(\mathrm{I})}}{Z^{(\mathrm{II})}} = 0 \quad \text{or} \quad Z^{(\mathrm{I})} = Z^{(\mathrm{II})}. \tag{A.3}$$

We now calculate the partition functions. The general formula for a system of N equal particles is

$$Q = \frac{1}{N!} \int dx\, dp\, e^{-E(x,p)/kT}.$$

Here x stands for all $3n$ position coordinates and p stands for all $3n$ momentum coordinates. $E(x, p)$ is the energy of the system which can be split into three parts:

$$E(n, p) = N\epsilon + P(x) + \kappa(p), \tag{A.4}$$

where ϵ is the energy per atom when the atoms are at rest, $P(x)$ is the potential energy, and $\kappa(p)$ the kinetic energy. The potential energy P is measured such that it is zero if all atoms are at their rest position. $P(x)$ also includes the potential energy pV of the gas container which keeps the pressure constant. Then the partition function can be factorized:

$$Q = \frac{1}{N!} e^{-N\epsilon/kT} \int dx\, e^{-P(x)/kT} \int dp\, e^{-\kappa(p)/kT}. \quad (A.5)$$

In our model the kinetic energy will always be

$$\sum_i p_i^2/2m,$$

so that the integral over the momenta will be the same in all phases:

$$\int dp\, e^{-\kappa(p)/kT} = (2\pi mkT)^{3N/2}. \quad (A.6)$$

We now introduce the definition of the "available volume" v_i of the phase i:

$$v_i = \left(\frac{1}{N!} \int dx\, e^{-P_i(x)/kT}\right)^{1/N}. \quad (A.7)$$

We then get from **A.2**

$$Z^{(i)} = e^{-\epsilon_i/kT} v_i (2\pi mkT)^{3/2},$$

and the equilibrium condition **A.3** becomes

$$\frac{v_I}{v_{II}} = e^{(\epsilon_I - \epsilon_{II})/kT}. \quad (A.8)$$

This is the expression that we have used.

It is easy to show that the definition **A.7** indeed yields the expression **3.7** for v_G and **3.5** for v_S. In the Einstein model for the solid the potential energy is

$$P(x) = \frac{1}{2} \sum_k m\omega_E^2 x^2$$

where the sum is taken over all particles coordinates. The factor $1/N!$ in **A.7** is cancelled by the fact that N atoms can be distributed over the N oscillator sites in $N!$ ways. Hence we get

$$v_S = \int dx^3 \exp\left(-\frac{1}{2} m\omega_E^2 x^2/kT\right) = \left(\frac{2\pi kT}{m\omega_E^2}\right)^{3/2}.$$

When we apply **A.7** to an ideal gas at a constant pressure p, the only potential energy entering is $P(x) = pV$, that is, the work exerted against the containers wall. Now $pV = NkT$ so that we get

$$v_G^N = \frac{1}{N!} \int dx\, e^{-P(x)/kT} = e^{-N} V^N.$$

We make use of the Stirling formula $e^N N! = N^N$ and obtain

$$v_G = V/N.$$

The determination of v_L from v_S is described in the section entitled "Liquids."

A FEW SPECIALIZATIONS OF THE GENERIC LOCAL FIELD IN ELECTROMAGNETISM AND GRAVITATION

John Archibald Wheeler

Physics Department
University of Texas
Austin, Texas 78712

How many narrowing-downs it takes to go from the generic local field to the special field for the special purpose (that is, what specialization makes certain electromagnetic and gravitational fields more interesting than others) is illustrated here by (1) the static magnetic field employed by Rabi and his group in their early molecular-beam experiments, (2) the corresponding electrostatic field used in measurements of electric moments, (3) the elementary magnetic mirror, (4) a "minimum-B" configuration of the type given by a "baseball" field winding, (5) the electric field associated with an atom bound in a crystal lattice, (6) the metric and gravitational field around a black hole or other compact mass associated in a lattice with 4, 7, 15, 23, 119, or 599 identical masses, regularly disposed to constitute a model "lattice universe" that is closed and follows closely the dynamics of the Friedmann universe, (7) the generic local electromagnetic field, reduced by a suitable "boost" of the observer to the canonical configuration, in which **E** and **B** are perceived as parallel, (8) the generic "part of the gravitational field that is of local origin," identical up to a factor with the stress-energy tensor, and reduced to diagonal form by a suitable Lorentz transformation, and (9) the generic "part of the gravitational field that is of non-local origin," composed of "electric" and "magnetic" parts each described by a 3 × 3 traceless symmetric matrix, these two matrices coming simultaneously to diagonal form in quite a different Lorentz frame.

Special Fields for Special Purposes

Not least of the attractions of the seminars organized jointly by Gregory Breit and I. I. Rabi to one participant of 1933-34 was contact with the nuts-and-bolts design of experiments for the deflection of atomic and molecular beams. Everyone by then was familiar with the issues raised by the Stern-Gerlach experiment and by the Ehrenfest-Einstein[1] discussion of the process of spin orientation: "The difficulties referred to show how unsatisfying are both ways discussed here to understand the results of Stern and Gerlach. Bohr's view—that in complicated fields no sharp quantization at all occurs—remains undiscussed [*unbesprochen*]." Discussed and developed and understood it had become by 1933. The central point had grown clear. Is the change in the magnetic field slow or fast, adiabatic or nonadiabatic? In the one case quantum jumps occur practically not at all. In the other case jumps occur with substantial probability. More specifically, an "adiabatic change" came to be understood as a change in field, all significant Fourier components of which have frequencies lower than the frequency of the lowest possible quantum jump of the atom or molecule in the field in question. Still in the future was the advance inspired by the attempts of C. Gorter[2,3] (see also his review[4]) later to be analyzed by Rabi[5] and achieved by Rabi, Zacharias, Millman, and Kusch[6] with the aid of microwave technology: make a clean separation between the orientation-producing component of the field and the transition-producing component, of tunable frequency. Already available, however, was the beautiful treatment by Zener[7] of the probability of a transition as a function of rate of change of field under conditions where the field change brings about a level crossing, prefiguring much to come in the years ahead.

In the arrangement of premicrowave days (see the volume of reminiscences edited by Estermann[8]; for the subsequent development of molecular beam research see for example Ramsey[9]), the magnetic lines of force typically converge from a long hollow "V" onto a long wedge running lengthwise parallel to each other and to the beam. The cross section at any point along the length has the shape of a wedge of pie partly removed from an otherwise complete pie. The magnetic field is essentially independent of the coordinate x in the direction of travel. In the region of the beam the magnetic potential is representable in the form

$$\Phi = -az - \tfrac{1}{2}bz^2 + \tfrac{1}{2}by^2, \tag{1}$$

with

$$\begin{aligned} B_x &= 0 \\ B_y &= -by \\ B_z &= a + bz \end{aligned} \tag{2}$$

and

$$\nabla \cdot \mathbf{B} = 0. \tag{3}$$

The force on an object of natural magnetic moment μ oriented in the z-direction is determined by the one independent coefficient that shows up in the terms in Φ of second order; thus,

$$F_z = -d(\text{energy})/dz = \mu \partial B_z/\partial z = \mu b. \tag{4}$$

What kind of specialization of field shows up here? More specifically, how many narrowing-downs does it take to go from the generic local field to the special field for the special purpose? And what community is there between the specialization seen here and the specialization that makes certain electromagnetic and gravitational fields more interesting than others? We focus on this question in the present brief report, not with the thought of bringing any important new point to light, but with the idea to bring together under a single simple and cheerful umbrella one of Rabi's favorite areas of investigation and other fields of endeavor to which he has also given his support.

Beam Deflector Field as a "Dimension 2" Specialization of the Generic Local Field

In the elementary example of a static magnetic field the generic local potential is well known to be characterized by one additive constant, three independent coefficients of x, y, and z, five independent coefficients of the five second-order solid harmonics, and so on. In the specialization to the beam deflector field the following changes were made: (1) all solid harmonics in Φ of order higher than the second have been neglected; (2) the five second-order harmonics, $2z^2 - x^2 - y^2$, zx, zy, $x^2 - y^2$, xy, have been narrowed to two, $z^2 - y^2$ and zy, by the demand for independence of x; (3) the three first-order terms x, y, z have been reduced to two by the same demand; (4) they have been further reduced to one, $-az$, by a suitable rotation of the coordinate axes in the (y, z)-plane; and (5) the zero-order term, an additive constant, has been dropped from Φ as irrelevant to the analysis of the field. Finally, as essential step in giving the deflecting field its characteristic form, (6) it has been demanded that no magnetic line of force passing through the beam shall curve; that is, it has been required that the direction z of the first-order term shall be a principal axis of the second-order term, thus killing the term zy and leaving expression (1). In summary, the generic local field with its infinite number of adjustable constants, or its "infinite dimensionality", or—in mathematical language—its "zero codimension," has been specialized to a field with only two adjustable constants: strength and inhomogeneity—a field thus with "dimension two."

When one turns from measuring the magnetic moments of an atom or molecule, both intrinsic and induced, to measuring the corresponding electric moments, one encounters a field identical to that already discussed, except for the replacement of the word "magnetic" everywhere by the word "electric." Again the generic local field, with its zero codimension, has been specialized to a field of dimension two.

The Elementary Magnetic Mirror as a "Dimension 2" Specialization

When one transfers attention from a static magnetic field that transmits—and deflects—a beam to one that makes a localized trap for a charged particle, one comes to another important instance of specialization.

The radius of gyration of the particle is idealized to be negligible in comparison with the region over which the magnetic field suffers a percentagewise appreciable change. The angle α of climb of the orbit,

$$\tan \alpha = v_\parallel / v_\perp \tag{5}$$

starts with a value α_0 at the center of the zone of trapping, where the magnetic field has the strength B_0. It diminishes in magnitude to zero, and the particle reverses its excursion, at the place where the field has climbed to a certain value B. This value is given by making use of the condition of adiabatic invariance, variously stated as "conservation of angular momentum" or "conservation of flux encircled" or "conservation of $v_\perp^2/B$"; thus,

$$(v_{\perp 0})^2/B_0 = v_0^2 \cos^2 \alpha_0/B_0 = \text{(adiabatic invariant)}$$
$$= v_\perp^2 \text{(at turnabout)}/B = v_0^2/B; \tag{6}$$

or

$$\cos^2 \alpha_0 = B_0/B;$$

and

$$\tan^2 \alpha_0 = B/B_0 - 1. \tag{7}$$

In the familiar textbook example of trapping, the central magnetic line of force runs straight from $z = -\infty$ to $z = +\infty$. The field is generated by circular coils centered on that line and located in the planes $z = -c$ and $z = +c$. Out of the generic local potential, the potential in the important region is obtained by these specializations: (1) drop all solid harmonics of order four and higher; (2) demand symmetry under rotation about the z-axis, thus killing all harmonics except 1, z, $3z^2 - r^2$, and $5z^3 - 3zr^2$; and (3) demand that Φ reverse sign on reflection in the plane $z = 0$. We end up with the "dimension two" trapping potential equal to

$$-az - (b/6)(2z^3 - 3\rho^2 z), \tag{8}$$

and the trapping field

$$B_z = a + bz^2 - \tfrac{1}{2}b\rho^2,$$
$$B_\rho = -b\rho z, \tag{9}$$
$$B_\varphi = 0.$$

This field, bulging most as it pierces the equatorial plane $z = 0$, is good for restraining the excursions of an individual particle. Thus B_z, evaluated near the central line of force, $\rho = 0$, increases both above and below the equator. However, the field is not good for confining plasma. Evaluated in the equatorial plane, it decreases in strength as ρ increases outward from the axis. In contrast, to withstand the particle pressure of a locally confined plasma, the magnetic pressure $B^2/8\pi$ should increase outward everywhere on the boundary of the region of trapping.

It is no help in achieving this goal to replace the two magnetizing coils at $z = \pm c$ by one coil at $z = 0$. That change, reversing the sign of b in Equation 9, would make the magnetic field increase with increasing distance from the axis, to be sure, but decrease along the axis, and would thus destroy altogether any chance for trapping.

A Minimum-B Configuration

For trapping particles it was a great step forward when the move was made from such a symmetric configuration to one with lesser symmetry but one in which B^2 increased in all directions from a central point. The work of Tamm (1958) and of Sakharov (1958), beginning as early as 1950, and also the investigations of Budker (1953) stimulated discussion of such novel configurations in the Soviet Union. Solov'ev,[10] in his chapter in the four-volume work of Leontovich[11] remarks that "The idea of a magnetic well [minimum field configuration] was introduced by Rosenbluth and Longmire[12]."

The state of thinking as it stood at the time of the Second Geneva Conference on Peaceful Uses of Atomic Energy was briefly but clearly summarized by Artisimovich.[13] Also at that conference Grad and his collaborators[19,20] reported their extensive investigations over the years at the Institute of Mathematical Sciences at New York University on improving the performance of a mirror by departure from symmetry.

It was an exciting moment, observers say, when at the fusion conference at Salzburg in September 1961, Ioffe—closely associated with Artsimovich—reported a remarkable improvement in magnetic confinement achieved by adding to the usual mirror device supplementary coils generating asymmetric field components.[14] The approach was empirical. The improvement was big. Any proper theory was lacking. Soon general asymmetric field configurations received intensive consideration. To be sure, "... it was shown [in 1963] by Jean Andreoletti (Fontenay) and Harold P. Furth (Livermore) that minimum-B stabilization could also be accomplished in axially symmetric magnetic configurations. [But among asymmetric arrangements] ... the coil configuration that was ultimately to become known as the baseball coil was evolved at Culham in the fall and winter of 1963. Informal conversations among members of J. B. Taylor's theoretical division and D. R. Sweetman's Phoenix group developed two different lines of thought leading to what later came to be known at Culham as the 'tennis-ball winding'" (quoted from Hiskes[15]; see also the article by Lubkin[16] that stimulated Hiskes' letter; and see also the article by Moir and Post[17] on the more recent Yin-Yang winding that has superseded the baseball winding). Real understanding of such configurations finally came with the work of J. B. Taylor.[18]

It is not necessary to enter into the subsequent development of asymmetric windings, with all their ramifications, to illustrate the central point in the language of "specialization of the generic local field." Retract part of the specialization enforced in the preceding example. Give up the requirement of symmetry under rotation about the z-axis. Or—to turn temporarily from how the field is to look locally to what kind of a farther-out coil will produce the field—deform the one coil most recently considered, so that with increasing azimuthal angle, φ, the wire rises up and down with respect to the equatorial plane, as suggested by the formulas

$$\rho = \text{constant}, \quad z = \epsilon \cos 2\varphi. \tag{10}$$

For an apter description of the deformation, imagine a hollow glass sphere that just fits inside the original flat coil, and let the wire, initially in the equatorial plane $\theta = \pi/2$, be slid on the surface of the sphere to

$$\theta = (\pi/2) - \delta \cos 2\varphi. \tag{11}$$

When the deformation is larger, so that so simple a description no longer applies, and $\theta(\varphi)$ even ceases to be single valued, one can still easily visualize the shape. Cup the hands together to imprison a maximum volume and look at the line of contact between them. Or pick up an American baseball and look at the seam that binds together the two separate hour-glass-shaped skins.

What about the field? The coil, subjected to a small deformation of the form (11), may be regarded as equivalent to (1) the original flat coil, creating a magnetic potential of the form (8) (with negative b), supplemented by (2, 3) a small pair of coils at $\varphi = 0$ and $\varphi = \pi$ ("x-coils") working in opposition to each other (lines of force approaching along the x-axis, only to spray out in the (y, z)-plane) and generating a magnetic potential whose leading term is

$$\Phi_{2,3} = (g/6)(2x^2 - y^2 - z^2), \tag{12}$$

and by (4, 5) a small pair of coils at $\varphi = \pi/2$ and $\varphi = 3\pi/2$ ("y-coils") likewise working in opposition to each other, but with current going around them in the opposite sense (lines of force coming in in the (x, z)-plane, and shooting away from each other along the y-axis) and generating a magnetic potential whose leading term is

$$\Phi_{4,5} = (g/6)(-2y^2 + x^2 + z^2). \tag{13}$$

Thus the generic local potential with its infinite number of adjustable constants, or zero codimension, is specialized in the baseball configuration to the "three-dimension" function

$$\Phi = \Phi_1 + \Phi_{2,3} + \Phi_{4,5} = -az - (b/6)(2z^3 - 3x^2z - 3y^2z)$$
$$+ (g/2)(x^2 - y^2). \tag{14}$$

Whatever modification of the coil it takes to produce exactly this potential, with no higher order terms, those modifications we assume and enforce. Moreover we assume the sign of b to be positive, as required for confinement of excursions in the z-direction, and as achieved by putting sufficient up-and-down "baseball ripple" into the shape of the coil.

The last term in Equation 14 generates transverse components of the magnetic field, B_x and B_y, which make B^2 rise in all directions outside the zone of trapping. The existence of one or more minima in B^2 within the zone of trapping is of course fully compatible with the theorem that Φ itself, as a solution of Laplace's equation, can have no minimum.

Where are the maxima, minima, and saddles of B^2? It economizes the analysis of this question to go to such units for distance, potential and field,

$$\begin{aligned} x &= (a/b)^{1/2} x_{\text{new}}, \text{etc.}, \\ \Phi &= (a^3/b)^{1/2} x_{\text{new}}, \\ B_x &= aB_{x\,\text{new}}, \text{etc.}, \\ g &= (ab)^{1/2} g_{\text{new}}, \end{aligned} \tag{15}$$

that the coefficients a and b take on unit values. We suppose this change made in what follows, but drop the subscripts "new" in the interest of simplicity. Then the components of the magnetic field have the values

$$B_x = -x(z + g)$$
$$B_y = -y(z - g)$$
$$B_z = 1 + z^2 - \tfrac{1}{2}(x^2 + y^2),$$

and the square of the field is

$$B^2 = 1 + (g^2 - 1)\rho^2 + 2z^2 + 2gz\rho^2 \cos 2\varphi + (\rho^4/4) + z^4. \tag{16}$$

All three components of the field vanish at the point,

$$x = 0, z = g, y = 2(1 + g^2)^{1/2}. \tag{17}$$

Not only there does B^2 vanish ("minimum-B configuration") but also at three other points,

$$x = 0, z = g, y = -2(1 + g^2)^{1/2};$$
$$x = 2(1 + g^2)^{1/2}, z = -g, y = 0;$$

and

$$x = -2(1 + g^2)^{1/2}, z = -g, y = 0. \tag{18}$$

There is a fifth and final minimum, $B^2 = 1$, at the origin itself, for values of the "baseball coefficient" g greater than unity. One gets from this minimum to the zero-field minimum (17) with the least increase in B^2 along the way by passing through a saddle. Consider for example a value of the baseball coefficient $g = 2$. Starting with $B^2 = 1$ at the origin, one comes to a saddle point with $B^2 = 3$ at $x = 0, y = 1.414\ldots, z = 1$ [more generally at

$$x = 0, z = (5g^2/4 - 1)^{1/2} - g/2, y = (4gz - 2g^2 + 2)^{1/2} \tag{19}$$

and of height

$$B_{\text{saddle}}^2 = (-1 + 7g^2 - 11g^4/2) + 4g(5g^2/4 - 1)^{3/2}] \tag{20}$$

before dropping down to $B^2 = 0$ at $x = 0, y = 3.162\ldots, z = 2$. From there, on the the way to still greater distances, B^2 steadily increases, whatever the direction of travel. The trapping is complete.

When one turns from confining a single charged particle to containing plasma, one meets new questions of stability (see, for example, Rosenbluth and Longmire,[12] Berkowitz et al.,[19,20] and the chapter by Solov'ev in Leontovich[10]) which lie outside the scope of this discussion.

Include not one, two, or three solid harmonics in the mathematical expression for the magnetic potential but go on including more and more terms and one will end up approximating arbitrarily closely any magnetic field. If we were to pursue this line of inquiry, we would be led from local questions to global questions, including for example the twisting of lines of force typical of the original stellarator design[21] and the winding numbers considered in quite other connections by Ulam.[22] Instead we shall continue to limit attention to fields locally.

Central Field Plus Lattice Perturbation

In all four cases considered so far, the potential, whether electric or magnetic, was regular at the origin, and was expanded in polynomials around that point. Nothing prevents consideration of a potential that (1) is singular at the origin, (2) is a spherically symmetric function that falls to zero with increasing distance, but that nevertheless (3) is once again supplemented by a polynomial perturbation with some small number of degrees of freedom or "low dimension." Two cases of this kind may be mentioned here, one having to do with electric fields, the other with gravitational fields.

In the realm of electric fields, symmetry considerations show up beautifully in the splitting of the energy levels of an atom located in a larger system, as treated by Ehlert[23] and especially by Bethe,[24] along the general lines laid down by Wigner[25] for dealing with a narrowing of symmetry. If the free atom is symbolized by a rotationally symmetric potential that falls off with distance $V = V(r)$, then the atom-in-the-crystal may be symbolized by such a potential supplemented by a polynomial in $x, y,$ and z that has the requisite symmetry. In Bethe's analysis, "For the case in which the individual electrons of the atom can be treated separately (interaction inside the atom turned off) the eigenfunctions of zeroth approximation are stated for every term in the crystal; from these follow a concentration of the electron density along the symmetry axes of the crystal which is characteristic of the term."

Later in the development of physics, in considering the wave function appropriate to describe a conduction electron in a crystal lattice, Wigner and Seitz[26,27] summarized in the book by Seitz[28] found it profitable to go in the opposite direction: to idealize the angle-dependent potential in which the electron moves as spherically symmetric, and to replace the actual lattice cell, with its many-faceted boundary, by a spherical cell of the same volume.

Following Wigner and Seitz, Lindquist and Wheeler[29] applied the method of the spherical lattice cell to deal with quite a different problem, one of gravitation physics, the dynamics of a model "lattice universe" composed of 5, 8, 16, 24, 120, or 600 identical point masses (black holes). The universe is closed but dynamic. Separations increase with time from "big bang" to a moment of maximum expansion, then contract to a catastropic "big stop," much as in the case of the Friedmann model universe with its uniform density of "dust." All of the dynamics is summarized in the law of rise and fall of a test mass, located at the interface between one lattice cell and the next. This motion is treated as taking place under the influence of the gravitational attraction, or Schwarzschild metric, of the one mass located at the center of the one representative cell, idealized as spherical. In this approximation the calculated dynamics agrees with the Friedmann dynamics to $\sim 5\%$ or 10%, depending upon the quantity considered. A more nearly accurate treatment would allow for the departures of the metric from spherical symmetry occasioned by the presence of the neighboring lattice cells, analogous to the perturbation of the electric potential in an atom by the field of the surrounding crystal. The machinery for such a perturbation analysis has been given by Regge and Wheeler[30] and Vishveshwara[31,32]; and the necessary tensorial spherical harmonics are given by Zerilli[33,34]; and the equivalent spinoral field components, by Bardeen and Press.[50] These harmonics correspond to the polynomials in x, y, z that one adds to the potential $V(r)$ in an atom to describe the perturbation induced by the crystal.

The Generic Local Electromagnetic Field and Its Specializations

Having considered local electric and local magnetic fields individually, we naturally turn to the generic local electromagnetic field—again in empty space—and its specializations. In this context it is natural to give up the restriction to a static field. However, as the price for considering a field that normally is dynamic, we change our definition of "local" from "local in space" to "local in spacetime." In the vicinity of the reference event we think of the four components of the 4-potential as being expanded as power series in the four coordinates t, x, y, z. It is the simplest specialization of the generic local field, and the only one we examine, to drop all terms in this expansion of second degree or higher in the coordinates. At the level of first degree terms, Maxwell's equations impose no conditions on the coefficients. From such potentials follow values of the field components themselves which are locally independent of position and perfectly arbitrary:

$$\mathbf{B} = (B_x, B_y, B_z),$$
$$\mathbf{E} = (E_x, E_y, E_z). \tag{21}$$

Only the choice of spacetime axes is at our disposition. Whatever else the general combination of boost and rotation accomplishes, it leaves untouched the values of the scalar

$$- F_{\alpha\beta} F^{\alpha\beta}/2 = \mathbf{E}^2 - \mathbf{B}^2 \tag{22}$$

and the pseudoscalar

$$F_{\alpha\beta} * F^{\alpha\beta}/4 = \mathbf{E} \cdot \mathbf{B}. \tag{23}$$

In the case of maximum specialization of the generic local field, all six components of $\mathbf{E}$ and $\mathbf{B}$ vanish ("0-dimensional field"). One step short of such extreme specialization is the case of a null field, so exhaustively analyzed by Synge,[35,36] in which $\mathbf{E}$ and $\mathbf{B}$ individually differ from zero, but $\mathbf{E}^2 - \mathbf{B}^2 = 0$ and $\mathbf{E} \cdot \mathbf{B} = 0$. The two field vectors stand perpendicular to each other and have the same magnitude. The Poynting flux, $\mathbf{E} \times \mathbf{B}/4\pi$, can be made to point in any direction one pleases, say the z-direction, by rotation of the space axes (2 parameters); and $\mathbf{E}$ can be made to point along the x-axis and $\mathbf{B}$ (equal in magnitude to $\mathbf{E}$) along the y-axis by a subsequent rotation about the z-axis (1 parameter). Finally, by a suitable boost (1 more parameter) in the direction of the Poynting flux—now the z-axis—that flux can be increased or decreased as much as one pleases. In other words, $\mathbf{E}$ and $\mathbf{B}$ can be raised or lowered until they reach a standard field magnitude, F_{standard} ("zero-parameter standard configuration"). Thus the generic null field is characterized by $2 + 1 + 1$ parameters; it is "4-dimensional"—as of course also follows from the imposition of the two conditions, $\mathbf{E}^2 - \mathbf{B}^2 = 0$ and $\mathbf{E} \cdot \mathbf{B} = 0$, on the six otherwise independent components of the field.

When the specialization is one degree less extreme, either $\mathbf{E}^2 - \mathbf{B}^2$ or $\mathbf{E} \cdot \mathbf{B}$ or both differ from zero. Then the Poynting flux $\mathbf{E} \times \mathbf{B}/4\pi$ is less in magnitude than the energy density $(\mathbf{E}^2 + \mathbf{B}^2)/8\pi$, as follows from the identity

$$(\mathbf{E}^2 - \mathbf{B}^2)^2 + (2\mathbf{E} \cdot \mathbf{B})^2 = (\mathbf{E}^2 + \mathbf{B}^2)^2 - (2\mathbf{E} \times \mathbf{B})^2.$$

Then according to Misner, Thorne, and Wheeler,[37] the Poynting flux is reduced to zero by viewing the field from a local inertial frame that is travelling in the direction of $\mathbf{E} \times \mathbf{B}$ with a velocity

$$v = \tanh \alpha, \tag{24}$$

where the velocity parameter α is given by the formula

$$\tanh 2\alpha = \frac{(\text{Poynting flux})}{(\text{energy density})} = \frac{2|(\mathbf{E} \times \mathbf{B})|}{\mathbf{E}^2 + \mathbf{B}^2} \tag{25}$$

(3-parameter boost; z-axis of new frame pointing in the direction of the boost). In this new frame $\mathbf{E}$ and $\mathbf{B}$ still stand perpendicular to the z-axis, but there is no energy flow. Therefore $\mathbf{E}$ and $\mathbf{B}$ have to be parallel to each other. By a suitable rotation around the z-axis (1 parameter) these vectors can be made to point along the x-axis, with magnitudes

$$E_x = \{[(\mathbf{E}^2 - \mathbf{B}^2)^2/4 + (\mathbf{E} \cdot \mathbf{B})^2]^{1/2} + (\mathbf{E}^2 - \mathbf{B}^2)/2\}^{1/2} \tag{26}$$

and

$$B_x = \{[(\mathbf{E}^2 - \mathbf{B}^2)^2/4 + (\mathbf{E} \cdot \mathbf{B})^2]^{1/2} - (\mathbf{E}^2 - \mathbf{B}^2)/2\}^{1/2}, \tag{27}$$

(one or two parameters, depending on whether one or both of the invariants $\mathbf{E}^2 - \mathbf{B}^2$, $\mathbf{E} \cdot \mathbf{B}$, is nonzero). Thus the nonnull local electromagnetic fields form a $3 + 1 + 1 = 5-$ or $3 + 1 + 2 = 6$-dimensional set according as one or both of the invariants have nonzero values.

The Generic Local Gravitational Field and Its Specializations

When we turn from electromagnetism to gravitation, the center of attention becomes the local value of the Riemann curvature tensor, the measure of the gravitational tide-producing acceleration [difference of acceleration per unit of difference in location; expressed in cgs units as $(\text{cm/sec}^2)/\text{cm}$, or in geometrical units as $(\text{cm}/\text{cm}^2$ of light-travel time)/cm or cm^{-2}], with its 20 independent components

$$R_{\alpha\beta\eta\delta}. \tag{28}$$

For example, in a space ship in an orbit of radius r about the earth [mass in geometrical units, $m = Gm_{\text{conv}}/c^2 = (6.67 \times 10^{-8} \text{ cm}^3/\text{gs}^2)(5.98 \times 10^{27} \text{ g})/(3.00 \times 10^{10} \text{ cm/s})^2 = 0.444$ cm] the tide-producing action of the earth draws apart at the relative acceleration

$$(\text{relative acceleration})_z = -R_{z0z0}\Delta z = (2m/r^3)\Delta z, \tag{29}$$

two test particles that have the separation Δz, along the direction of r. It draws together two test particles that have a separation in either of the two directions x or y, perpendicular to r:

$$\begin{aligned}(\text{relative acceleration})_x &= -R_{x0x0}\Delta x = -(m/r^3)\Delta x;\\ (\text{relative acceleration})_y &= -R_{y0y0}\Delta y = -(m/r^3)\Delta y.\end{aligned} \quad (30)$$

For fast-moving test particles, components of the curvature tensor come into play other than those of the form R_{i0j0}, much as components (magnetic) of the electromagnetic field tensor $F_{\alpha\beta}$ other than the electric component F_{i0} act on fast charged particles.

The relevant quantity is not the acceleration of a free uncharged test particle in the local Lorentz frame, for that is always zero, according to Einstein's principle of weightlessness in free fall, foundation of his "equivalence principle." The relevant quantity is the acceleration of the separation vector running from the one test particle to another that is moving on almost the same world line. In terms of proper time τ and covariant differentiation D, that separation follows the "equation of geodesic deviation"

$$D^2\eta^\alpha/d\tau^2 + R^\alpha{}_{\beta\gamma\delta}(dx^\beta/d\tau)\eta^\gamma(dx^\delta/d\tau) = 0. \quad (31)$$

Here we raise—and lower—indices with the help of the metric tensor (and its reciprocal) in the Landau-Lifshitz spacelike convention,

$$\begin{aligned}\text{all } \eta_{\alpha\beta} &= 0 \text{ except } \eta_{00} = -1, \eta_{11} = \eta_{22} = \eta_{33} = +1;\\ \text{all } \eta^{\alpha\beta} &= 0 \text{ except } \eta^{00} = -1, \eta^{11} = \eta^{22} = \eta^{33} = +1.\end{aligned} \quad (32)$$

In terms of laboratory time t (measured, like τ, in cm of light travel time) and velocity v^j ($j = 1, 2, 3$; more generally Latin letters are used here for 3-geometry and Greek for 4-geometry) relative to the laboratory Lorentz frame, the relative acceleration is

$$d^2\eta^i/dt^2 = (\text{tide})^i_j \eta^j. \quad (33)$$

Here the tide-producing tensor is a velocity-dependent quantity,

$$-(\text{tide})^i_j = R^i{}_{0j0} + R^i{}_{mj0}v^m + R^i{}_{0jn}v^n + R^i{}_{mjn}v^m v^n. \quad (34)$$

It is difficult to avoid comparing this expression with the Lorentz force-per-unit-charge,

$$F^i{}_0 + F^i{}_j v^j = (\mathbf{E} + \mathbf{v} \times \mathbf{B})^i. \quad (35)$$

In pursuance of this analogy, one is accustomed to follow Matte,[38] and use the six distinct components, R_{i0j0}, of the curvature to define a symmetric 3-tensor $\mathcal{E}$, the "electric part of the gravitational field,"

$$\mathcal{E}_{ij} = R_{i0j0}. \quad (36)$$

At this point the subject divides, according as the region of spacetime under consideration is or is not free of sources of mass-energy. Matte and Esposito limit attention to source-free regions. In this case the Riemann curvature tensor reduces to Weyl's tensor measure of conformal curvature,

$$C^{\alpha\beta}{}_{\gamma\delta} = R^{\alpha\beta}{}_{\gamma\delta} - 2\delta^{[\alpha}{}_{[\gamma} R^{\beta]}{}_{\delta]} + \tfrac{1}{3}\delta^{[\alpha}{}_{[\gamma}\delta^{\beta]}{}_{\delta]}R$$

$$(= R^{\alpha\beta}{}_{\gamma\delta} \text{ in source-free space}). \quad (37)$$

Here the bracket symbol indicates that one is to antisymmetrize with respect to the labels linked by paired brackets [ ]. Both tensors in the representation $C_{\alpha\beta\gamma\delta}$ or $R_{\alpha\beta\gamma\delta}$ have the symmetries:

(a) change of sign on interchange of α and β;
(b) change of sign on interchange γ and δ;
(c) symmetry with respect to interchange of the pair $[\alpha,\beta]$ with the pair $[\gamma,\delta]$;
(d) $R_{0123} + R_{0231} + R_{0312} = 0$, or, $\quad\quad\quad\quad\quad\quad\quad\quad(38)$
$C_{0123} + C_{0231} + C_{0312} = 0$.

The antisymmetries (a) and (b) have frequently led one to replace the double label $\alpha\beta$ by a single label A, taking on 6 distinct values:

$$01 = \text{I}, 02 = \text{II}, 03 = \text{III}, 23 = \text{IV}, 31 = \text{V}, 12 = \text{VI}. \quad (39)$$

Similarly one substitutes a single index B for the pair $\gamma\,\delta$. The resulting 6×6 tensor, because of the further symmetry

$$R_{AB} = R_{BA}$$

or

$$C_{AB} = C_{BA} \quad (40)$$

might seem to have $6 \times \tfrac{7}{2} = 21$ independent components, as compared to the 6 of the Maxwell field. However, one of the otherwise free parameters of the generic local tide-producing field is killed by the final symmetry,

$$R_{0123} + R_{0231} + R_{0312} = 0,$$

or

$$R_{\text{I IV}} + R_{\text{II V}} + R_{\text{III VI}} = 0. \quad (41)$$

Thus the gravitational field in the general case, regarded locally, has 20 independent components.

When no matter is present, and $R_{\alpha\beta\gamma\delta}$ reduces to pure conformal curvature $C_{\alpha\beta\gamma\delta}$, the symmetry (41) persists. However, there are now 10 additional conditions. They come from the vanishing of the 10 components of the stress-energy tensor $T_{\mu\nu}$ in Einstein's field equation. Alternatively, one can cite from the purely geometrical side the 10 symmetries possessed by the tensor of conformal curvature as compared to the Riemann curvature tensor. Whichever way one does the reasoning, the number of independent components of the gravitational field reduces in the source-free case from 20 to 10.

One can translate these words into formulas by writing down explicitly Einstein's field equation, with its 10 components

$$G_{\mu\nu} = 8\pi T_{\mu\nu}. \quad (42)$$

It is only necessary to express the Einstein curvature tensor, $G_{\mu\nu}$, in terms of the Riemann curvature tensor[39] to be completely explicit. The writing is simplified by going to a "mixed representation" ($R^{\alpha\beta}{}_{\gamma\delta}$, $G^{\alpha}{}_{\beta}$, etc.) with the help of an index-raising operation N^{AB} as applied to the 6 indices B of R_{AB}. Here N^{AB} is a diagonal matrix with the components $N^{I\,I} = N^{II\,II} = N^{III\,III} = -1$ and $N^{IV\,IV} = N^{V\,V} = N^{VI\,VI} = +1$. Specifically, one has

$$-8\pi \begin{pmatrix} \text{mass-energy per} \\ \text{unit volume} \end{pmatrix} = 8\pi T_0{}^0 = G_0{}^0 = \begin{pmatrix} t\text{-component of} \\ \text{"moment of} \\ \text{rotation" con-} \\ \text{tained in unit of} \\ 3\text{-volume} \\ \text{normal to } t\text{-axis} \end{pmatrix}$$

$$= -(R_{23}{}^{23} + R_{31}{}^{31} + R_{12}{}^{12})$$
$$= -(R_{IV}{}^{IV} + R_{V}{}^{V} + R_{VI}{}^{VI}); \qquad (43)$$

$$8\pi \begin{pmatrix} x\text{-component of} \\ \text{all nongravita-} \\ \text{tional forces} \\ \text{acting from side} \\ x - \epsilon \text{ to side} \\ x + \epsilon, \text{ across a} \\ \text{unit area} \\ \text{normal to } x\text{-axis} \end{pmatrix} = 8\pi T_1{}^1 = G_1{}^1 = \begin{pmatrix} x\text{-component of} \\ \text{"moment of} \\ \text{rotation" con-} \\ \text{tained in unit of} \\ 3\text{-volume} \\ \text{normal to } x\text{-axis} \end{pmatrix}$$

$$= -(R_{02}{}^{02} + R_{03}{}^{03} + R_{23}{}^{23})$$
$$= -(R_{II}{}^{II} + R_{III}{}^{III} + R_{IV}{}^{IV}); \qquad (44)$$

$$8\pi \begin{pmatrix} x\text{-component of} \\ \text{total momentum} \\ \text{per unit volume} \\ \text{of all kinds} \\ \text{other than} \\ \text{gravitation} \end{pmatrix} = 8\pi T_1{}^0 = G_1{}^0 = \begin{pmatrix} t\text{-component of} \\ \text{"moment of} \\ \text{rotation" con-} \\ \text{tained in unit of} \\ 3\text{-volume} \\ \text{normal to } x\text{-axis} \end{pmatrix}$$

$$= R_{12}{}^{02} + R_{13}{}^{03} = R_{VI}{}^{II} - R_{V}{}^{III}; \qquad (45)$$

$$8\pi \begin{pmatrix} x\text{-component of} \\ \text{all nongravita-} \\ \text{tional forces} \\ \text{acting from side} \\ y - \epsilon \text{ to side} \\ y + \epsilon, \text{ across a} \\ \text{unit area} \\ \text{normal to } y\text{-axis} \end{pmatrix} = 8\pi T_2{}^1 = G_2{}^1 = \begin{pmatrix} x\text{-component of} \\ \text{"moment of} \\ \text{rotation" con-} \\ \text{tained in unit of} \\ 3\text{-volume} \\ \text{normal to } y\text{-axis} \end{pmatrix}$$

$$= R_{20}{}^{10} + R_{23}{}^{13} = R_{II}{}^{I} - R_{IV}{}^{V}; \qquad (46)$$

and equations obtained from the foregoing by natural permutation of indices.

"Distant" versus "Local" Components of the Gravitational Field

Both mathematically and physically Equations 37 and 42 make a clear distinction between the parts of the full gravitational field, $R_{\alpha\beta\gamma\delta}$, that are of distant origin ($C_{\alpha\beta\gamma\delta}$) and of local origin ($G_{\alpha\beta}$). Both are normally present in any region that contains matter or any sources of mass-energy other than gravitation itself. Moreover, to reconstitute the full field out of the two separate contributions, one

TABLE 1

ELECTROMAGNETIC AND GRAVITATIONAL FIELDS COMPARED AND CONTRASTED WITH SPECIAL ATTENTION TO THE NEW COMPONENTS THAT COME INTO PLAY IN A REGION CONTAINING A CONTINUOUS DISTRIBUTION OF SOURCE
(**D** *versus* **E** and $R_{\alpha\beta\gamma\delta}$ *versus* $C_{\alpha\beta\gamma\delta}$).

	Electromagnetism	Gravitation or Tidal Action of Spacetime Geometry
Characterization in empty space	**E**, **B**, or $F_{\alpha\beta}$	$C_{\alpha\beta\gamma\delta}$
Independent components	6	10
Number of combinations of these invariant under Lorentz transformations	2	4
Characterization in region containing sources	**E, B, D, H**	$R_{\alpha\beta\gamma\delta}$
Independent components	12	20
Normal connection between the additional components and the components that are relevant in source-free space	Multiplicative (dielectric constant, magnetic permeability, etc.)	Additive (governed exclusively by local source density)
Is any component of the field at a point fixed exclusively by the source density at that point?	No (Example: $\nabla \cdot \mathbf{E} = 4\pi\rho$ gives derivative of field, not the field itself)	Yes (The 10 $G_{\alpha\beta}$ at a point are completely determined by the 10 $T_{\alpha\beta}$ at that very point)
Are any components of the field appropriately described as of "distant origin"?	Yes (All 12)	Yes (The 10 $C_{\alpha\beta\gamma\delta}$, as compared and contrasted to the 10 $G_{\alpha\beta}$. It takes all 20 of these quantities—"local" plus "distant"—to reconstitute the full gravitational field $R_{\alpha\beta\gamma\delta}$.)
Method to measure field	Acceleration of moving, charged, test particle	Relative acceleration of two nearby moving, uncharged, test particles
Distinction between new components—in region containing sources—and the components which suffice for source-free space	**H**: put test particle in needle-shaped slot cut out of medium parallel to field. **B**: penny-shaped slot normal to field. **D**: penny-shaped slot. **E**: needle-shaped slot.	$C_{\alpha\beta\gamma\delta}$: two test particles in a small hole cut out of medium, with their separation small compared to the size of the hole. $R_{\alpha\beta\gamma\delta}$: the two test particles are in two separate holes small in comparison to the separation between them.

employs simple addition. In contrast, the **D** and **H** fields in a region containing matter are usually connected to the **E** and **B** components of $F_{\alpha\beta}$ by multiplicative relations like $\mathbf{D} = \epsilon \mathbf{E}$ and $\mathbf{H} = \mathbf{B}/\mu$ (TABLE 1). In further illustration of the distinction between $R_{\alpha\beta\gamma\sigma}$ and $C_{\alpha\beta\gamma\delta}$, one can mention two examples of a source-free region of space, where the entire gravitational field is described by $C_{\alpha\beta\gamma\delta}$: (1) the Schwarzschild geometry outside a spherically symmetric distribution of mass-energy. In a local Lorentz frame with its x-axis pointing along the direction of increasing Schwarzschild r-coordinate, the spacetime curvature is all conformal, none of local origin, and is a purely diagonal tensor, with diagonal elements as indicated here:

$$R^A{}_B = C^A{}_B = \begin{pmatrix} \begin{array}{c|c} \begin{matrix} 01 \\ 02 \\ 03 \\ \hline 23 \\ 31 \\ 12 \end{matrix} & \begin{matrix} 2 & & & & & \\ & -1 & & & & \\ & & -1 & & & \\ \hline & & & 2 & & \\ & & & & -1 & \\ & & & & & -1 \end{matrix} \end{array} \end{pmatrix} \times (m/r^3), \qquad (47)$$

and (2) the geometry associated with a gravitational plane wave advancing in the z-direction. The wave is described by a "wave factor" $\beta = \beta(u) = \beta(t-z)$ and a "background factor" $L = L(u)$. The *effective* energy carried by the wave influences the background, according to the "equation of influence" (see Misner et al.[40] for details and references to the original literature)

$$d^2L/du^2 + (d\beta/du)^2 L = 0. \qquad (48)$$

With $\beta(u)$ zero outside of a limited region, and prescribed arbitrarily within that region, subject to natural conditions of continuity, the background factor is found from (48). The geometry is

$$ds^2 = -dt^2 + L^2(e^{2\beta} dx^2 + e^{-2\beta} dy^2) + dz^2. \qquad (49)$$

The components of the gravitational field vanish except as otherwise indicated in the following matrix:

$$R^{\hat{A}}{}_{\hat{B}} = C^{\hat{A}}{}_{\hat{B}} = \hat{A}\downarrow \begin{pmatrix} \begin{array}{c|c} \begin{matrix} 1 & & & 0 & -1 & \\ & -1 & & -1 & 0 & \\ & & 0 & & & 0 \\ \hline 0 & 1 & & & 1 & \\ 1 & 0 & & & & -1 \\ & & 0 & & 0 & \end{matrix} \end{array} \end{pmatrix} L^{-2}(d/du)(L^2 d\beta/du). \qquad (50)$$

(with $\hat{B}$ indicated across the top)

The carats over the indices emphasize that the components are given in a local Lorentz frame. Its axes are parallel to t, x, y, z but its "basis 1-forms" are dt, $Le^\beta dx$, $Le^{-\beta} dy$, and dz. One verifies that the Einstein curvature tensor, or "moment of rotation associated with the spacetime curvature", the quantity $G_{\alpha\beta}$ given in terms of the $R^{\hat A}{}_{\hat B}$ by Equations 43–46, vanishes throughout the region in question, as expected from the absence of any local sources, $T_{\alpha\beta}$, of stress, momentum, and energy.

In contrast is (3) the geometry inside a spherically symmetric cloud of mass-energy of uniform density—whether a complete Friedmann universe with the geometry

$$ds^2 = -dt^2 + a^2(t)[d\chi^2 + \sin^2\chi\,(d\theta^2 + \sin^2\theta\,d\varphi^2)], \qquad (51)$$

or a sector thereof, interior to the Schwarzschild geometry of (1), associated for example with a collapsing cloud of matter. The total gravitational field, the Riemann curvature tensor, referred to a local Lorentz frame with the "basis 1-forms" $\omega^0 = dt$, $\omega^1 = a(t)\,d\chi$, $\omega^2 = a\sin\chi\,d\theta$, $\omega^3 = a\sin\chi\sin\theta\,d\varphi$, is diagonal, with the diagonal elements listed here:

$$R^{\hat A}{}_{\hat B} = \begin{pmatrix} \ddot a/a & & & & & \\ & \ddot a/a & & & & \\ & & \ddot a/a & & & \\ \hline & & & a^{-2}(1+\dot a^2) & & \\ & & & & a^{-2}(1+\dot a^2) & \\ & & & & & a^{-2}(1+\dot a^2) \end{pmatrix}. \qquad (52)$$

The 3-sphere is conformally flat, being reducible to Euclidean 3-space by a simple position-dependent multiplicative rescaling of the metric that leaves all angles unchanged (conformal transformation). Either in this way or by direct calculation from Equation 37, one confirms that the whole of the "long distance" part of the gravitational field vanishes:

$$C^{\hat A}{}_{\hat B} = 0. \qquad (53)$$

Only the part of the curvature of local origin remains, $G_{\alpha\beta}$. All components of this Einstein curvature are zero except for the diagonal elements,

$$-G_{\hat t}{}^{\hat t} = G_{\hat t \hat t} = 3a^{-2}(1+\dot a^2) = 8\pi T_{\hat t \hat t} = 8\pi\rho \qquad (54)$$

and

$$G_{\hat \chi}{}^{\hat \chi} = G_{\hat \chi \hat \chi} = G_{\hat \theta \hat \theta} = G_{\hat \varphi \hat \varphi} = -2\ddot a/a - a^{-2}(1+\dot a^2) = 8\pi T_{\hat \chi \hat \chi} = 8\pi p. \qquad (55)$$

Here ρ and p are the density of mass-energy and the pressure of the homogeneous isotropic medium, both measured in geometric units:

$$\begin{aligned}\rho(\text{cm}^{-2}) &= (G/c^2)\rho_{\text{conv}}(g/\text{cm}^3),\\ p(\text{cm}^{-2}) &= (G/c^4)p_{\text{conv}}(g/\text{cm}\cdot\text{sec}^2).\end{aligned} \qquad (56)$$

If the medium has a boundary, and a nonzero pressure inside, it will blow off the surface. Conditions inside will cease to be isotropic. Therefore the present illustration is restricted to the starting instant, or to the case of zero pressure (collapse of a cloud of "dust"), unless the system is unbounded (closed universe).

In a final example (4) the geometry inside a rapidly rotating neutron star, no conformal transformation will reduce the geometry to flatness. Here both the long distance or conformal part, $C_{\alpha\beta\gamma\delta}$ of the full gravitational field, $R_{\alpha\beta\gamma\delta}$, and the part of local origin, the Einstein part $G_{\alpha\beta}$, have significant magnitudes.

If in case (1) or (2) we have one kind of specialization of the generic local gravitational field, where $G_{\alpha\beta}$ vanishes, and in (3) quite another, where $C_{\alpha\beta\gamma\delta}$ is zero, we are still far from having reviewed the simplifications that can be made in the presentation of this field by appropriate choice of the local Lorentz frame. As the first step in such a review it would be possible for us to list for gravitation all 14 algebraically independent invariants built out of the field components. They are the analogs of the two invariants, $\mathbf{E}^2 - \mathbf{B}^2$ and $\mathbf{E} \cdot \mathbf{B}$, of electromagnetism. They are known. The philosophy for finding them was given by Grace and Young[41] (see also Thomas[42]). A full list of invariants was given by Géhéniau and Debever[43] and independently by Petrov.[44] The latter list was simplified by Norden and Vishnevskii.[45] It is given in the book by Petrov.[46] The subject is further clarified in a paper of Penrose.[47] Nevertheless, it is not appealing to try to simplify the fully generic local field by Lorentz transformations that keep an eye on these 14 invariants. There is a physical reason for this renunciation, and also a mathematical one.

Two Lorentz Frames, Not One, Required for the Physically Appropriate Presentation of Gravitation in Simplest Form

From the mathematical point of view some of the 14 invariants are too complicated to be transparent, being built out of products of three or four components of the Riemann curvature tensor, as typified by an expression of the form

$$\epsilon_{\gamma\delta}{}^{\alpha\beta} R_{\sigma\tau}{}^{\gamma\delta} R_{\lambda\mu}{}^{\sigma\tau} R_{\alpha\beta}{}^{\lambda\mu}. \tag{57}$$

From the physical point of view it is not natural to try to simplify by one and the same Lorentz transformation two fields of very different origin and symmetry: the conformal part of the gravitational field, of distant origin, and the Einstein part of the field, of local origin. It is not an adequate counterpoise to this reasoning to say that the separation of two nearby test particles responds only to the total field, $R_{\alpha\beta\gamma\delta}$, as one sees it in the equation of geodesic separation (31). A closer look (TABLE 1) discloses idealized experiments where one cannot escape from the separate consequences of the two parts of $R_{\alpha\beta\gamma\delta}$. Let the acceleration of the separation of nearby test particles be observed (1) when they move through a common slot in the medium, and (2) when they move through separate slots. Any adequate bookkeeping of the motions is only then possible when one recognizes the separate contributions of $C_{\alpha\beta\gamma\delta}$ and $G_{\alpha\beta}$ to the geodesic deviation. Then let each part of $R_{\alpha\beta\gamma\delta}$ be cast in its simplest form by viewing it in the most appropriate local Lorentz frame. Let not the attempt be made to look at both within the straightjacket of a single inertial reference system. A particle is most naturally described in a Lorentz frame that is instantaneously comoving. The electromagnetic field takes its simplest form in quite another Lorentz frame. If two distinct frames already have to be called into service before one even comes to analyzing gravitation, one is not shocked to find two more Lorentz frames needed to make gravitation itself stand out most nakedly.

What difference does this philosophy make? Before articulating it, one could think of the generic local gravitational field as a single physical object, characterized by (1) 20 independent components, (2) 14 Lorentz-frame-independent invariants, and therefore (3) 20 − 14 = 6 meaningful parameters, the extraction of which is a worthwhile but most difficult and never yet achieved objective. Now, however, one may use in an unintended context Pauli's joking words, "Let not man join together what God hath put asunder." One recognizes $G_{\mu\nu}$ and $C_{\alpha\beta\gamma\delta}$, each with its 10 independent components, as separate and distinct physical objects. Each is to be presented in its own most appropriate Lorentz frame. Each frame, relative to the original laboratory frame, is characterized by 6 parameters. Each field has its own 10 − 6 = 4 recognized and well-known invariants. They, or quantities built on them (as E_x and B_x in Equations 26 and 27 are built on $\mathbf{E}^2 - \mathbf{B}^2$ and $\mathbf{E} \cdot \mathbf{B}$) provide the natural parameters for the characterization of the local gravitational field.

Is one to feel shortchanged by this reasoning? One had counted on 14 parameters. Instead, one ended up with only 4 (conformal) + 4 (local Einstein field) = 8. What has happened to the other 6? The answer is clear. They characterize the Lorentz transformation required to go from the frame in which the generic $G_{\mu\nu}$ looks simplest (i.e., takes on a prescribed canonical form) to the frame in which the generic $C_{\alpha\beta\gamma\delta}$ looks simplest. Nothing can keep anybody interested in the difficult task from distilling these 6 quantities out of the 14 invariants of the full $R_{\alpha\beta\gamma\delta}$. However, if one wholeheartedly accepts the split of $R_{\alpha\beta\gamma\delta}$ into separate and distinct physical fields, $G_{\mu\nu}$ and $C_{\alpha\beta\gamma\delta}$, one may be inclined to say that neither these 6 "relative-Lorentz-transformation" parameters, nor the 14 invariants themselves, have much to do with the physics of gravitation. Instead those 6 parameters and those 14 invariants might well be viewed as the right answer to the wrong question. The right question, on this view, is the characterization of $G^{\mu\nu}$ and $C_{\alpha\beta\gamma\delta}$ individually.

The Local Part of the Gravitational Field and Its Specializations

What frame makes the "moment of rotation" $G_{\mu\nu}$—or, equivalently (cf. Equation 42) the stress-energy tensor $T_{\mu\nu}$—take its simplest form? When the stress-energy originates solely from electromagnetism, we already have the answer in the frame that makes $F_{\mu\nu}$ itself look simplest. In the case of maximum specialization (see discussion following Equation 23), where both $\mathbf{E}^2 - \mathbf{B}^2$ and $\mathbf{E} \cdot \mathbf{B}$ vanish, the appropriate choice of Lorentz frame makes

$$E_x = F_{\text{standard}}, \quad E_y = 0, \quad E_z = 0,$$
$$B_x = 0, \quad B_y = F_{\text{standard}}, \quad B_z = 0,$$

(58)

and gives the local part of the gravitational field, or Einstein curvature, the value

$$G^\alpha{}_{\beta|} = 8\pi T^\alpha{}_{\beta} = 2F_{\text{standard}}^2 \overset{\beta}{\begin{pmatrix} -1 & 0 & 0 & 1 \\ 0 & 0 & 0 & 0 \\ 0 & 0 & 0 & 0 \\ -1 & 0 & 0 & 1 \end{pmatrix}}. \quad (59)$$

Here and below for simplicity, the fields have been taken to be expressed in geometrical units,

$$F(\text{cm}^{-1}) = (G^{1/2}/c^2) F_{\text{conv}}(\text{gauss or esvolts/cm}) \quad (60)$$

When the specialization is less extreme, and $E^2 - B^2$ or $E \cdot B$ or both differ from zero, then the appropriate simplifying Lorentz frame is given by (25). In it the electric and magnetic fields point along the x-axis with strengths given by (26) and (27). The Maxwell stress-energy tensor in its 4-dimensional form, multiplied by 8π, directly gives the Einstein curvature in its simplest version,[48]

$$G^\alpha{}_\beta = 8\pi T^\alpha{}_\beta = [(E^2 - B^2)^2 + 4(E \cdot B)^2]^{1/2} \begin{pmatrix} -1 & 0 & 0 & 0 \\ 0 & -1 & 0 & 0 \\ 0 & 0 & 1 & 0 \\ 0 & 0 & 0 & 1 \end{pmatrix}. \quad (61)$$

When the stress-energy arises from other fields either alone or in conjunction with electromagnetism, one is still open to have the case of maximum specialization (described by Equation 59 except for a change in the multiplicative constant before the matrix) in which the energy flows in a preferred direction with the speed of light. In a case of lesser specialization one expects from the physics that the energy must flow at a speed less than the speed of light. Then the appropriate Lorentz frame is one that moves in the direction of this flow with a speed sufficient to reduce the flow to zero. In this frame all G_{0i} components vanish. By a rotation of the space axes in this frame the space-space part of the Einstein curvature can be diagonalized. One ends up with the local part of the gravitational field, the Einstein curvature, given by an expression of the form

$$G_\beta{}^\alpha = \begin{pmatrix} -a & 0 & 0 & 0 \\ 0 & b & 0 & 0 \\ 0 & 0 & c & 0 \\ 0 & 0 & 0 & d \end{pmatrix}. \quad (62)$$

The 10 independent parameters of the generic local gravitational field are thus reduced to 4 meaningful parameters ("4-dimensional manifold of possibilities") with the help of the 6-parameter Lorentz transformation. The case of Friedmann geometry illustrated in Equations 54 and 55 was a specialization in that it contained only two disposable parameters (ρ and p), rather than four.

The Part of the Gravitational Field of Distant Origin and Its Specializations

The conformal curvature $C_{\alpha\beta\gamma\delta}$, the part of the gravitational field of distant origin, in the generic case is, like $G_{\alpha\beta}$, characterized by 10 independent components. In the past the components have been divided unabashedly into "electric" and "magnetic" components only in the special case where the full gravitational field $R_{\alpha\beta\gamma\delta}$ is of conformal origin exclusively (case of "empty space" and vanishing $G_{\alpha\beta}$). However, with the Einstein curvature and the conformal curvature now recognized as happily and independently coexisting entities, one has no reason to forego the convenience of this split—analogous to the split of $F_{\alpha\beta}$ into **E** and **B**—even when matter is present. Therefore let the tensor decomposition by Matte[38] of $C_{\alpha\beta\gamma\delta}$ into two 3-space symmetric tensors, $\mathcal{E}$ and $\mathcal{B}$, of zero trace, be taken over from the case of empty space (discussed especially clearly in the book of Zakharov[49]) to the general case:

$$C_{\alpha\beta\gamma\delta} = \left(\begin{array}{c|c} -\mathcal{E} & \mathcal{B} \\ \hline \mathcal{B} & \mathcal{E} \end{array}\right); \tag{63}$$

$$C_{\gamma\delta}{}^{\alpha\beta} = \left(\begin{array}{c|c} \mathcal{E} & \mathcal{B} \\ \hline -\mathcal{B} & \mathcal{E} \end{array}\right); \tag{64}$$

or, more concretely,

$$C_{0101} = -\mathcal{E}_{xx}$$
$$C_{0102} = -\mathcal{E}_{xy} = -\mathcal{E}_{yx} = C_{0201}$$
$$C_{2323} = \mathcal{E}_{xx}$$
$$C_{2312} = \mathcal{E}_{xy} = \mathcal{E}_{yz} = C_{1223}$$

and

$$C_{0123} = \mathcal{B}_{xx}$$
$$C_{0131} = \mathcal{B}_{xy} = \mathcal{B}_{yz} = C_{0223}, \text{ etc.} \tag{65}$$

Being traceless symmetric tensors, $\mathcal{E}$ and $\mathcal{B}$ have 5 linearly independent components each, adding up to the expected total count of 10. (The conventions in Equations 63–65 differ from Matte's because we use the $-+++$ metric and because we use for symbol $\mathcal{B}$ rather than his $\mathcal{H}$. In addition, our $\mathcal{B}$ has been reversed in sign from the H of Equation 14.46, p. 361 of Misner et al.[37] (a) to produce agreement in sign between the expressions for the flux in electromagnetism and gravitation and (b) to make the signs in the formulas for a Lorentz transformation run parallel in gravitation and electromagnetism.]

Do now the analog of what one does in the generic case of electromagnetism: go to a Lorentz frame in which the Poynting flux $\mathbf{E} \times \mathbf{B}/4\pi$ is zero. Such a choice of frame allowed one to align **E** and **B** simultaneously parallel to the x-axis. In gravitation, in the generic case, the corresponding choice of frame allows one to diagonalize simultaneously the 3 × 3 matrices $\mathcal{E}$ and $\mathcal{B}$:

$$\mathcal{E} = \begin{pmatrix} a \sin(\gamma - 2\pi/3) & 0 & 0 \\ 0 & a \sin(\gamma + 2\pi/3) & 0 \\ 0 & 0 & a \sin \gamma \end{pmatrix}; \quad (66)$$

$$\mathcal{B} = \begin{pmatrix} b \sin(\delta - 2\pi/3) & & \\ & b \sin(\delta + 2\pi/3) & \\ & & b \sin \delta \end{pmatrix} \quad (67)$$

Here each matrix, being traceless as well as diagonal, contains only two independent components. This circumstance is taken into account here by expression of the three diagonal elements in terms of a "magnitude parameter" and a quantity with the character of an angle, a "shape parameter". The resulting four quantities, a, γ, b, and δ, constitute the four interesting parameters of the generic "long-distance part", $C_{\alpha\beta\gamma\delta}$, of the local gravitational field.

Six other parameters specify the Lorentz transformation required to go from the laboratory frame to the "rocket frame" in which the Lorentz-transformed $\mathcal{E}$ and $\mathcal{B}$—call them $\mathcal{E}''$ and $\mathcal{B}''$—assume the canonical form (Equations 66 & 67). The part of the required Lorentz transformation of greatest interest is its velocity.

$$\mathbf{v} = \mathbf{n} \tanh \alpha. \quad (68)$$

Here $\mathbf{n}$ is a unit vector and α is the so-called "velocity parameter" of the "boost."

How does one tell whether one has the right boost (3 parameters of the Lorentz transformation) regardless of whether one yet has the right orientation of axes to put the "part of the gravitational field of distant origin" into the canonical form (Equations 66 & 67) (Petrov's canonical form[44,46])? How does one find the boost that makes $\mathcal{E}$ and $\mathcal{B}$ simultaneously diagonalizable as distinguished from the space rotation that actually diagonalizes them? The answer is simple. The three components v^i of the boost have to be such that the three components of the "flux of superenergy" as calculated for the transformed field are zero:

$$\begin{pmatrix} \text{"flux of} \\ \text{superenergy"} \end{pmatrix} = (T^{000i})'' = 2\epsilon_{ijk}(\mathcal{E}''\mathcal{B}'')_{jk} = 0. \quad (69)$$

The proof is due to Bel.[51] In this equation $\mathcal{E}\mathcal{B}$ denotes the ordinary matrix product of the two matrices in question; and the subscript jk signals that we are to take the jk element of this matrix product. The quantity ϵ_{ijk} is defined by the statements that it changes sign on the interchange of any two indices, and that $\epsilon_{123} = 1$. It is the same alternating symbol that enters the definition of the Poynting flux in electromagnetism,

$$(\text{"Poynting flux"}) = T^{0i} = 2\epsilon_{ijk}E_j B_k/8\pi. \quad (70)$$

Moreover, as the energy density in the one case is

$$T^{00} = (\mathbf{E}^2 + \mathbf{B}^2)/8\pi, \quad (71)$$

so the "density of superenergy" (into the physical interpretation of which only a beginning has so far been made; see Sejnowski[52]) in gravitation is

$$(\text{"density of superenergy"}) = T^{0000} = \text{Tr}(\mathcal{E}^2 + \mathcal{B}^2), \tag{72}$$

where "Tr" means "trace of the matrix."

It is one thing to have a criterion for the right boost. It is quite another to be in possession of that boost itself. A simple way to get that boost might be expected from the following unjustified and incorrect line of extrapolation:
(1) Given a 4-vector field, A^μ, find the boost,

$$\mathbf{v} = \mathbf{n} \tanh \alpha, \tag{73}$$

that reduces it to the canonical form of a pure timelike vector. Correct answer:

$$n^i \tanh \alpha = A^i/A^0. \tag{74}$$

(2) Same for the generic Maxwell field, $F^{\mu\nu}$. Correct answer:

$$n^i \tanh 2\alpha = T^{0i}/T^{00}. \tag{75}$$

(3) Same for the generic conformal curvature, $C^{\mu\nu}{}_{\rho\sigma}$. Wrong answer:

$$n^i \tanh 4\alpha = T^{000i}/T^{0000}. \tag{76, wrong!}$$

That this procedure fails is shown by an acutal numerical example in Wheeler[53] where also gravitational and electromagnetic wave flux are compared and contrasted. It is shown there that Equation 75 predicts a boost wrong in magnitude and direction; and in the example the correct boost is given, and a numerical procedure is outlined to determine the boost in other generic cases.

The field is no longer generic when the flux of superenergy is as big as the density of superenergy. This is the analog of a null field in electromagnetism. However, there is nothing physically exceptional about a point in space, near a point electric charge and a magnetic North pole, where the electromagnetic field is null. Likewise there is nothing physically exceptional about a point in space where the part of the gravitational field of distant origin is nongeneric in the sense that

$$(T^{0001})^2 + (T^{0002})^2 + (T^{0003})^2 = (T^{0000})^2. \tag{77}$$

It may be an inconvenience in the mathematical description of the field at such points that no finite boost enables one simultaneously to diagonalize the "electric" and "magnetic" parts of the conformal curvature. However, it is nothing more than an inconvenience. The true distinction between a gravitational field that is "radiative" and one that is not is not a local one. It is marked by global signs such as the integrated outward flux across a closed surface.

Acting on a gravitational-wave detector built of slow-moving components, a gravitational wave produces an effect governed practically exclusively by $\mathcal{E}(t)$ at the location of the equipment. In that neighborhood define a potential (a solution of Laplace's equation) by the equation

$$\Phi(t, x, yz) = (1/2)\mathcal{E}_{ij}(t)x^i x^j.$$

From it and its gradient one deduces the motions of the separate masses of the detector as straightforwardly (for example, Misner et al.[37] Sections 37.2–37.3) as one deduces the motions of charged particles from an electric potential—or the motion of molecules through Rabi's result-rich molecular beam apparatus.

Acknowledgment

I thank Polykarp Kusch for comments on molecular beam history; William Drummond, Richard Epstein, Harold Furth, Gloria Lubkin, and Richard Post for guidance into the literature of the "baseball" field configuration; and R. Debever, Paul Esposito, Charles Misner, Larry Smarr, and Louis Witten for discussions of the conformal tensor.

References

1. EINSTEIN, A., & P. EHRENFEST. 1922. Quantentheoretische Bemerkungen zum Experiment von Stern und Gerlach. Z. Phys. **11**: 31–34.
2. GORTER, C. J. 1936. Negative result of an attempt to detect nuclear magnetic spins. Physica **3**: 995–998.
3. GORTER, C. J. & L. J. F. BROER. 1942. Negative result of an attempt to observe nuclear magnetic resonance in solids. Physica **9**: 591–596.
4. GORTER, C. J. 1967. Bad luck in making scientific discoveries. Phys. Today **20** (January): 76–81.
5. RABI, I. I. 1937. Space quantization in a gyrating magnetic field. Phys. Rev. **51**: 652–654.
6. RABI, I. I., J. R. ZACHARIAS, S. MILLMAN, and P. KUSCH, 1938, "A new method of measuring nuclear magnetic moment", Phys. Rev. **53**: 318.
7. ZENER, C. 1932. Non-adiabatic crossing of energy levels. Proc. R. Soc. London **A137**: 696–702. For a pictorial summary, see Fig. 34 of HILL, D. L. & J. A. WHEELER. 1953. Nuclear constitution and the interpretation of fission phenomena. Phys. Rev. **89**: 1102–1145.
8. ESTERMANN, I. 1959. Molecular beam research in Hamburg, 1922–33. In Recent Research in Molecular Beams: A Collection of Papers Dedicated to Otto Stern on the Occasion of his 70th Birthday. I. Estermann, Ed.: 1–8. Academic Press. New York, N.Y.
9. RAMSEY, N. F. 1956. Molecular Beams. Clarendon. Oxford.
10. SOLOV'EV, L. S. 1963. Hydromagnetic stability of closed plasma configurations. In Reference 11: **6**: 239–331.
11. LEONTOVICH, M. A. 1963. Voprosy Teorii Plasmy. 4 volumes. Gosatomizdat. Moscow. [English translation, by H. Lashinsky. 1965–1972. Reviews of Plasma Physics. 6 volumes. Consultants Bureau. New York, N.Y.] Cited in Reference 10.
12. ROSENBLUTH, M. & C. LONGMIRE. 1957. Stability of plasmas confined by magnetic fields. Ann. Phys. (N.Y.) **1**: 120–140.
13. ARTSIMOVICH, L. A. 1958. Research on controlled thermonuclear reactions in the USSR. Uspekhi fiz. Nauk (USSR) **66**: 545–569. [English translation in 1958. Sov. Phys.—Uspekhi **1**: 191–207.] (1958).
14. GOTT, YU., M. S. IOFFE & V. G. TELKOVSKY. 1962. Some new results on confinement in magnetic traps. Nucl. Fusion (Suppl. Pt. III): 1045–1047.
15. HISKES, J. R. 1967. Who made the baseball (Letter to the editor). Phys. Today **20** (January): 9–10.
16. LUBKIN, G. B. 1966. Baseball magnetic field. Phys. Today **19**: 70–71.
17. MOIR, R. W. & R. F. POST. 1969. Yin-Yang minimum-$|B|$ magnetic-field coil. Nucl. Fusion **9**: 253–258.

18. TAYLOR, J. B. 1963. Some stable plasma equilibria in combined mirror-cusp fields. Phys. Fluids **6:** 1529–1536.
19. BERKOWITZ, J., K. O. FRIEDRICHS, H. GOERTZEL, H. GRAD, J. KILLEEN & E. RUBIN. 1958. Cusped geometries. *In* Proceedings of the Second United Nations Conference on Peaceful Uses of Atomic Energy. Vol. 31: 171–176. United Nations. Geneva.
20. BERKOWITZ, J., H. GRAD & H. RUBIN. 1958. Magnetohydrodynamic stability. *In* Proceedings of the Second United Nations Conference on Peaceful Uses of Atomic Energy. Vol. 31: 177–189. United Nations. Geneva.
21. SPITZER, L., JR. 1958. The stellarator concept. Phys. Fluids **1:** 253–264.
22. ULAM, S. M. 1960. A Collection of Mathematical Problems:107. Interscience. New York, N.Y.
23. EHLERT, W. 1928. Über das Schwingungs—und Rotationsspektrum einer Molekel vom Typus CH_4. Z. Phys. **51:** 6–33.
24. BETHE, H. A. 1929. Termaufspaltung in Kristallen, Ann. der Phys. **3:** 133–206. [English translation, no date, Splitting of Terms in Crystals. Consultants Bureau. New York].
25. WIGNER, E. 1927. Einige Folgerungen aus der Schrödingerschen Theorie für die Termstrukturen. Z. Phys. **43:** 624–652.
26. WIGNER, E. P. & F. SEITZ. 1933. On the constitution of metallic sodium. Phys. Rev. **43:** 804–810.
27. WIGNER, E. P. & F. SEITZ. 1934. On the constitution of metallic sodium. II. Phys. Rev. **46:** 509–524.
28. SEITZ, F. 1940. The Modern Theory of Solids. Chap. 9. McGraw-Hill Book Co. New York, N.Y.
29. LINDQUIST, R. W. & J. A. WHEELER. 1957. Dynamics of a lattice universe by the Schwarzschild-cell method. Rev. Mod. Phys. **29:** 432–443.
30. REGGE, T. & J. A. WHEELER. 1957. Stability of a Schwarzschild singularity. Phys. Rev. **108:** 1063–1069.
31. VISHVESHWARA, C. V. 1968. Stability of the Schwarzschild metric. Doctoral dissertation. University of Maryland.
32. VISHVESHWARA, C. V. 1970. Stability of the Schwarzschild metric. Phys. Rev. **D1:** 2870–2879.
33. ZERILLI, F. J. 1970. Gravitational field of a particle falling in Schwarzschild geometry analyzed in tensor harmonics. Phys. Rev. **D2:** 2141–2160.
34. ZERILLI, F. J. 1970. Effective potential for even-parity Regge-Wheeler gravitational perturbation equations. Phys. Rev. Lett. **24:** 737–738. [Correction, ZERILLI, F. J. 1974. Errata. Appendix A. 7. *In* REES, M., R. RUFFINI & J. A. WHEELER. 1974. Black Holes, Gravitational Waves and Cosmology: An Introduction to Current Research. Gordon and Breach. New York, N.Y.].
35. SYNGE, J. L. 1935. Principal null-directions defined in space-time by an electromagnetic field. University of Toronto Studies, Applied Math. Ser. 1.
36. SYNGE, J. L. 1956. Relativity: The Special Theory. Chap. 9. North-Holland. Amsterdam.
37. MISNER, C. W., K. S. THORNE & J. A. WHEELER. 1973. Gravitation.: 481. Freeman. San Francisco, Calif.
38. MATTE, A. 1953. Sur de nouvelles solutions oscillatoires des equations de la gravitation. Canad. J. Math. **5:** 1–16.
39. Reference 37:344.
40. Reference 37: 347, 957–959.
41. GRACE, J. H. & A. YOUNG, 1903. The Algebra of Invariants. Cambridge University Press. New York, N.Y.
42. THOMAS, T. Y. 1934. The Differential Invariants of Generalized Spaces. Cambridge University Press. London.
43. GÉHÉNIAU, J. & R. DEBEVER. 1956. Les invariants de courbure de l'espace de Riemann à quatre dimensions. Bull. Acad. R. Belgique Cl. Sci. **42:** 114–123.
44. PETROV, P. I. 1957. Second order invariants of the quaternary differential quadratic form. Dokl. Akad. Nauk SSSR **113:** 1214–1217.

45. NORDEN, A. P. & V. V. VISHNEVSKII. 1959. The complex representation of the invariants of a four-dimensional Riemannian space. Izvesti. Vuzov Ser Matematika 2(9): 176.
46. PETROV, A. Z. 1961. Pro Stranstve Einshteina. Fizmatlit. Moscow. [English translation, by Kelleher, R. F. 1969. Einstein Spaces. Pergamon. London.]
47. PENROSE, R. 1960. A spinor approach to general relativity. Ann. of Phys. **10**: 171–201.
48. RAINICH, G. Y. 1925. Electrodynamics in the general relativity theory. Trans. Amer. Math. Soc. **27**: 106–136.
49. ZAKHAROV, V. D. 1972. Gravitatsionnye volny v teorii tyagoteniya Einshteina, Izdatel'stvo "Nauka" (Glavnaya redaktsiya fiziko-matematicheskoi literatury) Moscow. [English translation in SEN, R. N. 1973. Gravitational Waves in Einstein's Theory. Halsted Press of Wiley. New York, N.Y.].
50. BARDEEN, J. M. & W. H. PRESS. 1973. Radiation fields in Schwarzschild background. J. Math. Phys. **14**: 7–19.
51. BEL, L. 1962. Cahiers Phys. **16**: 59.
52. SEJNOWSKI, L. 1974. *In* Gravitation Energy and Gravitational Collapse. C. DeWitt-Morette, Ed.: 103–104. D. Reidel. Dordrecht, Holland.
53. WHEELER, J. A. 1977. Gravitational and electromagnetic wave flux compared and contrasted. Preprint. University of Texas.